ELEMENTARY SCIENCE TEACHER EDUCATION

International Perspectives on Contemporary Issues and Practice

ELEMENTARY SCIENCE TEACHER EDUCATION

International Perspectives on Contemporary Issues and Practice

Edited by

Ken Appleton
Central Queensland University

ASSOCIATION FOR SCIENCE TEACHER EDUCATION

Co-Published with the Association For Science Teacher Education

 LAWRENCE ERLBAUM ASSOCIATES, PUBLISHERS
2006   Mahwah, New Jersey   London

Lawrence Erlbaum Associates, Inc., Publishers
10 Industrial Avenue
Mahwah, New Jersey 07430
www.erlbaum.com

Cover design by Kathryn Houghtaling Lacey

Library of Congress Cataloging-in-Publication Data

Elementary science teacher education : international perspectives on contemporary
   issues and practice / edited by Ken Appleton.
      p.   cm.
   Includes bibliographical references and index.
   ISBN 0-8058-4291-8 (alk. paper)
   ISBN 0-8058-4292-6 (pbk. : alk. paper)
      1. Science—Study and teaching (Elementary)—Cross-cultural studies.   2. Science
   teachers—Training of—Cross-cultural studies.   I. Appleton, Ken.

LB1585.E393   2005
372.3'5—dc22   2004056609
                                                                                       CIP

Books published by Lawrence Erlbaum Associates are printed on acid-free paper,
and their bindings are chosen for strength and durability.

Printed in the United States of America
10   9   8   7   6   5   4   3   2   1

Contents

Foreword

This is a wonderful work that looks at both the state of the art and the questions about the future in teaching elementary science teachers. The authors describe and question our ways of teaching elementary science teachers from the generalist to those who wish to specialize. These authors focus, from international perspectives, on current issues and their impact on the future elementary science teacher. Some of these issues contain technology impacts that include distance learning, national standards and science literacy, and field experiences prior to student teaching.

The authors write in a manner that is engaging and self-reflective. This is an important aspect of this monograph, in that the authors are urging their preservice teachers to be self-reflective on their own classroom practice. Therefore, the authors are practicing what they preach. It is only "fair" that the teachers of teachers practice and meet the same requirements that they demand of their students. By meeting such standards, the authors are able to give concrete examples to illustrate the topic of their chapter.

A basic feature of a monograph like this is to capture a slice of the variety of practice. This monograph does this in elementary science teacher education. By presenting this variety, the science teacher enterprise can prosper, grow, and build a greater body of knowledge about what works and what does not work, and identify questions that do not have answers. This is especially important in areas such as the early childhood and elementary grades, where science has been less emphasized.

Another aspect of this monograph that is commendable is the attention to the policy and politics of the various countries represented and how that impacts teacher education programs. Certainly, the U.S. could learn from Australia about the devastating effects of alternative licensure. One can also review the impact of including or excluding science in the early years of schooling on language development. Again, this book is an impressive collection of information from around the globe about teaching the elementary science teacher. One hopes that the book triggers many great discussions and instigates many research projects over the years.

—*Meta Van Sickle*
ASTE Publications Chair

Preface

Elementary Science Teacher Education: International Perspectives on Contemporary Issues and Practice was born out of discussions with a number of colleagues at annual meetings of the Association for Science Teacher Education (ASTE), and the National Association for Research in Science Teaching (NARST). There was strong agreement that a book focusing on contemporary issues and practices in elementary science teacher education was needed. After several discussions, the conceptual shape of the book emerged. I then approached colleagues whose work was pertinent about contributing to the project.

This book constitutes a significant collection of work from several countries where people are actively engaged in research and innovation in elementary science teacher education. It is impossible to separate the education of teachers from education systems in which they will serve, or the cultural attitudes and practices that frame those education systems. This clearly emerges in many of the chapters. The purpose of the book is to provide elementary science teacher educators, aspiring teacher educators, and policymakers with a greater understanding of the complexities of preservice and inservice teacher education in elementary science, and of some of the effects of current policies. More importantly, I trust that it provides a clearer direction for the future of elementary science teacher education, to the benefit of teachers, teacher educators, and most importantly, the millions of elementary school children who will benefit from improved practice.

—Ken Appleton
Editor

1

Framing Issues of Elementary Science Teacher Education: Critical Conversations

Sharon E. Nichols
University of Alabama

Thomas Koballa
University of Georgia

This book includes a wide range of perspectives on elementary science teaching, representing the complex and, at times, divergent stances taken among colleagues in this field. The approach to writing this chapter was to investigate the sorts of issues and literature the authors would draw on to represent their positions, strategies, and recommendations. There are places where authors' views intersect and conflict. Some assume positional writing stances, whereas others engage in writing as personal introspections, with diverging concerns to standardize or contextualize science teaching and learning. This range of perspectives characterizes the ongoing nature of debates about elementary science teaching and learning. Thus, this introductory chapter is an opportunity to engage readers in dialogue about complex issues that challenge elementary science education.

FRAMEWORKS FOR REFLECTION

Bolman and Deal (1991) argued that "too often they [leaders and managers] bring too few ideas to the challenges that they face . . . because they cannot look at old problems in a new light and attack old challenges with different and more powerful tools—they cannot *reframe*" (p. 4). The ability to reframe ideas and practices in elementary science teaching and research is critically important toward leading the profession in new directions. Accordingly, the approach taken within this introductory chapter hopefully will promote a more critical reading and discourse about ideas between the lines and across the chapters of this book. The next section presents four thematic connections to frame issues highlighted within this book, which include the following: elementary science teacher knowledge, disciplinary depth and breadth, the place of elementary science, and organizing the work of elementary science. Each theme is followed by key questions intended to provoke discussions about the ideas presented. Some chapters are featured more than once in cases where the content of the text readily connects with a different theme. It is important to note that authors may not have conceptualized their work according to the reframing of their ideas within this introduction. Readers may use the *thematic connections* and *key questions* presented in the following section to engage in conversations about issues raised by the text.

THEMATIC CONNECTION 1:
ELEMENTARY SCIENCE TEACHER KNOWLEDGE

Several authors have chosen to emphasize approaches to improve the education of elementary science teachers that have their foundation in the literature on teacher knowledge. Abell (chap. 5), in her writing about student field experiences, reminds of Fenstermacher's (1994) work in which he discussed four research programs that have greatly impacted the landscape of teacher education. His writing is particularly instructive in highlighting the research roots of various science teacher education models and strategies.

In contrast to past approaches, which assumed expert knowledge of teaching could be developed apart from teachers and the places where they enact teaching, much of the contemporary research on teaching and teacher education has emphasized the role of teachers in generating professional knowledge, and the highly contextual nature of teachers' knowledge and practices. Research that emphasizes what teachers know is based on the belief that teachers have a wealth of knowledge resulting from experience and it is this experience-based knowledge that drives their professional practice. Fenstermacher discussed the work of Clandinin and Connelly (1996; Clandinin, 1992) and Schön (1983, 1991) as examples of this research emphasis. Clandinin and Connelly's work on the *personal practical knowledge* stresses the usefulness of narrative and story in revealing how teachers think about their work and what knowledge they use to illuminate and inform their actions. At the heart of education research based on Schön's work is the notion that reflection focuses attention on how teachers construct knowledge in the context of action, the consequences of their knowledge for practice, and how practice leads to changes in teacher knowledge. Alternatively, Fenstermacher (1994) described research dealing with knowledge of teaching based on Shulman's (1986) categories of knowledge required for teaching, which is collectively referred to as *pedagogical content knowledge* (PCK). A final research focus described by Fenstermacher (1994) is the *teacher-as-researcher* characterized in the work of Cochran-Smith and Lytle (1993, 1999). This work stressed the importance of the contributions of teachers to knowledge about teaching and is based on the belief that teachers are best positioned to propose questions about teaching and learning and to investigate those questions.

The central role of teachers as generators of their professional knowledge is highlighted by the authors of chapters 13 and 14. van Zee (chap. 13) describes how she helped her elementary methods students develop understandings about scientific research by investigating their own questions about the moon before challenging them to research their own teaching. Similarly, Akerson and Roth McDuffie (chap. 14) recount how they mentored elementary teachers as they researched their own teaching. The authors of chapters 13 and 14 describe challenges associated with guiding elementary teachers in open inquiry and note the positive influence of former students who are able to speak about their own research experiences to reluctant first-time investigators. In a similar vein, Bryan and Tippins (chap. 16) describe an instructional program that combines the use of teacher stories, in the form of case-based pedagogy and reflection, to elicit and further develop teachers' knowledge. Their program is firmly based on the ideas that teachers construct and reconstruct knowledge from their experiences, and that narrative can serve as a vehicle for knowledge construction as well as knowledge expression. Whereas these authors focus on the means for generating teacher knowledge, other authors focus more on the content and contextual aspects of teacher learning.

In chapter 3, Appleton's description of pedagogical content knowledge as a framework for elementary science teacher preparation is clearly influenced by Shulman's (1986) categories of teacher knowledge. Appleton sees pedagogical content knowledge to be an essential knowledge base elementary teachers must have in order to construct and implement science lessons. Olson joins with Appleton (chap. 8) to explore the sorts of decisions that challenge science teacher educators as they work to design programs to prepare prospective elementary teachers for science teaching. Abell's (chap. 5) discussion of situated learning theory provides a theoretical rationale for creating meaningful science field experiences for elementary teachers. An important perspective shared by this group of authors is that those involved in designing teacher education programs should avoid the assumption that a teacher training (i.e., teacher-proof) curriculum can simply be developed in a fixed sequence in order to ensure the production of highly qualified teachers. Rather, they consider that planning for teacher preparation ought to represent teaching as a decision-making and contextually situated practice.

Key Questions

1. Besides the four research programs described by Fenstermacher, what others have led to substantial changes in science teacher education? Is there a place for the results of process-product teacher education research in the education of elementary science teachers?

2. What is the role of case-based pedagogy in the context of science teacher education? How should science teacher educators develop skills that will enable them to facilitate case-based discussions intended to promote teacher learning?

3. Which do you believe might have more detrimental impacts on elementary students' science learning—a teacher having limited content knowledge or a teacher lacking pedagogical knowledge in science?

4. What approaches might be most beneficial when *initiating* elementary teachers into inquiry about science teaching and learning in their own classroom?

5. What experiences might prepare those (e.g., university faculty, classroom teachers, prospective teachers) involved in elementary teacher education to negotiate productively, and learn from dilemmas arising in the midst of prospective teacher experiences (e.g., classes, field-based experiences)?

THEMATIC CONNECTION 2:
DISCIPLINARY DEPTH AND BREADTH

A wide range of perspectives is offered regarding the breadth and depth of science understandings needed by elementary school teachers. Whereas all seem to agree that a teacher's science knowledge directly affects students' opportunities to learn science, the authors express different opinions about what the nature of the science knowledge should be and the kinds of experiences that will facilitate the development of knowledge useful for science teaching. Some call for more authentic or accurate representations of the nature of science and scientific work, others contemplate the need for more science content preparation, and still others portray the interdisciplinary nature of elementary teaching and the placement of science with an interdisciplinary school curriculum.

Rather than settling the debate about the nature of the science knowledge needed by elementary teachers, the science standards documents prepared in many countries during the decade of the 1990s seem to have raised new questions and encouraged greater intellectual exploration of the knowledge base for science teaching. For example, the U.S. *National Science Education Standards* (National Research Council, 1996) called for elementary school teachers to "develop a broad knowledge of science content in addition to some in-depth experiences in at least one science subject" (p. 60). These standards stressed the need for all teachers of science to understand the nature of scientific inquiry, use scientific understandings to deal with personal and social issues, and make conceptual connections between science and other school subjects. They also challenged scientists and science educators to engage both prospective and practicing teachers in inquiry-based learning experiences that should serve as models for school science. And, like other guiding documents, the standards did not limit their focus to learning science, but went further to include guidance about how teachers learn to teach science and how they develop the understandings and skills of lifelong science learning. Regarding the issue of content breadth versus depth, however, the standards were less directive, merely recommending that breadth and depth of teacher science learning should be debated and decided at the local level.

Local leadership in deciding issues of disciplinary depth and breadth has resulted in elementary education program and professional development efforts that emphasize different dimensions of science subject matter knowledge. The focus might be traditional science facts, concepts, and theories, or on the substantive and syntactic knowledge of science (Grossman, Wilson, & Shulman, cited in Anderson & Mitchener, 1994). Grossman and her colleagues viewed substantive knowledge as including the organizing frameworks and dominant structures that guide inquiry and data interpretation in a field as well as those that prevailed in the past. Examples of substantive knowledge in modern science include the sun-centered solar system, the germ theory of disease, and continental drift. The embrace of these ideas by the scientific community led to the rejection of others considered to be traditional teachings of their day. According to the same authors, syntactic knowledge focuses on learners understanding how new knowledge is incorporated into a subject area, and in science this occurs through inquiry. In addition, some local decisions tied to educational reform initiatives emphasize teacher beliefs about subject matter and encourage teachers of science to think broadly about the interrelationships between science and technology and between science and other school subjects (Anderson & Mitchener, 1994).

The chapters prepared by several authors are associated with different dimensions of science subject matter knowledge. For instance, traditional science content knowledge is a central element in Appleton's discussion (chap. 3) about the development of elementary teachers' science pedagogical content knowledge. He looks at science content knowledge components of formal science knowledge, everyday knowledge, and perception of the nature of science as the foundation for building science PCK. Teachers learning substantive knowledge and syntactic knowledge are emphasized in chapters by van Zee (chap. 13) and Akerman and Roth McDuffie (chap. 14), with inquiry serving as the vehicle for learning about science phenomena and personal teaching practice. Elementary teachers' beliefs about science are addressed in Wieseman and Moscovici's discussion (chap. 10) about the place of science in interdisciplinary teaching approaches. They describe a *standards-infused curriculum* in which science is one of several subjects taught through problem- or theme-based curricula. Similarly, Prain and Hand (chap. 9) promote an interdisciplinary strategy for learning science. By showing overlap between science standards and language standards, they build a case for the greater involvement of students in using language as a way to construct science explanations. Jones (chap. 11) brings yet a different perspective of breadth to suggest that elementary teachers extend the content of the school curriculum to address technological literacy, differentiating between technology and science, and suggesting that the curriculum include investigations of social issues that involve both science and technology.

Authors highlighted under this second thematic section pose an exciting range of possibilities for engaging children in science learning—through the breadth of interdisciplinary learning experiences and using language as a medium for meaning making, and spanning beyond the classroom to make science relevant to the contexts of everyday life. The practices of today's teachers, however, are constrained by accountability policies that expect teaching to be aligned with standardized national and state curriculum frameworks, and for teachers to use assessment strategies to comparatively document the extent to which science learning is taking place for individuals, and across groups of students. In the mix of teachers' efforts to plan for contexts of science learning, spanning the sorts of "breadth and depth" experiences described earlier, they must simultaneously make challenging decisions about what, when, and how to assess students as individuals and/or whether to develop collective accounts of learning. Thus, this section concludes by calling attention to Harrison's (chap. 12) concerns about educators being able to balance practices of assessment in ways that promote meaningful learning for all students—for elementary children and prospective teachers alike.

Key Questions

1. How is science pedagogical content knowledge different for elementary teachers as compared to secondary science teachers? For elementary teachers, what experiences would help them develop *interdisciplinary* pedagogical content knowledge?

2. How does teaching science as part of an integrated curriculum affect the teacher's need for science content preparation?

3. What is the relation between science learning and language learning? How can teachers' understanding of the language standards help them teach science?

4. What sorts of social issues involving technology and science might concern elementary children?

5. What decisions might influence when an elementary teacher would place greater emphasis on *assessment for learning* or on *assessment of learning*?

6. What practices best support assessment of learning science—when science is taught as the sole content focus, as an interdisciplinary practice (e.g., language, technology), or with an intent to relate science to everyday contexts beyond the classroom?

THEMATIC CONNECTION 3: THE PLACE OF ELEMENTARY SCIENCE

Over the past four decades, researchers have persistently attempted to address problems unique to elementary science education. Much of the literature concerning elementary science education has focused on issues, including the lack of science teaching taking place in elementary classrooms, elementary teachers' lack of science content knowledge and understanding of the nature of science, and elementary teachers' tendency to have negative attitudes toward science. More recently, science education researchers have considered sociocultural, historical, and epistemological aspects that uniquely characterize elementary science teachers and teaching. Accordingly, the third theme explores "elementary" as having a distinctive place within the science teaching and learning community.

Koch (chap. 6) discusses ways learning theories and pedagogical approaches have translated into unique practices and dilemmas within the context of elementary science teacher preparation and classroom teaching. She stresses that whereas phrases such as *project-based science, inquiry science*, and *constructivist practice* have been introduced across all levels of science education, these notions hold unique implications in terms of preparing elementary teachers to adopt these as referents for their own science learning and classroom practices. Koch characterizes constructivist inquiry enacted in the elementary classroom as requiring "wiggle room"—an interactive space where teachers and students can "struggle" together in science meaning making. Encouraging prospective teachers to adopt such an approach to teaching requires intensive modeling, and a personal willingness to diverge from a traditional teacher-centered role. Assisting prospective teachers to change their images and practices of science teaching is challenging, and as McGinnis (chap. 15) points out, elementary science enthusiasts meet even greater challenges once they become classroom teachers.

McGinnis provides an opportunity to follow up the experiences of an elementary science teacher who has received the sort of constructivist inquiry orientation to science education as described by Koch (chap. 6). McGinnis begins his chapter by describing elementary schools as "sociocultural systems"—a workplace "which translates between the classroom and community contexts" (Page, 1991, p. 42). Drawing on the experiences of a new elementary teacher, "Ms. Susan Lee," McGinnis describes the successes and difficulties she experiences teaching inquiry-based science. Within the elementary teaching community, peer teachers and administrators hold particular views about the place of science in the elementary curriculum—overall there is little support for science teaching and learning at this level of education. Accordingly, he argues that researchers need to develop a more nuanced understanding of science education as a cultural practice within the elementary school workplace.

Fleer (chap. 7) appears to share McGinnis' perspective that alternative ways of representing and evaluating science education as a practice involving young children are needed. She maintains that learning theories espoused in teacher education courses and in the literature do not reflect the experiences of those who have worked with very young children. She attributes this problem to the unavailability of experienced early childhood teachers to provide instruction to prospective early childhood teachers. However, a larger problem stems from the lack of recognition that early childhood does, indeed, require science teaching approaches that are unique to this educational level. For example, individual "interviews about incidents" might inadequately allow children to express their ideas, whereas nonconventional approaches such as singing, artwork, or storytelling in the sand might be more conducive toward exploring young children's science ideas. These kinds of discrepancies create teacher learning borders for those involved in early childhood science education. Fleer calls for early childhood science educators to build their own community of practice to inform science teaching and research at this level of education.

Key Questions

1. In what ways might prospective elementary teachers react to the metaphor of science teaching and learning as a "struggle"? Would their reactions resemble those of a prospective secondary teacher? Why or why not?

2. How might we describe the place of science education as it might be situated within the "culture" of an elementary school? What would the ideal culture for elementary science education look like?

3. What sorts of "cultural discontinuities" might new elementary teachers perceive when they enthusiastically begin teaching science in their elementary school workplace? What approaches might be taken to prepare them to successfully navigate such cultural gaps?

4. Fleer argues that the current research base is limited in terms of addressing early childhood science education due to inadequate methodological and theoretical research orientations. What methodological and theoretical approaches might be more appropriate for exploring the science experiences of teachers and young children in classrooms?

THEMATIC CONNECTION 4:
ORGANIZING THE WORK OF ELEMENTARY SCIENCE

Several chapters address issues of elementary science education that could be interpreted as having *organizational* features. The rise of organizational conceptions underpinning public education, for the most part, accompanied the industrial revolution at the turn of the 20th century. The emergence of competitive marketing, with its need for mass production through efficient means, created tremendous interest in systems management and productivity. Results of production needed to be predictable, thus management evolved into a bureaucratic design to reinforce the division of labor, hierarchical authority, adherence to specified work procedures, and rewards based on technical competence. Accordingly, classical organizational theory assumed a highly mechanistic, technocratic view of the world. For example, Taylor (1911), whose time-and-motion research enabled greater efficiency in assembly-line production, reflected his desire to enable controlled and standardized productivity in manufacturing processes.

Organizational theorizing has shifted a great deal, especially as the role of human relations configured into understanding the workplace. In the 1950s, the human relations movement called attention to the ways human needs and behavior shaped business operations. Needs—such as worker appreciation, democratic processes of decision making, and group cohesiveness—were recognized as having significant impacts on worker motivation and performance. More recently, organizational theorists have developed sociocultural analyses that consider human lifeworlds as constitutive of organizations. Senge's (1990) conceptions of "learning organizations" and "systems thinking," for example, forged new directions in thinking about the workplace as a social system of human learning and change:

> Systems thinking makes understandable the subtlest aspect of the learning organization—the new way individuals perceive themselves and their world. At the heart of a learning organization is a shift of mind—from seeing ourselves as separate from the world to connected to the world, from seeing problems as caused by someone or something "out there" to seeing how our own actions create the problems we experience. A learning organization is a place where people are continually discovering how they create their reality. And how they can change it. (pp. 12–13)

"Systems thinking" has been adopted by education reformers; however, the systemic metaphor can readily be interpreted in mechanical as well as sociocultural terms. Thus, systemic reform, coupled with the language of "standards-based" education, can potentially bring a sense of incoherency in efforts to enhance elementary science education community (Donmoyer, 1995).

Many authors herein clearly describe challenges of elementary science education as complex human matters. Flick (chap. 2) and Olson and Appleton (chap. 8) present cases (fictional and nonfictional) to describe the nature of problems associated with designing elementary science teacher education programs. These cases holistically illustrate the challenges of program design as intricate and interrelated dilemmas—for which the authors offer no simple solutions. In a similar manner, Jones and Edmunds (chap. 17) portray complex situations that complicate efforts to improve elementary science teaching in their study of three models involving use of elementary science teaching specialists. The results of their research provide insights about the systemic ways elementary teachers' science expertise and attitudes influence science learning opportunities in schools. Finally, as Boone (chap. 4) shares technical insights concerning use of distance education technologies, he underscores the critical role of the science teacher as a mediator of student learning. Elementary science education is a human endeavor, and as such, requires those working within the "organization" to reframe and enact changes needed to enhance science learning for all.

Key Questions

1. Will standards and accountability approaches ensure productive elementary science teaching and learning? What alternative approaches might be used to motivate the work of teachers and students toward a more productive elementary science education?

2. Think about drawing an "organizational model" of professional elementary science education. What might this model look like? Would this model have the same features if drawn as an "elementary science teaching community"? What would an analysis of these model depictions suggest about underlying beliefs concerning who holds the knowledge and power to make changes in elementary science teaching and learning?

3. What would an ideal elementary science teacher education program look like? What are the qualities of an ideal elementary science teacher educator?

4. Should elementary schools have science specialists to ensure science is regularly taught at the elementary level?

5. Should distance education programs used for prospective elementary science teacher education have qualities that would distinguish them from secondary science teacher program of study? Why or why not?

EXPANDING CONVERSATIONS
ABOUT ELEMENTARY SCIENCE

The voices of teachers, teacher educators, and researchers represented in this book reflect a wide range of perspectives concerning issues and approaches taken toward enhancing elementary science education. In this introduction, we have attempted to bring together their various stances as conversations conjoined within four themes. There are reminders, however, that the challenges of making elementary science teaching meaningful are complex. Prospective and practicing elementary teachers need opportunities to rethink what it means to learn and do science inquiry, and to be willing to reposition themselves within the community of young science learners in a classroom. Elementary schools need to think deeply about the place of science within the curriculum—especially as more emphasis has been given to teaching reading, writing, and mathematics—nearly to the exclusion of all other subject areas. Although a plethora of research literature and curricular resources has been generated with the intent of improving elementary science education, it appears there are gaps in terms of our understanding contexts that challenge elementary science teacher education and practice. Also, as reiterated by several authors, there is the need to involve a broader community of practitioners (viz., elementary classroom teachers) in research to better inform studies of elementary science education. Ultimately, it is our hope that the themes and key questions we have presented here might help to expand the conversations about elementary science and bring new insights to benefit young science learners, their teachers, and families.

REFERENCES

Anderson, R. D., & Mitchener, C. P. (1994). Research on science teacher education. In D. Gabel (Ed.), *Handbook of research on science teaching and learning* (pp. 3–44). New York: Macmillan.

Bolman, L. G., & Deal, T. E. (1984). *Modern approaches to understanding and managing organizations.* San Francisco: Jossey-Bass.

Clandinin, D. J. (1992). Narrative and story in teacher education. In T. Russell & H. Munby (Eds.), *Teachers and teaching: From classroom to reflection* (pp. 124–137). Bristol, PA: Falmer.

Cochran-Smith, M., & Lytle, S. L. (1993). *Inside/outside: Teacher research and knowledge.* New York: Teachers College Press.

Cochran-Smith, M., & Lytle, S. L. (1999). The teacher research movement: A decade later. *Educational Researcher, 28*(7), 15–25.

Connelly, F. M., & Clandinin, D. J. (1996). Teachers' professional knowledge landscapes: Teachers stories–stories of teachers–school stories–stories of schools. *Educational Researcher, 25*(3), 24–30.

Donmoyer, R. (1995). The rhetoric and reality of systemic reform: A critique of the proposed national science education standards. *Theory Into Practice, 34*(1), 30–34.

Fenstermacher, G. D. (1994). The knower and the known: The nature of knowledge in research on teaching. *Review of Research in Education, 20*, 3–56.

National Research Council. (1996). *National science education standards.* Washington, DC: National Academy Press.

Page, R. N. (1991). Kinds of schools. In K. Borman (Ed.), *Contemporary issues in U.S. education* (pp. 38–60). Norwood, NJ: Ablex.

Schön, D. A. (1983). *The reflective practitioner.* New York: Basic Books.

Schön, D. A. (Ed.). (1991). *The reflective turn.* New York: Teachers College Press.

Senge, P. (1990). *The fifth discipline: The art and practice of the learning organization.* New York: Doubleday.

Shulman, L. S. (1986). Those who understand: Knowledge growth in teaching. *Educational Researcher, 15*(2), 4–14.

Taylor, F. W. (1911). *The principles of scientific management.* New York: Harper & Row.

I

CONCEPTUALIZING ELEMENTARY SCIENCE TEACHER EDUCATION

This section deals with different aspects of the conceptualization of elementary science teacher education. In chapter 2, Flick looks at the complex nature of working in academia in elementary science teacher education, raising key questions about how to equip prospective professors for their workplace situation. Appleton, in chapter 3, discusses science pedagogical content knowledge (PCK) for elementary teachers, considering its relation to other forms of teacher knowledge and how activities that work serve as both a focus for PCK, and a means of communicating it.

With the increasing availability of information technologies, there is an increasing trend toward delivery of teacher education courses by distance. In chapter 4, Boone recounts what he has learnt from his own experiences in using information technology to deliver inservice courses in elementary science, including some of the pitfalls and advantages.

Another increasing trend has been the association of field experiences with preservice methods courses in elementary science, often in addition to the normal practicum requirements of a program. Abell discusses in chapter 5 how she has incorporated field experiences into her courses, examining some of the different options available and their relative effectiveness.

Koch examines learning theories, another major component of both preservice and inservice courses. Chapter 6 summarizes the main thrust of current trends in thinking about constructivism in science education, and considers how prospective teachers need to grasp these ideas, as well as how they need to be incorporated into the pedagogy within a methods course. Fleer's final chapter in this section raises the issue of how science education for those intending to teach in the early grades has suffered from limited research and conceptualization, particularly in terms of pedagogies appropriate to younger children. This, in turn, causes problems in elementary science teacher education courses, because ideas and materials relevant to early grades tend to be limited.

2

Being an Elementary Science Teacher Educator

Lawrence B. Flick
Oregon State University

Being an elementary science teacher educator involves a complex mix of working with preservice and inservice teachers; institutional colleagues in science, mathematics, and education; representatives of state department bureaucracies; and doctoral students preparing to be elementary science teacher educators. The elementary science teacher educator functions as a liaison between faculty in departmentalized disciplines in higher education and teachers whose responsibility is curriculum that spans several subject areas. Functioning in this role and preparing doctoral students to become new colleagues carries a unique set of demands and responsibilities that differ from those of traditional science educators and high school teachers. This difference is underscored by the fact that the general term *science teacher educator*, by tradition, has meant someone who majored in a science and works with teachers in departmentalized settings. Someone who works with elementary teachers in self-contained educational settings is known as an *elementary* science teacher educator. To examine the demands on and requirements for being an elementary science teacher educator, this chapter uses the device of discussing the work of a hypothetical faculty member. Whereas elementary science teacher educators may work in a variety of contexts and have differing educational backgrounds, this person is placed in a traditional university setting with a teacher education program and doctoral program in science education.

Consider a hypothetical faculty member of a university elementary education program specializing in science, Dr. Gordon. According to Weiss

(2001, p. 8), 92% of Grade K–4 teachers and 77% of Grade 5–8 teachers are female. However, only 40% of the teacher educators surveyed by Goodlad (1990) were female. For the purposes of this discussion, consider Dr. Gordon to be a female, representative of the teaching population with whom she works.

This chapter outlines the kinds of responsibilities Dr. Gordon has when working with higher education faculties, teachers and prospective teachers, and doctoral students. It takes an in-depth look at the graduate education of elementary science teacher educators in terms of teaching experience, knowledge of and about science, human development and cognition, and research skills. A position statement of the Association for the Education of Teachers in Science (AETS) identifies six professional knowledge standards: knowledge of science; science pedagogy; curriculum, instruction, and assessment; knowledge of learning and cognition; research and scholarly activity; and professional development activities (AETS, 1997). These standards, along with Fey's (2001) analysis of doctoral studies in mathematics education, form complementary frameworks for the task. Fey argued that educators in the fields of science and math education find themselves in a wide variety of roles, but all graduates should be prepared to work with teachers. A similar argument can be made about the varied roles of elementary science teacher educators. Each role would require the individual to possess knowledge of and about science, as well as to have experience in and extensive knowledge of elementary teaching environments.

WORKING WITH FACULTY

The colleagues in her elementary education program will expect Dr. Gordon to promote interest, enthusiasm, and success in learning and teaching science. However, most of her preservice students have avoided taking science and math classes. About 90% of Dr. Gordon's undergraduate students are majoring in some education field other than science or math. Far fewer than 10% are pursuing a major in science or math (Weiss, 2001, p. 9). Faculty in the departments of science and mathematics on campus will expect Dr. Gordon to champion additional subject matter requirements from their departments despite the fact that the courses available to non-majors are often taught in lecture-dominated formats where content is unconnected to situations meaningful to elementary teaching. Meetings with science and math faculty often confront the fact that about one half of the prospective elementary teachers take fewer than six semesters of science and almost half of these individuals will not take any physics or chemistry. The math faculty are only mildly appeased by the fact that virtually all students (96%) take math for elementary teachers, but most will take no more math courses. Dr. Gordon, herself, is aware that only about one half of the class will meet the National Science Teachers Association's course background standards (NSTA, 1998).

Dr. Gordon, herself, faced many challenges in developing her own science knowledge. It was difficult finding science courses that were both appropriate for her knowledge needs and would apply toward her own doctoral degree. The National Science Foundation has only recently focused on faculty and course development to broaden the scope and appeal for prospective teachers (NSF, 1996, 1997). AETS has advocated that beyond coursework, science teacher educators should have "active inquiry/research experiences within his/her discipline preparation in at least one science discipline and a strong functional knowledge in several other science disciplines" (AETS, 1997). Science inquiry experiences are not a typical component of science courses, therefore opportunities are usually provided by externally funded projects that team scientists with teachers. Whereas teachers report these experiences as valuable, they are often of short duration and their long-term effects remain unexamined (Caton, Brewer, & Brown, 2000). Some projects of this type are examining more in-depth experiences spanning more time (Gummer, 2002; Schwartz, Lederman, & Crawford, 2000).

WORKING WITH TEACHERS

In addition to her university colleagues, Dr. Gordon must work with practicing teachers and school districts on critical issues such as implementation of state and national standards and confronting an increasing emphasis by federal and state governments on required multiple choice achievement tests. Teachers are also concerned about how to incorporate standards-based, inquiry-oriented instruction into an already crowded curriculum that is constrained by multiple reform activities and state assessment policy. Dr. Gordon is current with the research literature, indicating over 60% of practicing teachers are not aware of the National Science Education Standards and, of those who are, there is only weak agreement about their value (Weiss, 2001, p. 19). She not only has a challenge communicating the value of national standards to teachers, but also to administrators and parents whose attention is captured by news of failing schools and national testing agenda.

Professional colleagues expect Dr. Gordon to stay knowledgeable of the issues driving reform. Growing knowledge about how people learn

(Bransford, Brown, & Cocking, 2000; Lambert & McCombs, 1998) is having significant impact on how to analyze and structure classroom instruction and assessments. The direction to be taken, however, is rarely clear-cut and Dr. Gordon must evaluate and synthesize divergent implications from educational psychology. There are avid proponents that imply teaching in science (and other subjects) should be student centered, guided by student interest, motivation, and what they already know (Tobin, Tippins, & Gallard, 1994). At the same time, there are proponents who say that teacher scaffolding of instruction is critical for reaching all students through explicit teaching of learning strategies (Lambert & McCombs, 1998). Elementary teachers are faced with teaching multiple subjects and coping with shifting roles in different subject matter contexts. Dr. Gordon must navigate among these different points of view as she helps elementary teachers integrate content understanding while building an effective blend of pedagogy and content knowledge, what Shulman (1987) called "pedagogical content knowledge," in each of several content areas. This involves helping elementary teachers develop what Borko (1991) identified as an "overarching conception" of what it means to teach a (particular) subject. Dr. Gordon recognizes that this is an intellectual challenge different in kind but no less rigorous from single-discipline conceptions of teaching subject matter.

Developing a concept of what it means to teach science is a long-term, developmental process for elementary teachers. Because her interest in science led to a concentrated effort to develop science background, skills in teaching, and knowledge of relevant professional literature, Dr. Gordon has a concept of teaching science that she communicates to teachers in a variety of professional development activities. Her concept of teaching science has been informed by direct work with gifted and reflective teachers as well as her own scholarship. Dr. Gordon has studied a number of issues relevant to elementary science education and designs course content and professional development programs from this research base. For example, she is prepared to guide critical examination of the possible limitations of young learners in developing the epistemological commitments necessary to understand science as inquiry. She provides teachers with research-based examples of constructivist pedagogy that has demonstrated how elementary students can develop sophisticated understandings of how scientists add to scientific knowledge (Smith, Maclin, Houghton, & Hennessey, 2000). She has developed research-based practices that teachers can learn to better guide and develop student discourse in science. She demonstrates how research has helped teachers conceptualize the role of peer groups and teacher scaffolding to guide student development of high quality scientific explanations (Hogan, Nastasi, & Pressley, 2000). More broadly, she has examined the nature of scientific literacy as it applies to elementary classrooms. She helps teachers interpret research to aid them in facing significant challenges in supporting science learning for cultural and language minority students (Anderson, Holland, & Palincsar, 1997).

Working with teachers is one of the central roles of the elementary science teacher educator. The first five professional knowledge standards of AETS build support for the sixth and final standard that states the principle: Elementary science teacher educators know the work of elementary teachers, the trajectory of their development, and the major milestones of knowledge and skill toward becoming a master teacher (AETS, 1997). This principle poses significant challenges to doctoral programs that work with students with varying backgrounds: those with deep knowledge of teaching and those with deep knowledge of science. These two characteristics are not mutually exclusive but they are not regularly coincident.

BECOMING AN ELEMENTARY
SCIENCE TEACHER EDUCATOR

Operating between her professional activities with teachers and her engagement with a professional community of science educators is the responsibility for preparing new elementary science teacher educators. Dr. Gordon is always motivated by the appearance of an elementary teacher who has a strong interest in science and who is interested in an advanced degree. This does not happen often, but when it does, it raises important questions about the doctoral program and elementary science education. Elementary science teacher educators must fill a variety of roles in their university setting and broader professional responsibilities. Fey's (2001) analysis of doctoral programs shows that math education doctoral programs are challenged to combine sufficient knowledge of and experience in teaching with discipline-based knowledge. Elementary science teacher education programs must meet the converse of this challenge. The elementary science teacher educator must combine sufficient knowledge of science with teaching experience. Fey's (2001, p. 55) analysis of doctoral studies in mathematics education is a useful guide to considering recurrent issues in elementary science education, posed as the following questions:

1. What is the purpose of doctoral programs in elementary science education? What are the professional roles that doctoral students are preparing for?
2. How can students be recruited to appropriate programs and placed in appropriate professional positions when graduated?

3. What knowledge, abilities, and dispositions do those students need to acquire?
4. How can graduate programs in elementary science education provide the essential knowledge and personal development?
5. How can graduate programs assess the competence of their doctoral candidates?

One of the first questions Dr. Gordon asks of a new doctoral applicant is, "What do you intend to do with this degree when you are finished?" There are two responses to this question and each poses its own challenge to a doctoral program with respect to Questions 1 and 2. One response is that the applicant wants to expand on a teaching background at the elementary level to help teachers understand and teach science. This applicant often presents extensive teaching experience but minimal academic preparation in science. The second response is from an applicant who has a bachelor's or master's degree in science but has not taught at the elementary level. Their teaching experience may have been only at the college level. They would like to work on "special projects" in curriculum or professional development with elementary teachers to share their knowledge and enthusiasm for science. Both applicants have roles to play in elementary science teacher education, but both have different needs from a doctoral program.

Doctoral programs in education traditionally prepare students to work in research-oriented university education programs. Faculty members in these positions are expected to conduct original research, pursue external funding, and maintain professional involvement with national and international organizations in their field. However, there are many other jobs besides traditional research positions for people with a doctorate in education. These include working for a school district as a staff development or curriculum expert, working in a state department of education as a science specialist, leading or participating in program evaluation, or working at the classroom level as an instructional leader for the school or district (Abell, 1990). These diverse positions influence the options provided in doctoral programs and how students are recruited and eventually guided to employment.

Academic preparation of new doctoral students in elementary science teacher education must address two major areas of knowledge, abilities, and dispositions: knowledge and experience in teaching and knowledge and experience in science. They represent the two strengths and two weaknesses of the prototypical candidates described earlier. One was strong in teaching experience but relatively weak in science knowledge. The other was just the reverse. The problem of addressing these areas of knowledge and skill is exacerbated by constraints of expertise within the elementary education faculty and small numbers of students. The science and mathematics departments at most institutions are geared toward the development of their own doctoral students and therefore do not offer courses for the development of appropriate knowledge for an elementary science educator. Teacher education programs with doctoral tracks are similarly constrained by the kinds of teaching experiences they can offer. Doctoral students with elementary teaching experience can be qualified as supervisors for student teachers. Those without teaching experience have few options for getting exposure to the complex roles of an elementary teacher. Yet, developing such background is essential for anyone who will be qualified to work in elementary science education.

Teaching Experience

Discussions at the National Conference on Doctoral Programs in Mathematics Education concerning the role of teaching experiences offer ideas useful for this discussion. "The majority of attendees were adamant that *all mathematics educators must be prepared as teacher educators*—not only knowledgeable about teaching and learning issues, but also competent in communicating with school teachers and administrators and informed about how schools function" (Lambdin & Wilson, 2001, p. 79). This position seems even more important in elementary science education where knowing the intensity and complexity of elementary classroom teaching is prerequisite to understanding science education at this level. The general point made by discussants at the conference applies directly to the elementary context. The educational community will expect anyone with a degree in elementary science education "to be prepared as a teacher educator, regardless of whether the student plans to work as a teacher educator in the future or not" (p. 79).

Preparing all elementary science educators in teacher education will require that the doctoral program provide options for gaining teaching experiences. The kinds of experience deemed appropriate will depend on the student's career goals and their teaching background. Teaching in a major research university teacher education program is different from a program in a small college. However, preparing for teacher education means that a person with strong science credentials but sparse teaching background needs to engage in sufficient teaching experiences at the elementary level to meet the assumption by the education community that this person appreciates the work of elementary teachers. Even a doctoral student presenting some elementary teaching experience raises the question of whether or not their experience is appropriate or sufficient. Often graduate programs view any teaching as good enough as long as it meets some minimal amount. But, is elementary teaching experience in a private school with small classes, supported by educated and resourceful parents, the kind of background needed for a contemporary teacher educator? Ideally, a doctoral student would get some direct experience teaching elementary students under the supervision of an experienced and skilled elementary teacher. However, other experiences could include observing classrooms as part of a specially designed course or as part of a professional development or research program. All of these experiences must be embedded in a program of supervision, reflection, and assessment.

Knowledge of and About Science

The applicant who has elementary teaching experience but is in need of developing science content knowledge poses a different set of challenges. Finding science courses that develop science knowledge for elementary teachers is a perennial problem. Science departments tend to offer courses that favor students majoring in science, so their introductory courses, accessible to elementary educators, do a poor job of presenting the most valuable aspects of the discipline. The National Science Foundation has outlined many of the most glaring shortcomings (NSF, 1996). However, science and mathematics departments have begun to respond to demands to improve courses for non-majors in general and prospective teachers in particular. Over the last dozen years, the NSF has funded 32 Collaboratives for Excellence in Teacher Preparation. The purpose of the NSF Collaboratives was "to improve significantly the science, mathematics, and technology preparation of future K–12 teachers and their effectiveness as educators in these areas" (NSF, 1997). Projects used a variety of approaches that included collaborations among science, math, and education departments as well as among institutions within a state. One NSF collaborative developed a set of indicators to assist faculty in evaluating their course modifications with respect to their value for prospective teachers. The indicators listed in Fig. 2.1 were evaluated by science, mathematics, and education faculties of various institutions before they were employed as a self-evaluation tool for course modifications. These broad recommendations are consistent with recommendations for changes in science education at the collegiate level (Committee on Undergraduate Science Education, 1999).

Further, Dossey and Lappan (2001) recommended that the elementary math educator have a command of mathematics equivalent to the first 3 years of undergraduate work in mathematics, the equivalent of a strong undergraduate minor in mathematics. Their criterion was that the elementary math educator should be able "to engage in conversation with teachers over the full range of K–12" (Dossey & Lappan, 2001, p. 68).

Indicators for Selection of Mathematics and Science
Content Courses Appropriate for Future Teachers

Characteristics of the Course:

- National and/or state Standards are incorporated in course design. (National Council of Teachers of Mathematics Standards, National Science Education Standards, AAAS Benchmarks, and/or Oregon Content Standards)
- Hands-on activities (laboratory experiences and/or use of manipulatives) constitute an integral part of the course.
- Opportunities are provided for students to learn about and engage in inquiry.
- Instruction is designed to encourage conceptual development through the use of a variety of methods, activities, resources and educational technologies.
- Course content integrates relevant issues of science, mathematics and society.
- Lecture portion of course is closely coordinated with laboratory, discussion and/or recitation sections.
- Course grades are based on a variety of evaluation methods including authentic assessment.
- Opportunities exist for connections to the K–12 classroom environment.

Characteristics of the Instructor:

- Engages students interactively in instruction.
- Takes student prior knowledge into account when planning for instruction.
- Promotes a sense that all students can succeed in the course.
- Models thinking and study skills important for succeeding in the course.
- Emphasizes the value of science, mathematics and technology for all people of all ages.
- Models an enthusiasm for an inquiry orientation to learning.
- Is familiar with K–12 classrooms and teachers.

FIG. 2.1. Indicators developed for instructor self-evaluation by participants in the Oregon Collaborative for Excellence in the Preparation of Teachers (1999), National Science Foundation (DUE/9996453).

The aforementioned recommendation is appropriate but it is impractical for elementary science educators to know all fields of science taught through high school at this level. However, elementary science educators need to have sufficient knowledge in one science discipline that makes them able to discuss issues of school science content with teachers through the high school level.

The professional knowledge standards suggested by AETS (1997) devote the most verbiage to this issue. In terms of science content, the basic principle is that science educators should have more science knowledge than "mentioned in the reform documents and required for teacher licensure in a particular state . . . (and more knowledge than) the level at which their instruction is focused" (Standard 1.a). In addition, the science educator should have knowledge of science process skills and of the philosophy, sociology, and history of science "exceeding that specified in the reform documents" (Standard 1.c, 1.d). These broad prescriptions are qualified slightly for the elementary science educator by suggesting that their interests would be better served by more breadth than depth in subject matter content. Most colleges and universities will not have appropriate courses to support the graduate-level development of science content knowledge. However, innovative doctoral programs can support readings, seminars, and internships similar to programs for preservice teachers (Gummer, 2002; Schwartz et al., 2000). Until and unless science, mathematics, engineering departments, and schools of education can collaborate on the formal development of subject matter knowledge for elementary teachers and elementary science teacher educators, options will have to be created within the doctoral program.

Child and Human Development

Elementary science teacher educators understand learners whose development spans childhood to early adolescence. Learners in this age range are in a period of significant cognitive, social, emotional, and physical development. Keating's (1990) review of psychological research on childhood and adolescence, therefore, has implications for learning science and mathematics. A foundation in educational psychology prepares elementary science teacher educators to evaluate and reflectively respond to a range of issues within science education that cover curriculum, instruction, and assessment. Bybee (1991) commented on the importance of teacher educators maintaining an ongoing awareness of research so that they can critically review the latest report or research study. For example, grounding in major developments in the psychology of cognition and learning, as well as science education, was necessary for elementary science educators to evaluate the arrival of the National Science Education Standards (NSF, 1996) and fit it into a larger context of educational and psychological research.

A foundation in educational psychology is important for evaluating important, yet enduring, debates in the field. For example, the elementary science educator should help present the content and mediate discussion of the role of Piaget in the psychology of teaching and learning as debated by Kuhn (1997) and Metz (1995, 1997). Elementary science educators should be well versed on the background necessary to lead critical reflection on gender issues that continue to pervade science and mathematics education. These include knowledge about gender differences in intellectual performance, learning and memory, verbal performance, quantitative performance, visual-spatial performance, and creativity (Lips, 1993).

In order to teach subject matter effectively to children and adolescents, teachers have to be aware of numerous classroom interactions, anticipate problems, and negotiate solutions that promote learning. Elementary science educators need to understand the research behind the practices of effective teachers. Raizen and Michelsohn (1994) adapted the following guidelines from the National Board for Professional Teaching Standards to outline the qualities of an effective elementary teacher of science:

- Understanding the Learner-Centered Psychological Principles of the American Psychological Association (1997)
- Assessing fairness in student–teacher interactions
- Spotting bias and stereotyping in texts and other materials
- Providing a wide range of role models
- Analyzing school policies and practices that negatively impact students on the basis of gender, race, ethnicity, and religion
- Establishing a learning environment where all students can learn science

Elementary science teacher educators translate research from diverse fields into guidelines for educational practice. They are a primary means of making research accessible and understandable to teachers.

Research Skills and Knowledge

Scholarship at the doctoral level implies that there is a "field of study" that sets elementary science education apart from other fields such as elementary mathematics education. This is the issue raised by Fey's (2001) final two questions. The elementary science teacher educator stands at a crucial intersection between a rapidly developing learner and transformations of subject matter for instruction. This intersection has become of greater interest as reformers promote a deeper understanding of the empirical and psychological nature of scientific knowledge. Shulman and Quinlan (1996) noted the variable relation between psychology and subject matter but concluded that "we can anticipate the return of the psychology of school subjects to its former centrality in educational psychology" (p. 421). Psychology currently recognizes that problem solving and transfer are significantly affected by subject-matter-specific factors (Wittrock, 1998). A focus of research in teaching and learning is on the way teachers transform subject matter, as well as on the way learners interpret instruction and interpret subject matter (Mayer, 1998). Elementary science educators can reasonably identify a "field" of research and scholarship ripe with important questions for study.

However, preparation for research and scholarship, the mainstay of promotion and tenure in academe, is less clear for science educators than for scientists. Fey (2001) captured the issue in reporting from the National Conference on Doctoral Programs in Mathematics Education:

Much as we mathematics educators (or educators in general) would like respect as legitimate members of our university communities, with recognition of the worth in our scholarship, I believe that a professional field like ours is different from traditional academic disciplines. It requires different talents and different kinds of knowledge, and it makes different kinds of contributions to society. Ours is an *appropriately eclectic field*. It requires advanced study and skills comparable to those in other professional fields that award doctorates, but a narrow focus on research isn't the right thing. (p. 57)

Scholarly contributions of elementary science teacher educators are likely to span curriculum development, forming policy guidelines, teaching (transformed) science content to elementary teachers, teaching future elementary teachers, writing for the media and practitioner journals, and doing substantive empirical research.

Shulman and Quinlan (1996) predicted that psychology will blur the boundaries with other social sciences, such as sociology and anthropology. "It will be primarily a classroom- and community-based empirical science" (p. 421). A doctoral program in elementary science teacher education will prepare the student to work in this eclectic, scholarly environment. In addition to the topics already discussed in this chapter, the doctoral student will develop skills in contemporary forms of empirical research. This includes a grounding in both quantitative and qualitative methods. Faculty face many issues in delivering the kind of education outlined here. Elementary science teacher educators are spread throughout the country with few institutions having a critical mass that supports specialization of courses and faculty.

CONCLUSIONS

Elementary science teacher educators are scholars in a key position to develop knowledge about important questions of both an empirical and theoretical nature. What is appropriate science content for the elementary grades? What are appropriate cognitive skills to develop in elementary students that support learning in science? What kinds of and how much science should elementary teachers understand? What are appropriate relations between science and the rest of the comprehensive elementary curricula? What are appropriate assessments of child knowledge of and about science?

Elementary science teacher educators work with educational psychologists, scholars in the sciences and mathematics, colleagues in curriculum and instruction, master teachers, school districts, and state departments of instruction. College and university science education faculty must also be prepared to work collaboratively with the larger science, mathematics, and engineering communities utilizing innovative and complex models of professional development (Loucks-Horsley, Hewson, Love, & Stiles, 1998).

In addressing these challenges, there are many issues that Dr. Gordon deals with that have not been discussed here. For example, what is the role of an EdD degree? Her education faculty have been considering offering a professional doctorate for those educators who seek jobs where research is not the primary focus. This raises questions about coursework appropriate for two doctoral programs, the problem of staffing two programs, and considering what the dissertation or final project should look like and how it should be assessed (see Reys & Kilpatrick, 2001, part 2). Dr. Gordon also reflects on the value of offering opportunities for elementary science teacher educators to focus their work on critical age levels such as early childhood and early adolescence as well as the childhood years. Allowing such options puts further demands on an already limited faculty and expertise. Yet, the state is considering licensure options for early childhood, elementary, middle, and high school.

Being an elementary science teacher educator is uniquely challenging, standing between the subject matter demands of the science disciplines and the psychological and physical demands of working with young learners. The field of study is young and currently fueled by rapidly evolving knowledge in cognitive and developmental psychology. Preparing elementary teachers for teaching standards-based science requires attracting talented classroom teachers into doctoral programs who are motivated by and knowledgeable in science.

REFERENCES

Abell, S. K. (1990). A case for the elementary science specialist. *School Science and Mathematics, 90,* 291–301.

American Psychological Association's Board of Educational Affairs. (1997). *Learner-centered psychological principles: A framework for school redesign and reform.* Retrieved April 2003, from http://www.apa.org/ed/lcp.html

Anderson, C. W., Holland, J. D., & Palincsar, A. S. (1997). Canonical and sociocultural approaches to research and reform in science education: The story of Juan and his group. *Elementary School Journal, 97,* 359–383.

Association for the Education of Teachers in Science (AETS). (1997). *Position statement: Professional knowledge standards for science teacher educators.* Retrieved April 2003, from http://www.theaets.org/

Borko, H. (1991). The integration of content and pedagogy in teaching. In A. L. Gardner & K. F. Cochran (Eds.), *Critical issues in reforming elementary teacher preparation in mathematics and science* (pp. 25–45). Greeley, CO: University of Northern Colorado.

Bransford, J. D., Brown, A. L., & Cocking, R. R. (2000). *How people learn: Brain, mind, experience, and school.* Washington, DC: National Academy Press.

Bybee, R. W. (1991). The reform of elementary school science: Critical issues. In A. L. Gardner & K. F. Cochran (Eds.), *Critical issues in reforming elementary teacher preparation in mathematics and science* (pp. 13–23). Greeley, CO: University of Northern Colorado.

Caton, E., Brewer, C., & Brown, F. (2000). Building teacher–scientist partnerships: Teaching about energy through inquiry. *School Science & Mathematics, 100,* 7–16.

Committee on Undergraduate Science Education, Center for Science, Mathematics, and Engineering Education, National Research Council. (1999). *Transforming undergraduate education in science, mathematics, engineering, and technology.* Washington, DC: National Academy Press.

Dossey, J. A., & Lappan, G. (2001). The mathematical education of mathematics educators in doctoral programs in mathematics education. In R. E. Reys & J. Kilpatrick (Eds.), *One field, many paths: U.S. doctoral programs in mathematics education* (Conference Board of the Mathematical Sciences: Issues in mathematics education, Vol. 9, pp. 67–72). Providence, RI: American Mathematical Society.

Fey, J. T. (2001). Doctoral programs in mathematics education: Features, options, and challenges. In R. E. Reys & J. Kilpatrick (Eds.), *One field, many paths: U.S. doctoral programs in mathematics education* (Conference Board of the Mathematical Sciences: Issues in mathematics education, Vol. 9, pp. 55–62). Providence, RI: American Mathematical Society.

Goodlad, J. I. (1990). *Teachers for our nation's schools.* San Francisco: Jossey-Bass.

Gummer, E. S. (2002, January). *Reasonably rich environments in professional development experiences in scientific inquiry.* Paper presented at the annual meeting of Association for the Education of Teachers in Science, Charlotte, NC.

Hogan, K., Nastasi, B. K., & Pressley, M. (2000). Discourse patterns and collaborative scientific reasoning in peer and teacher-guided discussions. *Cognition and Instruction, 17,* 379–432.

Keating, D. P. (1990). Adolescent thinking. In S. S. Feldman & G. R. Elliott (Eds.), *At the threshold: The developing adolescent* (pp. 54–89). Cambridge, MA: Harvard University Press.

Kuhn, D. (1997). Constraints or guideposts? Developmental psychology and science education. *Review of Educational Research, 67,* 141–150.

Lambdin, D. V., & Wilson, J. W. (2001). The teaching preparation of mathematics educators in doctoral programs in mathematics education. In R. E. Reys & J. Kilpatrick (Eds.), *One field, many paths: U.S. doctoral programs in mathematics education* (Conference Board of the Mathematical Sciences: Issues in mathematics education, Vol. 9, pp. 77–84). Providence, RI: American Mathematical Society.

Lambert, N. M., & McCombs, B. L. (1998). *How students learn: Reforming schools through learner-centered instruction.* Washington, DC: American Psychological Association.

Lips, H. M. (1993). *Sex & gender: An introduction* (2nd ed.). Mountain View, CA: Mayfield.

Loucks-Horsley, S., Hewson, P. W., Love, N., & Stiles, K. I. (1998). *Designing professional development for teachers of science and mathematics.* Thousand Oaks, CA: Corwin Press.

Mayer, R. E. (1998). Cognitive theory for education: What teachers need to know. In N. M. Lambert & B. L. McCombs (Eds.), *How students learn: Reforming schools through learner-centered instruction* (pp. 353–377). Washington, DC: American Psychological Association.

Metz, K. E. (1995). Reassessment of developmental constraints on children's science instruction. *Review of Educational Research, 65,* 93–127.

Metz, K. E. (1997). On the complex relation between cognitive developmental research and children's science curricula. *Review of Educational Research, 67,* 151–163.

National Science Foundation (NSF). (1996). *Shaping the future: New expectations for undergraduate education in science, mathematics, engineering, and technology.* Washington, DC: Author.

National Science Foundation (NSF). (1997). Collaboratives for excellence in teacher preparation. Retrieved April 2003, from http://www.ehr.nsf.gov/ehr/due/awards/cetp.asp

National Science Teachers Association (NSTA). (1998). *Standards for science teacher preparation*. In collaboration with the Association for the Education of Teachers in Science. Retrieved April 2003, from http://www.nsta.org

Oregon Collaborative for Excellence in Teacher Preparation. (1999). *Indicators for selection of mathematics and science content courses appropriate for future teachers*. Retrieved April 2003, from http://www.mth.pdx.edu/ocept/evaluation.htm

Raizen, S. A., & Michelsohn, A. M. (Eds.). (1994). *The future of science in elementary schools: Educating prospective teachers*. San Francisco: Jossey-Bass.

Reys, R. E., & Kilpatrick, J. (2001). *One field, many paths: U.S. doctoral programs in mathematics education* (Conference Board of the Mathematical Sciences: Issues in mathematics education, Vol. 9). Providence, RI: American Mathematical Society.

Schwartz, R., Lederman, N. G., & Crawford, B. A. (2000, April). *Understanding the nature of science through scientific inquiry: An explicit approach to bridging the gap*. Paper presented at the annual meeting of the National Association of Research in Science Teaching, New Orleans, LA.

Shulman, L. E. (1987). Knowledge and teaching: Foundations of the new reform. *Harvard Educational Review, 57,* 1–22.

Shulman, L. E., & Quinlan, K. M. (1996). The comparative psychology of school subjects. In D. C. Berliner & R. C. Calfee (Eds.), *Handbook of educational psychology* (pp. 399–422). New York: Macmillan.

Smith, C. L., Maclin, D., Houghton, C., & Hennessey, M. G. (2000). Sixth-grade students' epistemologies of science: The impact of school science experiences on epistemological development. *Cognition and Instruction, 18,* 349–422.

Tobin, K., Tippins, D. J., & Gallard, A. J. (1994). Research on instructional strategies for teaching science. In D. L. Gabel (Ed.), *Handbook of research on science teaching and learning* (pp. 45–93). New York: Macmillan.

Weiss, I. R. (2001). *Report of the 2000 national survey of science and mathematics education*. Retrieved April 2003, from http://2000survey.horizon-research.com/reports/elem_science.php

Wittrock, M. C. (1998). Cognition and subject matter learning. In N. M. Lambert & B. L. McCombs (Eds.), *How students learn: Reforming schools through learner-centered instruction* (pp. 143–152). Washington, DC: American Psychological Association.

3

Science Pedagogical Content Knowledge and Elementary School Teachers

Ken Appleton
Central Queensland University

There has been ongoing discussion about the knowledge that elementary school teachers need in order to teach effectively. However, the knowledge that teachers draw on during teaching has been shown to be highly complex. Shulman (1986, 1987), for instance, suggested that teachers draw on seven types of knowledge, including their knowledge of subject matter (content knowledge), general pedagogical knowledge (knowledge about how to teach), and a special form of knowledge called pedagogical content knowledge. He saw pedagogical content knowledge as content knowledge transformed by the teacher into a form that makes it understandable to students. This proposition spawned considerable research into the proposed construct. This chapter focuses on elementary school teachers' science pedagogical content knowledge (PCK). It proposes a way of representing the development of science PCK, and how it is related to other forms of teachers' knowledge and to science teaching practice. A model, as an aid to its conceptualization, represents diagrammatically the development of elementary science PCK. A discussion of the model and its implications follows.

Science PCK has been the subject of a number of studies, many of which have focused on secondary science teachers (e.g., van Driel, Verloop, & de Vos, 1998; Veal & MaKinster, 1999). Secondary school teachers tend to be specialists in one area of science, so the content knowledge basis for their science PCK is clearly defined by their specialized science knowledge. However, the situation for elementary school teachers is considerably different from their secondary school counterparts. For example, in Australia, elementary school teachers usually take responsibility for teaching eight subjects, of which two, physical education and music, are sometimes taught by specialist teachers. Elementary school teachers, therefore, need to have a workable store of both content knowledge and PCK in each of the eight subject areas. In practice, teachers tend to have particular subject strengths and weaknesses based on their own personal interests and educational experiences. Most teachers have personal subject preferences that are reflected in the subjects they prefer to teach, and in how they teach each subject area. For instance, many elementary school teachers have had negative experiences in science during their own schooling, so in their senior high school and tertiary study they have tended to avoid the sciences (Skamp, 1997).

The tendency for elementary school teachers to prefer nonscience subjects has had a number of effects on science teaching that have been consistently documented over many years. These include teachers having limited science content knowledge—particularly in the physical sciences, low confidence to teach science, and perceptions of science as a body of facts and laws (Abell & Roth, 1992; Australian Foundation for Science, 1991; Department of Education, Employment, and Training, 1989; Harlen, 1997). So it is not surprising that many elementary school teachers avoid teaching science, and when they do teach it, they tend to use teaching strategies normally associated with other subjects rather than those more compatible with science.

The difficulty with science experienced by numbers of elementary school teachers has generated a corresponding focus on how to help them implement an effective science program in their classes, despite the difficulties confronting them (e.g., Bell & Gilbert, 1996). Part of this focus has been research into the nature and development of science PCK (Appleton & Kindt, 1999; Gess-Newsome, 1999). A particular dilemma arises from the question: Given that many elementary school teachers have limited science content knowledge and therefore limited science PCK, how do such teachers manage to teach science at all? Although research on classroom teaching has shown that some teachers resolve the problem by not teaching science, or by teaching social studies or language with some peripheral science content, some teachers make a resolute and concerted effort to teach science well (Smith & Neale, 1991). One explanation is that elementary school teachers develop a working set of science PCK that bridges the content gap for them, through the use of science activities that work.

Some studies (Smith, 1999; Smith & Neale, 1991; Zembal-Saul, Starr, & Krajcik, 1999) have reported how elementary teachers and neophyte elementary teachers developed science PCK within the context of formal science education instruction. Smith (1999) highlighted the role that curriculum materials and activities can play in helping teachers shift their teaching toward that advocated in recent reform movements. Studies such as these have led the way in our understanding of elementary science PCK and its development. They have also shown that, although there are similarities to the development of science PCK in secondary school teachers, there may be aspects unique to elementary school teachers.

SCIENCE PEDAGOGICAL CONTENT KNOWLEDGE

Gess-Newsome (1999) commented that there are a number of different views about the nature of PCK, so this section first outlines views of knowledge, then discusses the nature of science PCK, and finally proposes a model for the development of elementary school teachers' science PCK.

Knowledge as Construction

Consistent with the views described in chapter 6, this chapter views all knowledge as constructed. Knowledge is conceptualized as organized into clusters of similar experiences, often called schemata, or, using Claxton's (1990) term, mini-theories. Mini-theories are context specific, and experiential memories on which they are based include the context of the experiences as well as the experiences themselves. That is, each experience is inseparable from the physical and social/cultural context in which it occurred. However, new mini-theories can be constructed that traverse a group of similar mini-theories, creating more generalized abstractions. Hence the views of knowledge and of learning espoused herein are a mix of neo-construct theory (Kelly, 1963) and social constructivism (Wertsch, 1985). Therefore, each form of teacher knowledge proposed by Shulman (1987) is organized cognitively as sets of mini-theories, perhaps with numerous "higher order" mini-theories within each form of knowledge. Science PCK, in particular, can be viewed as mini-theories that cross the "boundaries" of other forms of teacher knowledge.

What Others Have Said About Science PCK

Cochran (1997, p. 1) defined PCK as "the integration or synthesis of teachers' pedagogical knowledge and their subject matter knowledge." According to Shulman (1986, p. 9), it also includes "the most powerful analogies, illustrations, examples, explanations, and demonstrations—in a word, the ways of representing and formulating the subject that make it comprehensible to others." So firstly, science PCK includes knowledge of the content to be taught, and how that content may be made understandable to students. Although Shulman's suggestions at first glance imply a teacher-directed pedagogy, he went on to include in PCK "an understanding of what makes the learning of specific concepts easy or difficult: the conceptions and preconceptions that students of different ages and backgrounds bring with them to the learning" (p. 9). So, secondly, science PCK includes a knowledge of children and their preconceptions, and implies a student-centered pedagogy that is perhaps constructivist in orientation. A constructivist-pedagogical aspect of science PCK was emphasized by Parker and Heywood (2000), who in an investigation of elementary school teachers' development of understanding of floating and sinking, claimed that science PCK includes knowledge of "how to ascertain and challenge learners' ideas in productive ways . . . and how to represent the subject matter in that teaching learning interface" (p. 109). That is, learning experiences that focus on ascertaining and challenging students' ideas, and the sequencing of those experiences to enhance understanding, are part of science PCK (Morine-Dershimer & Kent, 1999).

Cochran, deRuiter, and King (1993) also took a constructivist orientation to PCK, proposing that besides content knowledge and pedagogical knowledge, it includes knowledge of students (abilities, learning strategies, age/developmental levels, attitudes/motivation, and prior knowledge) and knowledge of the learning environment (social, political, cultural, and physical). Further work by Cochran (Cochran & Holder, 1998) showed how an even greater range of teacher knowledge was incorporated into a constructivist perspective of PCK. The interrelations between different forms of teacher knowledge for secondary school teachers were explored by Veal and MaKinster (1999). They suggested that science PCK is based on and incorporates a teacher's knowledge of content, students, context, environment, the nature of science, assessment, pedagogy, curriculum, socioculturalism, and classroom management. Similar relations have been suggested for elementary school teachers (Smith, 1999).

A final dimension of science PCK is the notion of the teacher's orientation to teaching science (Magnusson, Krajcik, & Borko, 1999). Magnusson et al. suggested that teachers hold different views and beliefs about teaching, learning, and science that lead them to view the teaching and learning of science in different ways. Possible orientations that they identified from the literature were process, academic rigor, didactic, conceptual change, activity driven, discovery, project-based science, inquiry, and guided inquiry. Such orientations, according to Magnusson et al. (1999), influence the teacher's choice of aspects such as learning goals, learning experiences, and teaching strategies, and thus exert a considerable influence on developing science PCK.

A View of Science PCK

Based on this literature and other research (e.g., Appleton & Kindt, 1999), the view of science PCK adopted in this chapter is the knowledge a teacher uses to construct and implement a science learning experience or series of science learning experiences. It is a dynamic form of knowing that is constantly expanding and being transformed from other forms of teacher knowledge, and through the experiences of planning, implementing, and evaluating science teaching and learning. It includes versions of science content appropriate to the students concerned, and ways of making that content understandable—not just in terms of analogies and examples to use in explanations, but also in the types of learning experiences in which students should engage, and the sequence in which these should occur. The activities and sequence are chosen partly on the basis of the teacher's knowledge of student difficulties in understanding and preconceptions held about the content, and partly on the teacher's view of teaching and learning (i.e., their orientation to teaching science), with other areas of knowledge identified by Smith (1999) and Veal and MaKinster (1999) also contributing. The outcome of science PCK being used is the enacted science curriculum. The creation of new science PCK that can be communicated to other teachers is another possible outcome.

Gess-Newsome (1999) categorized views about PCK as either integrative or transformative. The former view holds that PCK is not a form of teacher knowledge different from other forms, but rather an expression of other forms of teacher knowledge that are integrated during the act of teaching. In comparison, the latter view suggests that other forms of teacher knowledge are transformed into a new form of knowledge, PCK, for the purpose of instruction. The view previously outlined is clearly transformative in orientation. However, as suggested by Gess-Newsome, in some instances PCK could be integrative. Consider two examples of construction of science PCK in support of this position.

I recently worked with a teacher, Rose,[1] to help her implement a new science curriculum in her sixth-grade class. After we jointly planned a science unit of work, I attended a number of the science lessons and occasionally contributed when Rose was troubled or struggling. During one lesson, I suggested a few activities for the students "off the cuff" that might help them to grasp an idea with which they were working. This challenged the teacher. As she explained later, she was "an experienced teacher so I should be able to do that too [come up with good ideas to help the students]. If Ken can do it, so can I." Rose consequently created a model to help explain the body's use of food for energy, and used it in a

---

[1]Teachers' names are pseudonyms.

lesson the next day. It drew on her knowledge of the human body, work the students had done on healthy eating, and her knowledge of the type of explanation that would make sense to the students. It involved a demonstration with a balloon.[2] It was her own idea completely, and provides an example of science PCK transformed from other forms of knowledge. This was new knowledge that did not exist before, and was not an integration of other forms of knowledge. In Claxton's (1990) terms, it constituted a new mini-theory derived from several mini-theories. Further, this new knowledge could now be communicated to others, as a useful idea for teaching.

However, science PCK does not consist solely of models and analogies. For instance, it includes making on the spot decisions about how to respond to a student's comment—weighing up the comment, envisioning possible responses, and choosing one. I saw Rose do this many times. In one instance, a student had explained that a partially filled jar of water, when struck, made a sound because it vibrated. However, it was unclear what she thought was vibrating: the water, the jar, or the air above the water. Rose's response was a query to ask her to clarify what she meant, leading to a discussion and exploration of the different pitch of sounds made with different levels of water in a jar. Specific inferences about pitch and what was vibrating followed. It is this sort of science PCK involving on the spot decision making that may be integrative, because there is no clearly identifiable new knowledge as in the previous example: The activities with different levels of water in a jar were those that Rose had prepared for the students after she found them in some teacher resources. In this case, I could argue that Rose was applying an existing mini-theory (Claxton, 1990) about general pedagogy to a new situation, so no new mini-theory was generated. However, despite the lack of teacher-generated new science PCK, this could still be seen as transformative science PCK, in that Rose generated a teaching sequence from a particular context, transforming it from other aspects of knowledge that she held into a new mini-theory. In these types of teaching incidents, there is simply insufficient data to be able to clearly categorize the event as transformative or integrative. Consequently, from the previous accounts and the point about science PCK being "higher order" mini-theories, I hold the view that science PCK is predominantly transformative, but there may be instances of integrative science PCK.

_____

[2]Rose inflated, then partially deflated, the balloon several times. The analogy she used was that the air put into and released from the balloon represented energy, and the balloon the human body. Energy consumed by the body as food (inflating the balloon a little more) is used up during activity (deflating the balloon slightly). If there is an imbalance, the balloon grows slowly larger if energy intake is higher than expenditure, or smaller if energy intake is too low. She took care to discuss the limitations of the analogy.

Science in the elementary school has its own set of contextual factors that influence the creation and use of science PCK, so these are outlined in the next section.

THE PARTICULAR SITUATION FOR ELEMENTARY SCHOOL TEACHERS

As mentioned earlier, many elementary school teachers have difficulties teaching science, however, there is a significant but smaller group of teachers who teach science regularly and effectively. Of those who experience difficulty, some avoid teaching science altogether, some deal with token science content in reading and social studies, and others do the best that they can with their limited knowledge and the resources at hand. One way that many of this last group of teachers cope is to base their science program around a number of "activities that work" (Appleton & Kindt, 1999). These are activities with which they feel comfortable, that they have taught before or have had recommended to them, for which they can readily assemble the necessary equipment, and that have fairly predictable outcomes that will provide the students with some science knowledge (Appleton, 2003). For instance, Rose relied almost exclusively on activities that work for her contributions to the science unit that we planned together. Appleton and Kindt suggested that activities that work may be a substitute for science PCK, or at least supplement it (Smith & Neale, 1991).

Any model for elementary school teachers' science PCK needs to be able to accommodate those teachers who teach little science as well as those who do so enthusiastically. It also needs to explain how activities that work relate to teachers' science PCK.

A PROPOSED MODEL FOR ELEMENTARY TEACHERS' SCIENCE PCK

Figure 3.1 portrays a representation or model showing the development of science PCK for elementary school teachers. The model has arisen from work with teachers over 3 years, and has undergone constant change as new research data emerge and further insights occur. Therefore, it cannot be considered a final or definitive model, but rather a working position based on what is known, or perhaps more accurately, what is known at this point in time. There has also been a struggle with how to represent the model to best convey the development of science PCK. The final decision was to use a flow diagram (see Fig. 3.1). In order to simplify the model, many possible cross-links have been omitted.

Each component of the model in Fig. 3.1 is discussed next.

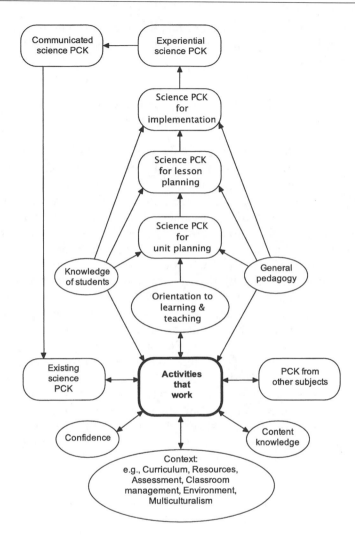

FIG. 3.1.   A model of science PCK development for elementary teachers.

Activities That Work

Earlier work outlined the characteristics of activities that work (Appleton, 2003) and also tentatively suggested that they may be a substitute for science PCK for some teachers (Appleton & Kindt, 1999). More recent work has led to the conclusion that activities that work assume a central role in the development of elementary science PCK. Neophyte elementary teachers who have little or no existing science PCK rely heavily on activities that work (Appleton & Kindt, 2002), as do experienced teachers working in a new area of science. Experienced teachers who have a store of science PCK represent it to themselves and others as activities that work (Appleton, 2003). When teachers talk or write about their science PCK, they describe activities that work, so these are essentially communications of aspects of teachers' science PCK. For instance, consider the case of Rose, a teacher who created a model to help her students understand the body's energy use and food. If she were to write down this model, and how to use it in a lesson, then it would become a description of an activity that works. But, at the same time, it is also a description of aspects of her science PCK. Note that it represents only *aspects* of her science PCK: She has chosen to communicate those components of her science PCK that she considers to be important, working on the assumption that what is omitted will be filled in by someone reading it.

The support material for the new science curriculum recently introduced in Queensland provides a further example. The organization that prepared the science material, the Queensland School Curriculum Council, invited members of the Queensland Science Teachers Association to write supporting modules (units of work with suggested activities, pedagogy, and science background information). After an orientation program, the participating teachers drafted modules based on their own teaching of science. That is, they drew on their science PCK, and attempted to communicate it within the prescribed format. Sally, another teacher with whom I recently worked, was able to make ready use of one of these modules in preparing her own unit. It provided her with an initial set of science PCK ideas that she was able to transform into her own, using her existing science PCK as well as other forms of teacher knowledge. Further, with assistance and encouragement, she readily extended the ideas provided, and generated new science PCK that she incorporated into her written plan for the unit. Her science PCK was further developed as she implemented her planned unit. When she came to aspects that troubled her, I stepped in temporarily to teach those sections. However, she readily assimilated her observations of my teaching into her science PCK repertoire. These included some activities that work, but most were the detailed aspects of pedagogy typically omitted from a description of an activity that works.

Loughran, Gunstone, Berry, Milroy, and Mulhall (2000) conducted a study in which they attempted to identify and document secondary science teachers' science PCK. They found that the best way of doing this was by the use of cases, where the main ingredient was teachers' descriptions of their practices. This is consistent with the view that activities that work are a form of communicated elementary teachers' science PCK.

This can be put in a different way, in that communicated science PCK in the form of activities that work can be seen as a representation of teacher knowledge in the form of narrative. Lyons (2002) emphasized how narrative forms of teacher knowledge are validated through a process of teachers trying for themselves, aspects of narratives from other teachers that they have come across. It is the process of exploring aspects of a narrative in different contexts that validates the knowledge embedded in the narrative: a process of personal reconstruction of knowledge using shared knowledge within a community of elementary school science teachers. This is exactly the process that occurs with activities that work. Teachers communicate through a narrative, an account of an activity that worked for them. Other teachers hear or read the narrative and try it out in their own teaching situation, modifying it where necessary, and perhaps in turn communicating it to colleagues. Curriculum resources containing activity suggestions are consistent with this view of activities that work being a narrative version of science PCK.

For teachers to use effectively another's communicated science PCK presented as a narrative form of an activity that works, they must transform what is written (or spoken) into their own, personal, science PCK. To do this, they draw on related science PCK, as well as other knowledge forms, such as general pedagogical knowledge and their knowledge of students. They also assimilate any science content described in the activity that works into their own science content knowledge framework, and use that to aid construction of their personal science PCK. However, it is essential that teachers can fill in the gaps where aspects of science PCK are omitted from the activity that works narrative. If teachers are unable to do this, then there is a high probability that they will begin to experience problems during implementation.

An inability to fill in the science PCK gaps in activities that work also explains why the beginning teachers in other studies (e.g., Appleton & Kindt, 2002) did not find the available teacher resources helpful, compared to more experienced teachers. Because the beginning teachers had limited science PCK and limited general pedagogical knowledge, it was difficult for them to fill in the gaps: They were unable to transform the written versions of science PCK into their own personal science PCK.

Further, activities that work conceived as communicated science PCK explains why teachers who rely on textbooks tend to have a limited science program unless they deviate considerably from the text. A textbook usually consists of descriptions of content, with some student activities suggested to supplement the content. However, the activities in textbooks are not normally written in a form that is useful science PCK for teachers, making it difficult for them to transform the activities into personal science PCK. Consequently, many suggested activities are glossed over, and the science program focuses almost exclusively on the content in the textbook.

Although activities that work play a central role in the development of elementary school teachers' science PCK, other forms of teacher knowledge also play significant roles.

Existing Science Pedagogical Content Knowledge

Experienced teachers who have previously taught elementary science would have developed a store of science PCK on which they could draw for future teaching events. As mentioned earlier, they represent this to themselves as activities that work, but when their ideas are probed, it is revealed that their science PCK incorporates much more than a description of activities in which students might be engaged. Much of this knowledge is expressed as narrative, in accounts of past teaching events with multilayered levels of detail and purpose. Even so, much remains unsaid as implicit knowledge that is either inexpressible, or is assumed to be common knowledge among teachers.

Beginning elementary school teachers, however, would have little if any existing science PCK. This creates especial problems for them in teaching science, and together with limitations in other forms of teacher knowledge illustrated in Fig. 3.1, may well be why many beginning teachers seem to fall into the habit of not teaching science (Appleton & Kindt, 1998).

Content Knowledge

An essential contributor to any PCK development is the teacher's knowledge of subject matter or content to be taught (Cochran & Holder, 1998; Shulman, 1986; Veal & MaKinster, 1999). Although elementary school teachers tend to have limited formal science content knowledge such as might be obtained in senior high school and university, they frequently have some informal everyday science knowledge arising from hobbies, interests, and experiences. This is usually a practical knowledge centered round everyday events. It may incorporate common misconceptions, may be domain specific, and may or may not include conceptual understanding similar to those held by scientists. Most teachers do not recognize such general knowledge as being useful for teaching science because they are uncertain of its accuracy, and it does not usually fit their view of what science is. For instance, in a discussion with a teacher a few years ago, I discovered that he had an amazing knowledge of Australian native flora; he did not consider this to be science—to him it was gardening. He did not specifically use this knowledge in his science teaching, although he felt more comfortable with topics on plants. He was initially surprised when I suggested to him that he was doing science by acquiring and using this knowledge—especially when applying it by harvesting seeds and growing them.

Teachers who attempt to teach science readily acknowledge their limited knowledge and may use resources such as printed descriptions of activities that work (Smith & Neale, 1991), or a science resource (Prinsen, 2001) to enhance their science knowledge.

A number of studies (e.g., van Driel et al., 1998) have shown that teachers tend to hold a positivistic view of science; that is, science is seen by elementary school teachers as a collection of numerous and complicated laws and definitions obtained from definitive and accurate sources such as books and scientists, and from following a formulaic version of the "scientific method." A teacher's view of the nature of science is, therefore, an important component of content knowledge. But, according to Veal and MaKinster (1999), its influence on secondary teachers is no greater than that of other forms of knowledge such as curriculum and classroom management. Prinsen (2001, p. 24), in following up two elementary school teachers' implementation of a science unit subsequent to a professional development orientation to the new Queensland science curriculum, commented that "Monica and Wendy . . . believe that the substantive content knowledge of science is fixed." The teachers' view of science contributed significantly to difficulties they had in implementing the science unit observed by Prinsen, because it inhibited their ability to generate appropriate science PCK for the unit.

Science cannot be taught without science PCK, because without science PCK the teacher has no idea of what to teach or how to do it. One of the foundational bases of science PCK is content knowledge, so teachers who have little science content knowledge have limited options, especially if they lack confidence to choose activities that work from science topics about which they know little, or to acquire new science content knowledge for themselves.

Teacher Confidence

Another critical influence on science PCK is an attitude cluster, rather than what is traditionally considered as knowledge. One of the most documented difficulties facing the majority of elementary school teachers is their low level of confidence in teaching science (e.g., Abell & Roth, 1992; Appleton, 1995; Goodrum, Hackling, & Rennie, 2001; Symington, 1974). They also tend to lack confidence in the adequacy of their own science knowledge, and in their ability to do and to learn science for themselves.

This general lack of confidence is directly attributable to limited science content knowledge, but is also related to a commonly held positivistic view of science. Because some teachers feel more confident about science and some less so, confidence level is a major influence on their development of science PCK. For instance, teachers who have extremely low levels of confidence may avoid teaching science altogether, and consequently develop little or no science PCK.

Recommendations from reports into elementary science teaching have advocated including more science in teacher preparation programs, and requiring science content during inservice programs (e.g., Department of Employment, Education, & Training, 1989). However, there is substantial evidence to suggest that unless teachers' perceptions of the nature of science, as well as science PCK, are addressed alongside the science content, more science per se does not necessarily lead to increased confidence, more science teaching, or better science teaching (e.g., Garet, Porter, Desimone, Birman, & Yoon, 2001; Parker & Heywood, 2000; Skamp, 1997; Skamp & Mueller, 2001a, 2001b).

Context

A further influence on elementary school teachers' development of science PCK is a broad group of other forms of teacher knowledge: classroom management, assessment, socioculturalism, curriculum, environment, and resources. The role played by each varies according to the situation. Most of these have been discussed elsewhere in the literature (e.g., Morine-Dershimer & Kent, 1999), so they are not elaborated on here. However, there is another one that is addressed specifically: resources.

Resources.   The issue of resources may be considered as a subset of curriculum, but it assumes such importance for elementary school teachers in science that special consideration is necessary. Elementary school teachers tend to refer to two types of resources with respect to science (Appleton, 2003). The first type of resource is equipment for the students to use during activities. This may be print and electronic resources, or manipulative equipment. Typically, manipulative equipment is in poor supply in the majority of schools, or is deployed in a way that effectively makes it unavailable for most teachers. The notion of manipulative equipment is sometimes complicated by the teachers' view of science: If science is seen as being done by specialist scientists with complicated equipment, then specialized equipment is seen as necessary for doing science in school. Hence teachers consistently complain about inadequate equipment resources. Even when a teacher attempts to use everyday equipment, the logistics of purchasing and assembling the material requires a time commitment that many teachers are unable or unprepared to make.

The second type of resource is teacher resources, usually in the form of printed versions of activities that work. What is in descriptions of activities that work determines whether the teacher considers them useful. A full description of what teachers want in activities that work is outlined in Appleton (2003). If an activity that works is considered to be useful by teachers, then it will enhance their science content knowledge and provide an outline of science PCK for the activity. When preparing to teach a new science topic, elementary school teachers typically search through teacher resources, looking for brief synopses of science content and summaries of activities that work (Smith & Neale, 1991). Activities are judged in terms of whether they apply to the particular context in which the teacher works, using multiple values and views to arrive at that judgment (Appleton, 2003; Appleton & Asoko, 1996). For instance, classroom management issues are considered; or, if the suggested pedagogy is based on a view of learning different from that held by the teacher (see later), then the activity may be rejected or modified (Prinsen, 2001); or, if the classroom physical environment does not readily allow multiple small group activity, then the activity may again be rejected or modified.

PCK From Other Subjects

As indicated earlier, some elementary school teachers have little existing science PCK and lack the confidence to acquire it. However, they are also sensitive to their responsibility to teach science to their students. One way that some teachers cope is to draw on pedagogical content knowledge developed in other subject areas like Studies of Society and Environment or English. For example, instead of using books to find and summarize information about a historical event, they would have students investigate a science topic using books. In doing so, they would draw on similar teaching strategies used during social studies investigations. Even though there are obvious limitations to doing this, such as no planned scaffold of student learning in science, and especially when it constitutes the majority of the science program, it does provide a means whereby some teachers can teach a form of science without existing science PCK. As they do so, however, they begin to develop a store of science PCK that differs little from the subject PCK from which it originated. The science PCK thus developed may not result in a science program consistent with recent reform movements, but to the teachers concerned they feel that they are discharging their responsibility to teach science in the best way that they can.

Teacher's Orientation to Learning and Teaching

The teacher's view of how learning occurs has a critical influence on that teacher's development of science PCK (Morine-Dershimer & Kent, 1999), and is related to the teacher's orientation to teaching (Magnusson et al., 1999). Something that teachers want from an activity that works is that students should preferably learn what is expected from engaging in the activity (Appleton, 2003). That is, there is an implicit view that learning occurs by doing: If the doing is done correctly, then the expected learning should occur with little or no further action by the teacher required. This is also related to the teacher's view of science as progressing by experiment. This view was evident in the two teachers, Monica and Wendy, observed by Prinsen (2001). If the expected learning does not occur, only then might the teacher tell the students "the answer," or perhaps refer them to a book or similar source containing "the answer." This seems to be a perverted version of discovery learning (one of the orientations suggested by Magnusson et al., 1999), and is a predominant view about learning in science held by many elementary school teachers that I have encountered. By contrast, a teacher working from a constructivist orientation to learning would recognize that the pedagogy employed prior to, during, and after an activity is critical to students' learning. Such a teacher would ensure that an appropriate scaffold of experiences and teacher–student interaction occurred to maximize opportunities to achieve understanding of the desired science concepts. This would be consistent with inquiry or guided inquiry orientations to teaching (Magnusson et al., 1999). Such a view would result in a very different type of science PCK, even though the same activity may be used in both examples.

Knowledge of Students

According to Veal and MaKinster (1999), the teacher's knowledge of students is a major contributor to secondary school teachers' science PCK. Similarly, it plays a major role in elementary school science. For instance, some teachers consistently weigh up potential activities that work in terms of what they know about their students. That is, teachers use what they know of their students' interests, ability to work in groups, and so on, to decide finally whether to use a particular activity, and what pedagogy to use when implementing it. A key consideration is whether the teacher thinks the activity would be fun for the students to engage in (Appleton, 2003; Appleton & Asoko, 1996). Knowledge of student preconceptions, a component of knowledge of students, also plays a part when a teacher decides how to scaffold learning around and within a selected activity

(Smith, 1999); but, this is usually only considered specifically by teachers holding a constructivist view of learning. Knowledge of students may also involve other forms of knowledge such as classroom management. For instance, if a teacher considers that a class is unruly, then small group work may be curtailed; but, with a well-behaved class, the same teacher may have the students fully engaged in small group activities.

In Fig. 3.1, Knowledge of Students is shown interacting with other components contributing to science PCK development in different ways. For instance, initially it is used to aid selection of suitable activities that work based on knowledge of students' existing knowledge and interests. Later, it is used to decide on an appropriate scaffolding pedagogical sequence for a unit of work based on knowledge of students' preconceptions; then to aid in selecting particular lesson teaching strategies such as KWL, based on knowledge of (say) the students' preference for ownership of their work; and finally when deciding which student to choose to answer a particular question, based on knowledge of each student's capabilities and contributions to the lesson.

General Pedagogy

The critical role played by this form of teacher knowledge in the development of elementary science PCK has been well documented by others (e.g., Smith, 1999). An important consideration is that elementary school teachers frequently rely on their general pedagogical knowledge to fill in gaps in their science PCK. For instance, teachers who are unaware of students' preconceptions and scientists' views about a science topic might use a general pedagogic strategy such as having students raise questions for a guest speaker to answer—a solution that allows them to deal with their lack of knowledge, and subsequently becomes built into their science PCK for that topic.

As shown in Fig. 3.1, General Pedagogical Knowledge interacts with developing science PCK when selecting activities that work, planning a unit, planning lessons, and when actually teaching.

THE ENACTED SCIENCE CURRICULUM

So far, the discussion has focused on the creation of science PCK from the other forms of teacher knowledge, shown in the lower half of Fig. 3.1. Next consider science PCK as it is played out in the enacted elementary science curriculum, illustrated in the upper half of the figure. The enacted science curriculum is the actual teaching and learning dealing with science content that occurs in the classroom. That is, it is what happens when teachers use their science PCK in practice. For the term to be fully comprehensive, it includes science taught using whatever pedagogy teachers choose to use even though some pedagogies are perhaps inappropriate for science or the circumstances. Similarly, what passes for science PCK for some teachers may be more like Studies of Society and Environment PCK with bits of science content; but, if those teachers believe that they are teaching science, then to them they are using their "science PCK,"[3] and what is taught is their enacted science curriculum.

Science PCK for Unit Planning

Aspects of teachers' science PCK are used to structure a unit of work focusing on a particular area of science. This includes the sequencing of activities that work and other learning experiences, and selection of a scaffolding pedagogy that enhances students' conceptual learning as they move from one learning experience to another, such as the learning cycle or 5Es approach,[4] as it is sometimes called. These choices are influenced by the teachers' orientation to learning and teaching.

Other forms of teacher knowledge are used at this time. For instance, Knowledge of Students and General Pedagogy are shown in Fig. 3.1 as contributing. However, elements of other forms of teacher knowledge, particularly context ones such as socioculturalism, assessment, and classroom management would also be drawn on during unit planning. This may mean that teachers construct different goals and purposes for teaching the same science topic to different classes, because the classes are perceived as differing in significant social or cultural ways (Friedrichsen & Thomas, 2002). The different goals and purposes determined by the teacher for different classes may result in very different enacted curricula in those classes, even if the science PCK drawn on by the teacher is substantially similar in each class (but there would be differences as different aspects of science PCK are drawn on).

This relation has not been incorporated into Fig. 3.1 simply because it would become too complex a diagram. I also acknowledge that some elementary school teachers do not draft unit plans in science.

Science PCK for Lesson Planning

During lesson planning, science PCK is used to determine the specific teaching strategies used during each activity or learning experience. For

---

[3]As this does not reflect the syntactic nature of science, calling it science PCK—even limited PCK—is problematic, but I do so here, with the qualifying quotes, for expediency.

[4]The 5Es approach includes the steps of Engage, Explore, Explain, Elaborate, and Evaluate (Australian Academy of Science, 1994).

instance, this might include a discovery component or a jigsaw activity. Other details of lessons, such as resources and classroom management, are also determined at this time by teachers. In doing so, they may draw on other forms of teacher knowledge such as Knowledge of Students, General Pedagogy, and contextual forms (not shown in Fig. 3.1). Choices again are influenced by the teachers' orientation to learning and teaching.

Science PCK for Implementation

Different aspects of science PCK come into play when the science is actually taught. Here, teachers' decisions based on science PCK determine the flow of interaction with students—the questions to ask, whether to respond to a student's answer, question wait time for particular questions, examples or metaphors to use in an explanation that had not been planned for, the responses to students' questions or comments, and so on. Other forms of teacher knowledge are also drawn on at this time.

"Untried" Science PCK

The first two planning components of science PCK discussed earlier (unit and lesson planning) can be considered as science PCK that is untried, if it has been generated from other teachers' suggested activities that work but has never before been taught by the teacher. That is, it is science PCK created by personally constructing knowledge from science PCK communications from other teachers. When planning a new science topic for the first time, teachers generate creatively from their own knowledge, or from communications from others (e.g., an activity that works in the teacher resource material), a form of science PCK that is as yet untried. It is theoretical in the sense that it is an untried mental plan.

Experiential Science PCK

In contrast, experiential science PCK is created by the act of teaching science. The mental plan embodying the untried science PCK is put into practice in the classroom. The teacher reflects on the enacted science curriculum and evaluates its effectiveness. The untried science PCK is transformed through this process into experiential science PCK. Changes may consequently be made to the science PCK, depending on the judgments about effectiveness made by the teacher.

Further, experiential science PCK can be generalized across classroom contexts. It may be used to arrive at some generalized aspects of science PCK, particularly after teaching a number of lessons. This generalized science PCK, which is consistent with a higher order mini-theory (Claxton, 1990), can be topic specific (Veal & MaKinster, 1999). For example, "The best way to start a grade 6 unit on magnetism is to . . ." It could alternatively apply across topics as, for example, "Always start a unit with an activity that has some 'WOW' factor." Cross-topic generalizations may border onto general pedagogical knowledge.

Teachers Communicate Their Science PCK

Generalized topic-specific science PCK can be communicated to others as activities that work. This revisits a point discussed earlier. When a teacher wishes to share with a colleague an activity or a unit that went particularly well, the essential features of the science PCK associated with the activity or unit are communicated, usually as a form of narrative. Teachers commonly refer to such narratives as activities that work. Sometimes a narrative may be mainly confined to a description of the learning experience the students engage in, with minimal comment about teacher actions. But, those narratives found most helpful by teachers include many other characteristics (see Appleton & Kindt, 2002).

Summary

Science PCK is necessary to teach science. For teachers to teach any science, they must have some form of science PCK. Although this sounds like common sense, it has an inherent contradiction: How does anyone teach science for the first time? Obviously, a neophyte teacher has to generate a small amount of initial science PCK to get started. If not, the lesson is likely to deteriorate rapidly and the neophyte will revert to other known, safer forms of pedagogy. This initial science PCK is generated from science PCK communicated by others in the form of activities that work.

Science PCK is not automatically generated from science content and other forms of teacher knowledge. It might be implied from the model in Fig. 3.1 and the literature that science PCK will naturally emerge from the other forms of teacher knowledge, particularly science content knowledge. Although this may happen, there is an equal chance that it may not happen. For instance, Akerson and Abd-El-Khalick (2002) reported how a teacher developed content knowledge about the nature of science, but could still not teach about the nature of science because she lacked the appropriate PCK. Only after observing Akerson teaching about the nature of science was she able to generate the science PCK that enabled her to teach it. That is, she observed a form of an activity that works being implemented. Similarly, Alonzo (2002) reported how elementary school teachers who had received instruction on science content did not necessarily make changes to their pedagogy when teaching the same content area to their students. Thus, the model portrays the interrelated component forms of knowledge from which science PCK is developed, but does not portray the process by which it is generated. However, Akerson and Abd-El-Khalick's study highlighted the central role played by activities that work in teachers' development of science PCK.

Teacher self-confidence in science is an integral part of elementary science PCK development. This aspect must be considered during both preservice and inservice teacher development. Science PCK is developed from activities that work and other forms of teacher knowledge that are mediated by teachers' orientation to learning and teaching. Science PCK can exist as "untried" knowledge when a science topic is planned for the first time, represented by unit and lesson plans. When planning and teaching science, teachers draw on their science PCK, but also use other forms of teacher knowledge, such as knowledge of the environment and knowledge of students, so that the enacted science curriculum is drawn from a variety of forms of teacher knowledge.

Experiential science PCK is created by the act of teaching science. Untried and existing science PCK are transformed through this process into experiential science PCK. Experiential science PCK can be generalized across classroom contexts and may be communicated to others as activities that work.

IMPLICATIONS FOR TEACHER EDUCATION

Preservice Elementary Science Teacher Education

Whereas it is self-evident that for preservice teachers to be prepared to teach elementary science they need the opportunity to develop science PCK, given the scope of knowledge needed this is problematic. The problem has several dimensions, most of which have been subject to research and comment in the literature. The first dimension is in developing science content knowledge, given that the majority of preservice teachers have self-selected nonscience courses during their high school and university studies: That is, there is an impossible amount to cover during required science components of a preservice program. The second dimension is the low level of confidence in science felt by many preservice teachers. The third is the range of other forms of teacher knowledge that are needed in order to be transformed into science PCK. For instance, by the end of their preservice program, few preservice teachers would have developed a comprehensive knowledge of general pedagogy or knowledge of students in general, let alone of a particular group of students. Further, few would have had the opportunity to understand comprehensively constructivist ideas about learning and the implications of these ideas for science pedagogy, or to have gained a different view of science from the commonly held positivist one. The fourth is that elementary school teachers are supposed to be able to teach all elementary grades, so for them to develop untried science PCK for even a few topics for each grade, as well as PCK for other subjects, is an overwhelming task. The fifth and final dimension is that the opportunity for preservice teachers to create experiential science PCK is limited, given the nature of the usual practicum experiences available to them.

To address adequately all of these dimensions within the time available during preservice teacher preparation programs is well nigh impossible. Some success has been reported in programs where all of these aspects have been addressed in an integrated way (e.g., Zembal-Saul et al., 1999). However, it is beyond the scope of this discussion to elaborate how teacher educators have attempted to achieve this. But, the evidence provided by ongoing surveys of teaching practices in elementary science suggests that, apart from a few exceptions, attempts have been mostly unsuccessful. Perhaps the best that can be hoped for is to prepare prospective teachers to be ready and willing to try teaching science, and trust that ongoing professional development will help them get past the difficulties they will face in their early teaching assignments. Further, it may be possible to target those most likely to be successful in teaching science, and prepare them to be a cadre of science enthusiasts in schools who can help their colleagues.

Inservice Elementary Science Teacher Education

The task in providing inservice professional development support for elementary school teachers in science is equally overwhelming, given the number of teachers in schools, and the nature of the support that we are now coming to understand is necessary in order for teachers to effect change. This usually comes down to the funds that education systems are prepared to invest in science teacher professional development. Consider a few characteristics of science teacher professional development that have a greater likelihood of success (see Bell & Gilbert, 1996, for a comprehensive discussion), as well as their relation to science PCK development.

Professional development for teachers in science should include the following:

- Workshops with a multiple-pronged focus:
  - Selected science content
  - The nature of science as revealed through studying the science content (and historical contexts)

— Science PCK associated with the science content

— Key ideas of constructivist learning theory on which pedagogical elements of the science PCK are based

— Positive personal encouragement and success in doing and teaching science

• Purposeful interaction between teachers to facilitate exchange of aspects of science PCK.

• Mentoring from someone with science PCK in the areas of the teacher's weakness. The mentoring needs to occur in both planning and implementation.

CONCLUSIONS

This chapter proposes a model for the development of science PCK in elementary school teachers, and explains its relation to the enacted science curriculum. Obtaining a better understanding of science PCK development in elementary school teachers again highlights the difficult task for both elementary school teachers and elementary science teacher educators. It is encouraging that there are some elementary school teachers who teach science often and teach it effectively. They, it would seem, have overcome considerable obstacles and developed sufficient science PCK to make a difference in their own teaching and their students' learning.

ACKNOWLEDGMENTS

I wish to thank the many teachers and colleagues who have contributed to my thinking over the years. The research on which this chapter is based was funded by Central Queensland University and the Rockhampton Diocese Catholic Education Office.

REFERENCES

Abell, S. K., & Roth, M. (1992). Constraints to teaching elementary science: A case study of a science enthusiast student. *Science Education, 76,* 581–595.

Akerson, V. L., & Abd-El-Khalick, F. (2002, April). *Teaching elements of nature of science: A year long case study of a fourth grade teacher.* Paper presented at the annual meeting of the American Educational Research Association, New Orleans, LA.

Alonzo, A. (2002, April). *Effects of a content-based professional development program in elementary school inquiry science instruction.* Paper presented at the annual meeting of the American Educational Research Association, New Orleans, LA.

Appleton, K. (1995). Student teachers' confidence to teach science: Is more science knowledge necessary to improve self-confidence? *International Journal of Science Education, 19,* 357–369.

Appleton, K. (2003). Science activities that work: Perceptions of primary school teachers. *Research in Science Education, 32*, 393–410.

Appleton, K., & Asoko, H. (1996). A case study of a teacher's progress toward using a constructivist view of learning to inform teaching in elementary science. *Science Education, 80*(2), 165–180.

Appleton, K., & Kindt, I. (1998, April). *Teaching elementary science: Practices of beginning teachers.* Paper presented at the annual meeting of the National Association for Research in Science Teaching, San Diego, CA.

Appleton, K., & Kindt, I. (1999, March). *How do beginning elementary teachers cope with science: Development of pedagogical content knowledge in science.* Paper presented at the annual meeting of the National Association for Research in Science Education, Boston, MA.

Appleton, K., & Kindt, I. (2002). Beginning elementary teachers' development as teachers of science. *Journal of Science Teacher Education, 13*(1), 43–61.

Australian Academy of Science. (1994). *Primary investigations.* Canberra, Australia: Australian Academy of Science.

Australian Foundation for Science. (1991). *First steps in science and technology: Focus on science and technology education No 1.* Canberra, Australia: Australian Academy of Science.

Bell, B., & Gilbert, J. (1996). *Teacher development: A model from science education.* London: Falmer.

Claxton, G. (1990). *Teaching to learn: A direction for education.* London: Cassell.

Cochran, K. F. (1997). *Pedagogical content knowledge: Teachers' integration of subject matter, pedagogy, students, and learning environments.* National Association for Research in Science Teaching: Research Matters—to the Science Teacher. Retrieved May 24, 2000, from http://narst.org

Cochran, K. F., deRuiter, J. A., & King, R. A. (1993). Pedagogical content knowing: An integrative model for teacher preparation. *Journal of Teacher Education, 44*, 263–272.

Cochran, K. F., & Holder, K. C. (1998, April). *The development of pedagogical content knowing of prospective secondary teachers.* Paper presented at the annual meeting of the American Educational Research Association, San Diego, CA.

Department of Education, Employment, and Training. (1989). *Discipline review of teacher education in mathematics and science.* Canberra, Australia: Australian Government Publishing Service.

Friedrichsen, P., & Thomas, D. (2002, April). *Exploring the science teaching orientation of highly regarded secondary biology teachers.* Paper presented at the annual meeting of the American Educational Research Association, New Orleans, LA.

Garet, M. S., Porter, A. C., Desimone, L., Birman, B. F., & Yoon, K. S. (2001). What makes professional development effective? Results from a national sample of teachers. *American Educational Research Journal, 38*(4), 915–945.

Gess-Newsome, J. (1999). Pedagogical content knowledge: An introduction and orientation. In J. Gess-Newsome & N. G. Lederman (Eds.), *Examining pedagogical content knowledge* (pp. 3–17). Dordrecht, The Netherlands: Kluwer Academic.

Goodrum, D., Hackling, M., & Rennie, L. (2001). *The status and quality of teaching and learning of science in Australian schools.* Canberra, ACT: Commonwealth of Australia.

Harlen, W. (1997). Primary teachers' understanding in science and its impact in the classroom. *Research in Science Education, 27*, 323–337.

Kelly, G. (1963). *A theory of personality: The psychology of personal constructs.* New York: Norton.

Loughran, J. J., Gunstone, R., Berry, A., Milroy, P., & Mulhall, P. (2000, April). *Documenting science teachers' pedagogical content knowledge through PaP-eRs.* Paper presented at the annual meeting of the American Educational Research Association, New Orleans, LA.

Lyons, N. (2002, April). *Valuing narrative as a valid form of knowledge in teaching, teacher education, and research.* Paper presented at the annual meeting of the American Educational Research Association, New Orleans, LA.

Magnusson, S., Krajcik, J. S., & Borko, H. (1999). Nature, sources, and development of pedagogical content knowledge for science teaching. In J. Gess-Newsome & N. G. Lederman (Eds.), *Examining pedagogical content knowledge* (pp. 95–132). Dordrecht, The Netherlands: Kluwer Academic.

Morine-Dershimer, G., & Kent, T. (1999). The complex nature and sources of teachers' pedagogical knowledge. In J. Gess-Newsome & N. G. Lederman (Eds.), *Examining pedagogical content knowledge* (pp. 21–50). Dordrecht, The Netherlands: Kluwer Academic.

Parker, J., & Heywood, D. (2000). Exploring the relationship between subject knowledge and pedagogic content knowledge in primary teachers' learning about forces. *International Journal of Science Education, 22*(1), 89–111.

Prinsen, M. (2001). *Teaching the dog to whistle: Case study exploring the professional development needs of teachers implementing a new constructivist-based science syllabus.* Unpublished honors thesis, Central Queensland University, Rockhampton, Australia.

Shulman, L. S. (1986). Those who understand: Knowledge growth in teaching. *Educational Researcher, 15,* 4–14.

Shulman, L. S. (1987). Knowledge and teaching: Foundations of the new reform. *Harvard Educational Review, 57,* 1–22.

Skamp, K. (1997). Student teachers' entry perceptions about teaching primary science: Does a first degree make a difference? *Research in Science Education, 27,* 515–539.

Skamp, K., & Mueller, A. (2001a). A longitudinal study of the influences of primary and secondary school, university and practicum on student teachers' images of effective primary science practice. *International Journal of Science Education, 23*(3), 227–245.

Skamp, K., & Mueller, A. (2001b). Student teachers' conceptions about effective primary science teaching: A longitudinal study. *International Journal of Science Education, 23*(4), 331–351.

Smith, D. C. (1999). Changing our teaching: The role of pedagogical content knowledge in elementary science. In J. Gess-Newsome & N. G. Lederman (Eds.), *Examining pedagogical content knowledge* (pp. 163–197). Dordrecht, The Netherlands: Kluwer Academic.

Smith, D. C., & Neale, D. C. (1991). The construction of subject matter knowledge in primary science teaching. *Advances in Research on Teaching, 2,* 187–243.

Symington, D. (1974). Why so little primary science? *Australian Science Teachers Journal, 20*(1), 57–62.

van Driel, J. H., Verloop, N., & de Vos, W. (1998). Developing science teachers' pedagogical content knowledge. *Journal of Research in Science Teaching, 35,* 673–695.

Veal, W. R., & MaKinster, J. G. (1999). Pedagogical content knowledge taxonomies. *Electronic Journal of Science Education, 3*(4). Retrieved July 31, 2000, from http://unr.edu/homepage/crowther/ejse/vealmak.html

Wertsch, J. V. (1985). *Culture, communication, and cognition: Vygotskian perspectives.* London: Cambridge University Press.

Zembal-Saul, C., Starr, M. L., & Krajcik, J. S. (1999). Constructing a framework for elementary science teaching using pedagogical content knowledge. In J. Gess-Newsome & N. G. Lederman (Eds.), *Examining pedagogical content knowledge* (pp. 237–256). Dordrecht, The Netherlands: Kluwer Academic.

4

Teaching by Distance

William John Boone
Indiana University

Elementary science teacher education can include preservice preparation of elementary teachers to teach science, as well as informal inservice professional development sessions and formal inservice courses for university credit. A recent trend has been to offer courses by distance education. Is it possible for the different types of elementary science teacher education to be offered by distance education? This chapter documents aspects of issues that have arisen when offering a science education course by distance, and suggests possible ways to deal with them.

DISTANCE EDUCATION

It is tempting to assume that when a new concept such as distance education appears on the horizon it would bring new issues along with it. It is important to note that science teacher education has used a variety of distance education technologies over many decades. Some of the earliest delivery modes of science teacher distance education have involved the use of correspondence courses that primarily relied on texts, articles, and written tests. When taking a correspondence course, the student might occasionally drive to a university and meet with the instructor either individually or as part of a larger group, if they were close enough to the campus. With the advent of relatively inexpensive videotape players, some distance education courses have included the exchange and viewing of tapes. Sometimes science teachers might work at their own speed with a single set of tapes. The purpose of such tape usage was, in theory, to bring a more realistic "classroom" to students. For science teacher educators, such correspondence and videotaped courses required some new teaching and assessment techniques, in light of the non-real-time interaction of the science educator and student that would normally take place.

With decreasing costs of technology and telephone line charges in the last decade it has become easier for science educators to organize interaction between an instructor and students at a distance using a speakerphone. With the use of the speakerphone, an initial attempt was made to create a classroomlike experience for students in which science educators can use teaching techniques they know to be effective. Clearly, even for a science educator who has not led or participated in a class using speakerphones, it should be obvious that the situation may be lacking in personal interaction.

A Critical Question

A critical question in teaching by distance is whether the instructor should attempt to simulate aspects of normal classroom transactions, or adopt a totally different set of pedagogies. Central to this question, as alluded to earlier, is the nature of instructor–student interaction. In general, the issues impacting science education and the use of various distance education technologies are influenced by the same factors that concern educators in other fields. However, because hands-on interactive experimentation is such an integral component of the work that science educators carry out, there may be an inclination to explore and push the boundaries of distance education more so than those in other disciplines. For those who may think that the use of distance education technology is just the commonsense application of what science educators think they know about classroom teaching—beware. It is true that most of what science educators view as important classroom considerations transfers to what should take place at a distance, but the transfer is often not an easy one. Speak to science educators who have attempted to utilize the technology and they will quickly reveal that it is easy to imagine how the technology might be used, but it is actually much more difficult to use the equipment. Seemingly simple teaching techniques often are used only after a science educator attempts and fails with a wide variety of teaching techniques. In this work, the term *science classroom* will be used to describe any setting in which science educators may work—be it with elementary students, preservice teachers, or active teachers.

INFORMATION TECHNOLOGIES

The use of print material, audio- and videotapes, Web sites, CDs, and DVDs all allow communication from the instructor to the student. Students can attend to the communication how and when they choose. That is, these are asynchronous forms of communication. Communication from students to the instructor is more problematic. Mail, fax, and e-mail are common asynchronous means at a student's disposal, and standard telephone offers an easy, real-time, two-way form of interaction. Teleconferencing, Internet chat sessions, and videoconferencing allow real-time, two-way interactions with the option of multiple participants. However, these need to be organized in advance and require specialized information technology equipment.

There may well be emergent pedagogies that have the potential to be equally effective using information technology in distance education, compared to face-to-face teaching. However, because elementary science teacher education is so dependent on classroom interaction (see chap. 8), this is convincing evidence that instructor–student and student–student interaction are both essential; and as much as is possible in distance education, these should approximate the type of interaction experienced in a normal classroom. This poses particular challenges in both the choice, and operation, of information technology. The obvious technology to achieve this is videoconferencing.

This chapter is based on a decade of distance education teaching by the author. It takes into consideration only science teacher education at a distance utilizing two-way audio/two-way video technology. Asynchronous distance education (using e-mail or other similar course delivery methods) is not discussed.

The remainder of the chapter focuses on synchronous distance education via videoconferencing, which represents a distance education innovation of use to science educators. In recent years there has been growth in the utilization of Web-based modes of distance education. Those techniques are evolving and can easily be considered in light of the logistical issues presented in this chapter. The distance education classroom of the future used by elementary science teacher educators will probably be one in which a variety of techniques are used, many of which are still in their infancy. However, the key issues raised in this chapter will be useful whatever the form of the equipment used to teach at a distance.

A number of studies have been carried out considering a range of technologies that have allowed students at a distance to attend a class. Only a few of these studies involve science educators. This is because little work has been done specifically with the unique issue of science educators utilizing distance education. However, some past studies do provide a useful overview of relevant research. Beare (1989) reported on the use of videotape, audiotape, and telelecture for teacher education. Dallat, Fraser, Livingston, and Robinson (1992) and many others evaluated videoconferencing issues. There have always been some researchers interested in the use of distance education technology for science education purposes (e.g., Charron & Obbink, 1993; Hoefkens, Diening, Berez, & Erdelyi, 2001; Jackson, 1998; Mose & Manley, 1993; Tresman, Thomas, & Pindar, 1988; Whitelegg, Thomas, & Hodgson, 1990), but these efforts do not appear to have involved highly interactive classes. This probably has to do with the fairly recent ability to cost-effectively interact using two-way audio/two-way video. A number of authors have also attempted to evaluate the achievement and/or attitude of those taking distance education classes—although those studies did not exclusively consider the topic of science (e.g., Martin & Rainey, 1993; Outwit & Nelson, 1991). Researchers have, of course, considered distance education issues with regard to other subject areas (e.g., Hummel & Smit, 1996; Sharp, 2001).

Multiple Site Videoconferencing Technology

Since the early 1990s, the decreasing costs of videoconferencing technology have made it feasible to offer science education classes utilizing so-called two-way audio/two-way video technology. This technology, ever evolving, allows video and audio interaction to take place between numerous remote sites and a studio or classroom. It is this technology that will be briefly reviewed to provide the reader with a framework that will more easily allow consideration of the ways distance education technology can be used in science education.

When two-way audio/two-way video technology is used, an optimal site would include a television, television speaker, a separate speakerphone, and a control panel of some sort. Other key pieces of technology are a document camera (a vertically mounted, adjustable television camera), a computer, and a video recorder/player (VCR). The document camera can be used as an overhead projector of sorts, allowing small objects to be more easily shown to distant sites. The computer allows transmission of digital pictures as well as e-mail communication. The VCR can be used to share video footage of classrooms that has been recorded by participants in the days between broadcasts, or to broadcast other film or video footage. Less expensive equipment may utilize a computer that can record video images and provide audio. The television should ideally display all other remote sites and all participants should be able to receive audio from all sites. The VCR will need to be replaced with a DVD player/recorder as the availability of DVDs increases. A combination VCR/DVD player may be most appropriate during the transition period of technology changeover.

There are many permutations regarding the technology available for science educators, and the type of technology used has considerable impact on the quality of science education delivery. To understand the variations, it is helpful to outline some common differences that may exist from one distance education system to another. For instance, some technology may only show selected remote sites and often may not display the studio site. Thus, sometimes a science educator may lead a class from University B to Remote Sites 1, 2, and 3. But all three remote sites may only be able to see University B, although they can hear the other two sites, as well as the studio. Other configurations of the technology may be systems in which any site can be displayed, but only one site at a time. Thus, video images from University B might be displayed at all three remote sites, or video images from one site (e.g., Site 2) can be displayed at all other sites (University B, Site 1, Site 3). Audio technology also varies at times from broadcast system to broadcast system. For example, when some technology is used it is only possible for one person to speak at a time, whereas other systems allow audio from any and all sites to be broadcast simultaneously. The system that permits simultaneous audio and video interaction between all sites lends itself to the creation of a classroomlike environment for science educators.

A relatively new system, brand-named *Poly Com*, uses a simple hardware interface and an Internet connection to allow up to four sites to view and speak to each other at the same time. This system also allows for one site to be displayed at a time. The camera is controlled from a remote site, and site switching is based on who is speaking. The advantages of this system are its low cost (no dedicated line connection is needed) and ease of connection and use. It also allows the distant sites to be located anywhere on the earth where there is access to the Internet, as long as all sites have an inexpensive *Poly Com* unit (personal or room-type). The disadvantage is that it requires a fair amount of bandwidth on the line connected to the Internet. A lack of bandwidth can result in a breakdown in video signal, audio signal, or both, as the videoconferencing traffic competes with other traffic on the line. In addition, there is a significant lag time (2–3 seconds) between one site transmitting and the other receiving.

PROBLEMS AND PROMISE FOR SCIENCE EDUCATORS UTILIZING DISTANCE EDUCATION TECHNOLOGY

Does the technology allow educators to practice what they preach? A goal of science educators is that effective science education involves meaningful interaction between science students and science educators. This means no "sage on the stage" performances, the use of cooperative learning groups, and the careful consideration of many topics such as the quality of questions posed and wait time. Acknowledgment of the importance of such issues suggests that distance education technology, when functioning optimally, should facilitate all of the strategies already noted and more. Thus, there should be easy interaction among participants no matter their location. This requires that a distance education technology should be in place that allows a science educator to not only facilitate learning, but also to allow for the opportunity to easily evaluate and improve a course. Presently, two-way audio/two-way video technology optimally (among distance learning configurations) supplies participant–instructor interaction that approximates true classroom interaction. This may well change over the next decade due to technological developments, economics, and convenience.

Key Issues

Considering those aspects of a normal elementary science teacher education classroom that may need to be transferred to distance education teaching helps greatly when assessing issues impacting the use of any distance education technology. For instance, when prospective elementary teachers participate in a science methods class, they can see and hear the science educator as well as other participants. Likewise, the science educator can see and hear the participants. In a real classroom, everyone can look easily about the room and hear one another. The instructor can walk around the room, ask the entire class questions, interact with small groups of participants, and direct questions to individuals. Everyone can move about the room. Projects can be carried out and then easily be shared with all involved. Transferring these requirements to distance education led to the earlier assertions about instructor–student and student–student interactions, because these same conditions for interaction are needed in a distance education classroom. The following are key issues of importance to science educators using this technology.

Video of All Sites. Optimal science teaching at a distance takes place if all participants from all sites can see and hear one another throughout a class. This allows them to interact and reflect on that interaction. What happens when all sites are not visible and may not be heard? First, sites with people who have questions, for instance, about the procedures for an experiment, may not be able to pose such questions in a timely manner. When all sites cannot see and hear one another, an instructor must check in with each site while other sites wait until the check is complete. Without two-way audio/two-way video, remote sites are also less able to demonstrate to other sites. For instance, participants at one site might wish to demonstrate how they interpreted the development of a model with hands-on supplies. If multiple site two-way audio/two-way video is not used, then participants at remote sites will not be able to interact with those at other sites. Another issue is classroom management. It is critical for an instructor to be able to view what is taking place at all sites, just as a good instructor would review what is taking place within a regular classroom. For a science educator considering the use of this type of technology, it is critical to work toward being able to see all sites, and to be seen by all sites simultaneously. Some individuals claim that being able to display a site on cue is enough, but experience attests that not having all sites viewable makes work very difficult.

Quality of Video and Audio. The size of video images is also a consideration for those interested in using distance education technology for science education delivery. When many participants and/or multiple sites need to be displayed, it is very helpful to utilize a large-screen television. If too many separate images (many sites) are displayed on a small screen, then it becomes increasingly difficult to manage such a class. If too many participants need to be displayed (e.g., a large class), then the images of individuals can be very small. The use of a large television screen allows larger images of multiple sites to be displayed (and also allows individual participants to be viewed more clearly). Also a large-screen display provides other benefits—for instance, participants can more easily show what they are doing. If, for example, laboratory work is being conducted or if specific items have been developed as part of an activity, laboratory materials and such items can be more easily displayed to the instructor and other participants. In sum, when viewing multiple sites, a large screen in the studio allows the instructor to see larger images of remote sites, facilitating interaction and better classroom management.

Different broadcast techniques are used to transmit video. Some methods result in crisp images, which do not blur when a person moves quickly. In general, science educators can adapt to blurred or slowly moving images. For science education delivery, video is critical, but it does not have to be perfect. The quality of audio, however, is very important. If audio is garbled, then it makes a transmission almost worthless. For science educators, therefore, both audio and video are important. Sometimes video and/or audio will cut in and out of a site, which is manageable. However, if a connection is to be temporarily lost, then it is better to lose video rather than audio.

ADAPTING DELIVERY TO DISTANCE EDUCATION

Distance education technology presently allows science educators to develop virtual science classrooms, but many logistical issues (which are well understood in a typical science classroom) must be addressed, and often confronted anew, in a distance learning science classroom.

Publicizing the Science Education Course

Confronting the issue of public relations is not new for science educators as full-time teachers are encouraged to participate in professional development, be it a summer grant or a course. Presenting a hands-on inservice course to full-time science teachers using this technology requires the instructor to carefully think about how to best publicize the course. Advertising is more difficult than in a traditional course because for many participants, there would be the automatic assumption that an interactive laboratory-based course would not be possible with distance education. Science educators must develop a number of techniques to explain the nature of the distance education offering, in terms of both content and organization. If a science education distance education offering (e.g., a formal course or an informal inservice session) has been previously presented, then it is helpful to create a short videotape (perhaps 5 minutes in length) that shows the science education effort in action.

A technique that can be used to publicize the offering of a science education experience by distance requires the identification of people who can work at "the remote site level" to increase the level of participation. Most often, such people oversee the technology at the remote site and/or are heavily involved in science education issues. Enlisting such people allows them to knowledgably publicize the effort with regard to both the science education delivery and the technological innovation. Potential participants may be located at a considerable distance from the instructor's institution, and may not think that the university or college would offer a science education course in their geographical area, simply because of the distance involved. Again, it is important that people at remote sites be enlisted to find out how best to publicize an offering in those areas.

Other public relations considerations involve how to describe the offering to those who may not have previously taken a distance education course, and how to attract those who may not have a predisposition to enroll in a science education course. Unless both of these issues are taken into account when publicity is designed, there is an increased chance that participation will be low. Science educators are already familiar with some of these issues with efforts to ensure enrollment in inservice contexts, but distance education provides a greater challenge requiring a creative response in order to let others know what is being offered.

Setting Up Remote Sites

There are a number of issues to be considered when setting up remote sites. The first is who initiates the setting up of the site. Often this is a university "offering" to make a school a remote site. In this case, the university is attempting to convince those at the remote location that they need the course. This is probably the most difficult to set up unless some sort of incentive can be offered to course takers, such as free or greatly reduced tuition costs. However, if a remote site is requested by a school administrator, such local backing can be used to ensure that some level of organization and public relations are in place. The best scenario is when both teachers and principal of a local school request that their school serve as a distance education site.

The second issue is whether a planned site is viable, which is related to its geographic location. A viable site is one where a critical mass of participants finds it convenient to attend. In practice, this means that a number of teachers from the school hosting the site must enroll in a course and teachers from other schools must also find it convenient and worthwhile to attend at the remote site. I have found that the geographic location of the site is perhaps the most important consideration for teachers outside the school. Ease of access is the main issue. If teachers in a district tend to live and/or work in another part of the city, their commute time will determine whether they are willing to attend. Related to this is ease of parking for those from other schools attending. If parking is difficult, or if there is a perception that a neighborhood is dangerous, then that can affect the viability of a remote site. Finally, consideration needs to be given to whether or not there is a competing course offered at a local university, and the location of that university in relation to the remote site.

A third consideration regarding the selection of a remote site is the accessibility of equipment and the availability of technical support before, during, and after broadcasts. Equipment is commonly located in a school library or in a room dedicated to technology. Because of security issues, these rooms are often locked after school hours. However, the most successful courses for teachers are those that start at the end of the school day and are offered once or twice per week. This means that a course will often run past the end of the work day. Some schools and districts have technical support staff, but often such people are on call (they may even be at other locations), and at the end of the work day they may even be unavailable. As a result, it is almost a necessity to have at least one site participant from the school serving as a remote site coordinator. The site participant must have access (keys) to the room with the equipment, as well as access to phones and fax for incoming and outgoing calls and transmissions. If at all possible it is best to enroll a teacher from the host school (who is familiar with the technology) in the course. That teacher might be charged at a reduced tuition rate, or may receive the course for free.

Finally, establishment and ongoing costs need to be considered. The costs of equipment and line charges can be high. Whoever sets up the site will determine who pays these costs, and whether or not they are factored into course fees. Often, schools in a district may initially be given equipment, or purchase it at a low cost, and then be provided with an amount of free or reduced cost broadcast time. It is understandable that at this point, districts and administrators are keen to utilize the equipment. Consequently, when costs are low it is possible to have lower student enrollment than when the full costs must be covered. This helps when a course is being offered for the first time. When charges are at a full rate, then it becomes very important to ensure that agreements can be developed to reduce costs for the school, attending participants, as well as the organization providing the course. At some universities, selected student charges can be waived when students are not attending at the main campus.

Preparing Future Participants for the World of a Virtual Classroom

Another issue that science educators must confront is that those participants who decide to take part in a science education course using distance education must be prepared for the differences between distance education and a traditional face-to-face offering. It is essential to do this so participants' expectations are realistic and potential difficulties can be minimized. First, almost certainly there will be technological difficulties, both minor and major, associated with the technology-mediated interaction. Participants must be prepared for the range of problems that may occur. It is essential that science educators leading a distance education class review what to do if the technology fails (e.g., if audio disappears from one or more sites, or if a link to a site never comes up during a broadcast). Second, science educators must explain how the science distance education delivery will differ from a traditional face-to-face class, especially to explain protocols to follow with multiple site discussions, and the limitations of interaction, field of view, and level of visible detail. It is often helpful to ask past participants to attend a new class and provide their own perspective.

It is also very important that both participants and the instructor are prepared to deal with fewer body language and gesture cues that are normally used to guide interaction between people speaking with one another. Procedures for orderly verbal interaction between participants must be instituted so that all know how, and when, to interact productively.

Clear expectations are critical. If a science educator is presenting a course, then key documents (e.g., class syllabus, assignments, activities, materials, and course assessments) should all be provided at the beginning of the course. Internet-based course management software, such as Web CT or Blackboard, may be used to accomplish this as a supplement to the videoconferencing. These materials are accessible to students on a 24-hour/7-day-a-week basis. If Internet-based support is not available, then other forms of printed information and communication procedures need to be established. If these procedures are followed, much of the confusion and uncertainty associated with not being able to interact with the instructor in the same way as if you were physically present will be greatly diminished. Further, if a broadcast to a particular remote site is not possible for a time due to technological problems, then participants at that site can still carry on to a degree.

Supplies

The types of supplies needed for a distance elementary science teacher education offering will vary according to the relative emphasis given to discussion, workshops, hands-on activities, and science laboratory work. Obviously, laboratory work must be limited to what is safe, and can be done outside of a science laboratory. The main constraint on supplies will be the ease or difficulty with which materials can be made available at remote sites and, in particular, whether necessary supplies can be cheaply and safely transported to remote sites. Sites should have similar, if not identical, equipment.

The adequate distribution of science supplies can significantly affect the quality of a course. Although specialized equipment may have to be delivered or mailed, the best option is to develop a syllabus in which readily available materials are utilized. For instance, if elementary teachers take a science education course by distance, then supplies at a typical school and/or those that are easily purchased locally should be used. Instructors need to make sure they provide ample warning to participants regarding the materials needed for a future broadcast. Organizing future supplies can be greatly impacted by technological failures if, for example, a link to a site can go down before an instructor has been able to review what is to be brought to the next session. For this reason, equipment requirements should preferably be outlined in a class schedule included in the syllabus. This might seem to be common sense, but science teacher educators usually only organize materials in detail a few weeks in advance, and rarely plan that far ahead with respect to supplies.

Class Size

There is a limit to the number of sites that can be concurrently handled by an instructor and there is a limit to the number of participants who can be enrolled. The first key issue has to do with the technology. If groups of students attend at specific sites, then there is a limit to the number of participants who can be present if they are to be easily heard and seen on a consistent basis by other sites and the studio. So that participants can feel a link to the studio as well as with other remote sites, there needs to be a limit on the number of sites that can be used. This has to do both with the mental challenge of remembering who is at which site, and the consequent size of video images resulting from simultaneously displaying many sites. The more sites one attempts to show at once, the smaller the image of each site. However, there are some techniques that can be used to maximize the number of sites and participants. For instance, with many remote sites, it is possible to limit the number of times all sites are displayed at any one time. Also, pictures of participants can be distributed in paper copy to each participant. This allows participants to memorize more easily the people at different sites. The technology described in this chapter makes it possible to manage approximately 10–15 participants at each remote site for a total of up to five sites. If fewer participants are present at remote sites, then around eight remote sites is a manageable number.

SERVING AS A SCIENCE EDUCATOR AT A DISTANCE

An elementary science teacher educator launching into distance education for the first time must consider how to make the offering comparable to a face-to-face class. For instance, the National Science Education Standards (National Science Teachers Association, 1998) must apply equally to traditional and distance education settings. This monograph provides detail of a range of themes that influence elementary science teacher education (e.g., science pedagogical content knowledge and elementary school teachers, dilemmas and directions for field experiences in elementary science teacher preparation, and inquiry). It is important to consider how such topics should be dealt with in a distance education classroom. For instance, issues regarding field experiences, such as the recruitment of sites, and the monitoring of quality from a distance, can and must be used when designing distance education classrooms. Aspects of inquiry presented in chapters 13 and 14 could readily be adapted to distance education.

Bumps in the Road

As mentioned earlier, technological problems can and will occur for those using videoconferencing, especially the more complex multiple site two-way audio/two-way video technology. The most crucial problems are the failure of audio and/or video to one or more sites. Such failures can occur at any time, and may take place prior to the start of a class. It is best to have a range of alternative techniques that can be used to contact a site. Such methods include fax machine, phone line, e-mail, and cell phone. Unfortunately, these are not always available in the room housing the videoconferencing equipment at a local remote site. Without the ability to communicate from studio to remote site, each group affected by the failure of the audio and/or video links has no way of knowing what is being done to rectify the problem. A related issue to the loss of links is the degradation of audio and/or video quality during a broadcast. For instance, the audio signal from one site may be reaching all other remote sites as well as the studio. But it could be the case that static from one site may compromise communication from all other sites to the studio. In such cases, it may be to the instructor's advantage to sever part of the link with a site. For example, the audio link may be cut off intentionally, but the video link may be retained. This is done to salvage quality interaction with other sites. Thus science educators utilizing this technology may find that they have to limit the extent of the technology used at one remote site in order to maximize success at other remote sites during a broadcast. Both participants and instructor need to have preplanned arrangements about how to deal with such events. Improvisation is just as necessary in the distance education classroom as it is in traditional classrooms. For example, all sessions may be videotaped, and the tape sent to any sites that experience problems. The presence of a local coordinator makes local improvisation much easier.

Evaluation

As in any study within science education, there are a range of techniques (qualitative and quantitative) that can be used to monitor delivery. Because the number of teachers who commonly can be accommodated for a set of broadcasts should be limited to 30 or fewer (due to the issue of screen image size), the efficacy of quantitative data gathering tools such as tests and attitudinal surveys is greatly dependent on the robustness of the measurement device used. This is simply because the low number of students does not allow for extensive statistical manipulation. Tried-and-true measures that have been validated in other studies are best. A robust attitudinal survey can allow distinctions to be made as a function of subgroups. However, if a measurement device is not well targeted, then the sample size of teachers participating is often too small to monitor changes. If a course is offered a number of times, then it is possible to collect quantitative survey data for a number of semesters, and then evaluate the data set.

I have used other quantitative techniques to monitor the quality of broadcasts and to improve distance education courses. For instance, I have recorded each broadcast and later monitored the amount of time the instructor spends at each site, how often sites interact, and a number of other issues. Such investigations provide a wealth of information, but the analysis of data is time consuming. It is likely that in the future some automated techniques to collect data, such as time spent with particular remote sites, might become available.

Qualitative techniques—such as interviews, open-ended surveys, and the review of broadcast tapes—can be revealing. Follow-up interviews with past participants can prove particularly helpful. Several articles (Boone, 1996, 2001) describe a variety of tools that can be used for evaluation of distance education classes in science education.

Perhaps the best way of evaluating the delivery of distance elementary science teacher education courses is to compare the distance education version with what can take place in a traditional classroom. In a productive science education classroom, a range of sharing and interaction should occur. How did this differ in the distance education classroom? Why?

SOME FINAL THOUGHTS

In a distance education classroom, an instructor cannot easily change the room arrangement, where participants are positioned, or how tables are arranged. Moreover, the television monitor cannot be quickly moved. As a result, participants at remote sites must be positioned so they are easily visible on the screen that presents their site. Participants must not move off the screen during the course of a broadcast. This often requires the instructor to remind those at remote sites to sit so that they are easily visible to the instructor and the other sites. The size of the image of each participant is also very important. The larger the image, the more lifelike can be the interaction of students and teachers. To accomplish a large image, two techniques can be used. One is to have students sit grouped as close to the camera as possible. Another technique is to use the camera remote control to best frame the group by tilting, panning, and zooming. Some distance education technology allows for so-called dynamic quad. This is a technology that allows a particular site to be broadcast full-screen to all remote sites and a studio. This results in larger video images being provided. What seems to work best? All of these techniques, to some degree, help foster interaction across sites and with the studio. The key responsibility falls, however, on the shoulders of the instructor to do what needs to be done so that all participants are involved in the broadcast.

The dynamics of a science education/distance broadcast can also be shaped greatly by the verbal interaction that occurs. In a distance science education setting of the sort described in this chapter, verbal discussion can clearly take place in a traditional manner among those participants attending a particular single site. The level of the interaction between sites and the studio is dependent on the location, quality, and use of microphones. Sometimes microphones are portable so that they can be passed from participant to participant at a particular site. Sometimes microphones are located in the ceiling or in the broadcast unit, which requires participants to direct their voices toward the microphone. In all of these scenarios, it is up to the instructor to guide the use of the microphones so that all sites can hear students, but the verbal guidance must be made in such a way as not to disrupt the flow of the class. In a well-functioning distance education class, participants learn when and how to move microphones and cameras to create a classroomlike environment without continual guidance from an instructor.

In the case of hands-on science distance education classes, often presented in a room that lends itself to videoconferencing, but not necessarily science teaching, there are times when participants will have to carry out activities off camera and beyond the range of microphones. If participants have to conduct activities out of microphone or camera range, then there must be a method devised to ensure that they check with the instructor periodically.

This chapter has not attempted to summarize the types of distance education methods that can be used for delivery of elementary science teacher education courses. However, it has tried to present some selected nuts and bolts issues that need to be addressed by science educators when one type of distance education technology is used. This chapter should help the reader understand that optimal distance education technology should provide for many of the same interactions seen in a traditional setting. Different techniques, however, may have to be used by the instructor to encourage interaction, monitor hands-on experiments, and carry out all the other logistics of a science classroom. Science education with distance education technology is a challenge because some tasks that are quite simple in a traditional setting become much more difficult in a distance education format. Two research studies involving many years of effort to utilize distance education for science education efforts are a good starting point for those interested in further exploring this issue (Boone, 1996, 2001).

Distance education holds the promise of helping elementary science teacher educators work with many groups, in particular, full-time teachers. Full-time teachers may be interested in learning about new science teaching techniques, but they have so little free time. Distance education technology that allows teachers to attend a virtual class at local sites (often their own school) can be potentially helpful. The reduction or elimination of commuting time can make it much more attractive to both participants and instructor to take a class, and with thoughtful teaching techniques and planning, a useful and productive experience can take place. A danger is that distance education science methods courses can be of dubious quality. Such courses can have little group interaction, and some may not provide hands-on, minds-on activities. This may be due to factors such as poor technology, inadequate planning, or merely the inexperience of the instructor with distance teaching. It is up to each elementary science teacher educator considering using distance education to consider carefully and address the many logistical and pedagogical issues raised in this chapter. If those issues are confronted, then a quality course can be the result. Distance education technology has the potential to make good courses available to a larger population, significantly increasing the number of elementary teachers who are using research-based science pedagogy in their classrooms.

REFERENCES

Beare, P. L. (1989). The comparative effectiveness of videotape, audiotape, and telelecture in delivering continuing teacher education. *American Journal of Distance Education, 3*(2), 57–66.

Boone, W. J. (1996). Evaluating selected aspects of quad screen technology used in a hands-on science, distance education course. *International Journal of Educational Telecommunications, 2*(1), 29–44.

Boone, W. J. (2001). A qualitative evaluation of utilizing dynamic quad screen technology for elementary and middle school teacher preparation. *Journal of Technology and Teacher Education, 9*(1), 129–146.

Charron, E., & Obbink, K. (1993). Long-distance learning: Continuing your education through telecommunications. *The Science Teacher, 60*(3), 56–59.

Dallat, J. G., Fraser, R., Livingston, R., & Robinson, A. (1992). Teaching and learning by video-conferencing at the University of Ulster. *Open Learning, 7*(2), 14–22.

Hoefkens, J., Diening, L., Berez, M., & Erdelyi, B. (2001). The WebCOSY system for course management in distance education. *Journal of Computers in Mathematics and Science Teaching, 20*(3), 307–321.

Hummel, H., & Smit, H. (1996). Higher mathematics education at a distance: The use of computers at the Open University of the Netherlands. *Journal of Computers in Mathematics and Science Teaching, 15*(3), 249–265.

Jackson, M. D. (1998). A distance-education chemistry course for nonmajors. *Journal of Science Education and Technology, 7*(2), 163–170.

Martin, E. D., & Rainey, L. (1993). Student achievement and attitude in a satellite-delivered high school science course. *American Journal of Distance Education, 7*(1), 54–61.

Mose, D., & Manley, T. (1993). An experiment in distance learning in geology. *Journal of Computers in Mathematics and Science Teaching, 12*(1), 5–18.

National Science Teachers Association (NSTA). (1998). *Standards for science teacher preparation.* In collaboration with the Association for the Education of Teachers in Science. Retrieved November 30, 1998, from http://www.nsta.org

Outwit, B. J., & Nelson, K. R. (1991). Meta-analysis of interactive video instruction: A 10 year review of achievement effects. *Journal of Computer-Based Instruction, 18*(1), 1–6.

Sharp, J. (2001). Distance education: A powerful medium for developing teachers' geometric thinking. *Journal of Computers in Mathematics and Science Teaching, 20*(2), 199–219.

Tresman, S., Thomas, J., & Pindar, K. (1988). The potentiality of distance learning. *School Science Review, 69,* 687–691.

Whitelegg, L., Thomas, J., & Hodgson, B. (1990). The Heinekin principle re-visited: Collaborative presentation of retraining courses for physics teachers. *Physics Education, 25*(3), 158–162.

5

Challenges and Opportunities for Field Experiences in Elementary Science Teacher Preparation

Sandra K. Abell
University of Missouri–Columbia

Field experiences are considered the *sine qua non* of teacher preparation programs. They come in many shapes and sizes. They occur in schools with informal connections to the university, or in more formal collaborative partnerships, such as laboratory schools and professional development schools (Book, 1996; Byrd & McIntyre, 1999; Darling-Hammond, 1994; Goodlad, 1991; Holmes Group, 1990; Shroyer, Wright, & Ramey-Gassert, 1996; Teitel, 2000). Field experiences typically take place at the beginning of teacher preparation programs to initiate students into thinking like a teacher and to help them decide if teaching is the right career choice. They continue in conjunction with methods courses,[1] where preservice teachers examine how children learn subject matter and have the opportunity to test specific instructional strategies. They culminate in student teaching and other intensive forms of internship, where beginning teachers take primary responsibility for all classroom instruction. For future teachers of elementary science, these field experiences involve teaching reading, mathematics, and social studies in addition to science. This chapter focuses on field experiences over which science teacher educators can

---

[1] I have chosen to use the term *methods* course throughout this chapter for expediency, although some have commented on the mismatch between the term and constructivist views of teacher learning (V. Lunetta, personal communication, January 10, 2002). Because the science teacher education community does not have a suitable agreed-on substitute, I have opted for simplicity of terminology over political correctness.

leverage the most control in elementary teacher preparation—field experiences associated with a science methods course.

Most science teacher educators have heard an all too frequent lament from preservice and practicing teachers with whom they have worked: These teachers claim to have learned little in their teacher preparation programs outside of the "real" experiences they had in field experiences in elementary classrooms. Perhaps with this view in mind, education reform documents in the United States have, for many years, recommended an increased emphasis on field experiences in teacher preparation (Carnegie Forum, 1986; Committee on Science and Mathematics Teacher Preparation, 2001; Holmes Group, 1990; National Commission for Excellence in Teacher Education, 1985), often tied to stronger university–school partnerships. More recently, in the advent of teacher shortages (especially in science and mathematics), many states in the USA have developed new teacher licensure rules that permit teachers to teach with little or no teacher education preparation (e.g., Missouri Department of Elementary and Secondary Education, 2002). This is called "learning on the job"—the ultimate form of field experience.

Yet, what is the basis for continuing and expanding the role of the field experience in science teacher preparation? What is the potential for student learning in the field? What are the constraints? Is it viable to continue to invest in field experiences for teacher education programs? If so, what should these field experiences look like and where should they take place? These questions frame this chapter. The first part of the chapter lays out the empirical and theoretical foundations for thinking about the science education field experience. The next section tells the story of developing and enacting field experiences in elementary science teacher education, and then discusses both the challenges to field experiences and particular models that may provide opportunities to overcome these challenges. The final section formulates recommendations for research and practice.

THE RESEARCH BASIS FOR FIELD EXPERIENCES

The use of field experiences in elementary teacher preparation programs appeals to commonsense views of quality teacher education. Field experiences seem right to teacher educators, students, and school-based colleagues: Everyone works hard to develop field experiences that they think will be effective. But will increased practice in the field lead to increased professional growth? Despite the widespread use of field experiences in association with elementary teacher preparation, the empirical support for field experiences is weak. McIntyre, Byrd, and Foxx (1996) reviewed the research on field experiences in teacher education through an analysis of variables, including context, length, supervision, and administration. They reported no clear findings in any of these areas, and concluded that "research on field experience has not been conducted in a systematic fashion . . . and falls short of providing the profession with answers about what works best in field and laboratory experiences" (p. 188). Recent reviews of the professional development school (PDS) literature (Byrd & McIntyre, 1999; Teitel, 2001) offer some insights about field experiences as embedded in the context of the school–university partnership model.

Turning to the science teacher education literature specifically, the research on field experiences is even more spotty. The largest share of what might be classified as field experience research has taken place in the context of student teaching (e.g., Nichols & Tobin, 2000). Studies concerned with the science methods field experience are rare. Earlier publications about the science methods field experience focused on description (e.g., Andersen & Gabel, 1981), or on measuring specific teaching attitudes and behaviors in quasi-experimental settings where field experience was the treatment variable (Harty, Anderson, & Enochs, 1984; Strawitz & Malone, 1986; Sunal, 1980). More recently, Huinker and Madison (1997) examined preservice teacher learning in an elementary methods block that included science, mathematics, and social studies with associated fieldwork in urban schools. Although the researchers claimed that the fieldwork component was "an integral part of the methods courses in this study" (p. 124), their data collection methods limited their ability to interpret the role of the field experience in student learning. Tippins, Nichols, and Tobin (1993) found that the field experience associated with a methods course for prospective middle and high school teachers did not significantly change student concerns about teaching science, but did improve their attitudes toward science and science teaching. Placements in a school where science teachers were excited about enacting a new science curriculum saw the greatest increase in attitudes. Beyond these studies, the review of the science education literature turned up little empirical evidence surrounding the science methods field experience.

The dearth of evidence concerning the field experience in teacher education may be due to the lack of a well-conceived theoretical base for field experience research (McIntyre et al., 1996). The following section provides a brief history of thinking about teacher learning, and proposes a framework for examining field experiences in elementary science teacher preparation, building from the ideas of Putnam and Borko (2000). I believe this framework can provide guidance in developing model field experiences and an associated research agenda.

A THEORETICAL FRAMEWORK
FOR TEACHER LEARNING

Teacher learning as a subject for research has found its voice over the past 25 years. Early views of teacher learning treated teachers as recipients of knowledge created by researchers. Such a view led to the common teacher education practice of presenting future teachers with the results of process–product research so that they could reproduce successful strategies in their own teaching. The emphasis was on teaching as technology (Schön, 1983), not on teaching as an intellectual pursuit. More recent notions of teacher knowledge have recognized teachers as active participants in generating knowledge *in situ*. Fenstermacher (1994) outlined four research programs that have changed the face of teacher knowledge research: Clandinin and Connelly's work on personal practical knowledge through teacher narrative (Clandinin, 1992; Clandinin & Connelly, 1996); Schön's notions of reflective practice for professional development (Schön, 1983, 1987, 1991); Shulman's research program on teacher knowledge types (Grossman, 1990; Shulman, 1986; Wilson, Shulman, & Richert, 1987); and Cochran-Smith and Lytle's leadership in the teacher researcher movement (Cochran-Smith & Lytle, 1993, 1999a, 1999b). These research programs have shifted the perspective from teacher knowledge produced by others to teacher knowledge residing within teachers, from teachers as objects of research to teachers as coresearchers. They also have generated a different set of instructional strategies for teacher education, including reflection on practice, action research, case-based instruction, and a focus on developing pedagogical content knowledge.

Alongside this empirical work on teacher knowledge, a new theory of teacher learning is being fashioned. Current learning theories propose that learning is situated in authentic contexts, which allow learners to participate in communities of practice (J. S. Brown, Collins, & Duguid, 1989; Lave, 1988; Lave & Wenger, 1991; Rogoff, 1990). Although learning theorists may disagree on some of the details, Druckman and Bjork (1994) characterized four common aspects of the situated approach:

1. Action is grounded in the concrete situation in which it occurs. A potential for action cannot fully be described independently of the specific situation, and a person's task-relevant knowledge is specific to the situation in which the task has been performed.

2. Psychological models of the performer in terms of abstract information structures and processes are inadequate or inappropriate to describe performance. A task is not accomplished by the rule-based ma-

nipulation of mediating symbolic representations. Certain task-governing elements are present only in situations, not representations.

3. Training by abstraction is of little use; learning occurs by doing. Because current performance will be facilitated to the degree that the context more closely matches prior experience, the most effective training is to act in an apprenticeship relation to others in the performance situation.

4. Performance environments tend to be social in nature. To understand performance, it is necessary to understand the social situation in which it occurs, including the way in which social interactions affect performance. (p. 33)

One implication of the situated approach is that learners learn in authentic apprenticeship contexts (Lave & Wenger, 1991; Rogoff, 1990), although proponents are light on details of how this process occurs.

The situated learning perspective makes a great deal of sense when applied to teacher learning (see Putnam & Borko, 2000). The authentic context is the elementary classroom, where students of teaching take part in apprenticeships in which they join, peripherally at first, and more fully as time goes on, the community of practice of teaching. Knowles and Cole, with Presswood (1994) proposed a spiral apprenticeship model of teacher learning through field experience. In their model, "experience is the foundation for learning" (p. 10), supported by information gathering, reflection, analysis, and informed action. Furthermore, they believed that experience leads to an inquiry stance for teacher learning, supported by narrative methods of inquiry. Their model attempts to link theory and practice in a situated context. Thus, the situated learning perspective provides theoretical support for the field experience from which to build models of practice; it may be the strongest rationale available for including field experiences in science teacher preparation programs.

However, the situated perspective also creates the ultimate paradox for science teacher preparation. In methods courses, preservice teachers are asked to think reflectively about best practice in science teaching, and they are familiarized with the latest policy documents. The goal is to get them to observe, create, and enact reform-minded practice in real elementary classrooms, because the field experience is the definitive form of legitimate peripheral participation (Lave & Wenger, 1991). However, the desirable kinds of classrooms in which students should serve their apprenticeships quite often do not exist. Because science teacher educators are limited in the field placements that can be provided, students are often placed in classrooms where reform-minded practice is not occurring. This brings up new questions: Into what sort of community of practice are students being inducted? To what ideals and forms of practice are students being apprenticed? How does the apprenticeship theory play out when the context is less than ideal?

CHALLENGES TO THE ELEMENTARY
SCIENCE FIELD EXPERIENCE

I have been involved in teaching future elementary teachers of science in U.S. universities since 1986, and I have witnessed many challenges to the science methods field experience. Some of these are logistical, requiring patience, time, and commitment to solve. Others are philosophical, requiring new ways of thinking about and supporting teacher learning.

Logistical and Institutional Challenges

The immediate challenges that face science teacher educators trying to establish field experiences in association with the science pedagogy course are logistical. Finding enough sites in which to place all of the students enrolled in the course and making sure students have a diversity of field placements throughout their program (e.g., rural, urban, and suburban; placements at different grade levels) are major issues. Once there are enough appropriate sites, then there is the difficulty of scheduling. Elementary teachers in the United States teach science for only a small fraction of the day (if at all), only on certain days, and typically only in the afternoon. Furthermore, science instruction is often interrupted by standardized testing, field trips, assemblies, and the like. University students, on the other hand, take courses throughout the day and week, their course and work schedules tend to be unyielding, and their science methods instructors require that field-based assignments be completed on a given timeline. Even when placement and scheduling challenges have been overcome, there are transportation issues—how to get all of the students and science equipment to the school site safely and efficiently, and then get them back to campus in time for students' next courses. In institutions like mine, where the elementary science methods course has multiple sections or classes, these problems are exacerbated. Communications among course instructors, partner teachers, and students about schedules and expectations become critical. Logistical challenges can appear as insurmountable obstacles to enacting field experiences. Heaven forbid that snow forces the partner school to close on a scheduled field experience day!

Furthermore, institutional barriers await faculty attempting to establish field experiences. These faculty members face realistic concerns about jeopardizing tenure and promotion by devoting time and energy to creating and carrying out field experiences. They face bureaucratic challenges of working with elementary education faculty in other disciplines to institutionalize their work. I have seen such challenges dissuade faculty from dealing with the field experience altogether. Yet strategies have been tried that ease the burden. For example, some teacher preparation institutions hire field experience directors who coordinate placement, scheduling, and transportation. Assigning preservice teachers to classrooms in teams of two to three reduces the need for as many placement sites. Distance technologies help support communication among all participants, but do not replace the need for face-to-face interaction between instructors and school personnel. More radical and long-term strategies involve building partnerships with schools and building field experience blocks into students' schedules. Such changes require faculty leadership and demand faculty time to ensure quality. Educators must continually ask themselves if the benefits of providing apprenticeship opportunities for students of teaching are worth the costs.

Supervision Challenges

For the sake of discussion, assume the logistical challenges have been addressed and the field experience for science methods students is in place. Now comes the challenge of how to operate the field experience. That is, how will students be supervised, by whom will they be supervised, and how will their grades be determined? These questions lead back to the situated perspective on learning. Professors of science education, graduate teaching assistants, and partner school teachers have been inducted into the community of practice of elementary science teaching. However, to be a supervisor demands thinking, not only about teaching science to children, but also about teaching teachers. How do supervisors move from thinking like a teacher, to thinking like a teacher educator? How do they help graduate students and partner teachers make this transition? The answers to these questions will help direct the ways in which they supervise students in field experiences.

The challenge of supervision relates not only to who supervises and how the supervision takes place, but also involves what aspects of teaching these individuals choose to focus on and how they provide feedback to students. Students typically come at this issue by asking, "How will we be graded?" No matter who asks the question, the underlying issue is the same: What is expected of students in a field experience and how should this be communicated clearly to students, other instructors, and supervisors alike? How can a shared discourse be developed in this community of preparing future teachers of science? These are not easy questions to answer. It is understood that students of teaching are novices; they are not expected to excel, they are expected to learn. In field experiences associated with my methods courses, classroom teachers have given immediate feedback to teaching teams, using their own background and prompts that I provide. Because I observe teams on a more limited basis, I give informal verbal and written feedback about practice, but I formally grade only their written work (e.g., lesson plans and reflective writing). However, accreditation and licensure agencies expect performance evidence to be part of the evaluation of teacher candidates. Thus, we must plan for supervision and evaluation deliberately, and ensure that all participants understand the process.

Challenges to Teacher Learning

Once logistical and supervision challenges have been addressed, students are out in the field, teaching science in small and large group settings alongside a veteran teacher. What are they learning and how are they learning it? What can be done to better facilitate learning? Experience shows that practice does not necessarily lead to learning, expertise, or improvement. For example, I practiced tennis for years, but never became worthy of an above average opponent. Furthermore, there are teachers who have taught for many years who appear not to have learned from their experience. Clearly, practice does not make perfect.

Munby and Russell (1994) explained the issue of teacher learning in field settings as one of authority, and raised the question, how can teachers learn from the authority of experience? A body of research on the role of reflective thinking in teacher education (Grimmett & Erickson, 1988; LaBoskey, 1993, 1994; MacKinnon, 1987; McIntyre, 1993; Richert, 1992; Russell, 1993; Russell & Munby, 1992; Valli, 1992; Wildman, Niles, Magliaro, & McLaughlin, 1990) has attempted to demonstrate that teachers will learn from experience if they are involved in reflection activities. It is clear from this research that tools such as autobiography, journals, video analysis, and action research (Abell, Bryan, & Anderson, 1998; Gore & Zeichner, 1991; Knowles et al., 1994; Smith, 2000; van Zee, Lay, & Roberts, 2000) can help students make sense of their field-based experiences and develop as professionals. The practice of placing methods students in teams in their field experiences creates a social setting consistent with the situated framework where students of teaching can plan and debrief together, thus providing opportunities for increased learning.

However, we must return to the question of the quality of the field experience. How and what do students learn when placed with a teacher who does not teach science in reform-minded ways? Such apprenticeships often serve to undermine the efforts of campus-based science educators. Consider whether the trade-offs are reasonable—if student time in field experiences is decreased, then will there be a concomitant drop in learning? Currently, there is little empirical evidence to support or refute this conjecture.

Although this discussion has painted a bleak picture of the challenges facing science methods instructors in instituting field experiences, I am not at all pessimistic about the potential of such endeavors. Instead, I would like to see us continue to design the best field experiences possible given our theoretical framework and work contexts, and then develop a research agenda that will help us learn from our own experience. The next section examines various models of field experience, and is followed by a proposal for a concomitant research agenda.

SCIENCE EDUCATION FIELD EXPERIENCE MODELS

In more than 15 years of teaching elementary science methods, I have experienced and invented a variety of field experience models in conjunction with the methods course. I have also read about and interacted with instructors of other elementary methods courses in an attempt to generate new field experience ideas. It is not the intent here to provide a comprehensive compendium of field experience efforts in elementary science methods. Instead, I outline some models that could become the foundation for further thinking about science teacher education field experiences.

Model 1: The Virtual Field Experience

The virtual field experience addresses logistical challenges to field experiences by avoiding school-based placements altogether. Microteaching (Yeany, 1977, 1978) is the most typical example. Science methods students plan and teach lessons to each other, pretending to be children and their teachers. Reflecting on the situated model of teacher learning, educators recognize that the learning of future teachers may be jeopardized by placing them in an apprenticeship context so unlike what they will encounter in their future teaching.

Other types of virtual field experiences have been expedited via various kinds of distance technology, for example, classrooms linked via video to university courses (Boone, 2001) and e-mail and web-based communities of students and preservice teachers (Abell & Julyan, 1994). Another type of science methods virtual field experience has been accomplished via case-based instruction (e.g., Tippins, Koballa, & Payne, 2002). Cases provide windows into classrooms that are discussed at a distance. In my own teaching of elementary methods, I created interactive video-cases (Abell, Bryan, & Anderson, 1998; Abell, Cennamo, & Campbell,

1996; Abell et al., 1996; Cennamo, Abell, George, & Chung, 1996) to present examples of reform-minded practice to methods students. The virtual field experience has numerous advantages: Instructors can select examples of practice consistent with their beliefs, focus student observations on particular aspects that illustrate specific learning goals, and avoid many of the logistical hassles associated with real-time field experiences. The major disadvantage of most virtual field experiences is that students do not have opportunities to legitimately participate inside the teaching community; rather, they watch from afar.

Model 2: The Add-on Field Experience

When faced with prospects of having no field experience associated with their science methods course, many science teacher educators have attempted adding on a field experience, either bit by bit or as an integrated part of the course. For example, at Queens University in Canada, a field experience was added to the program immediately prior to the science methods course (Chin & Russell, 1996). Through the field experience, students generated problems of practice as they were inducted into the teaching community. These self-generated problems then guided their work in the methods course.

A more typical add-on occurs simultaneously with the methods course itself. My experience as an elementary methods instructor is representative. I began adding field experiences slowly but deliberately to the campus-based methods course at my institution. At first I asked students to complete two assignments in the field. The first, interviewing science learners about their science ideas, was coordinated by the students themselves. They found the children and scheduled the interviews outside of class time. I coordinated the second, a 3-day teaching experience, by asking teachers I knew to volunteer to accept a team of students into their classrooms. I spent hours fitting together schedules and teams as best as I could. Although this was a logistical nightmare, I managed to grow the experience to include four or five course sections each semester, to partner with a large group of local teachers, and to involve students in several classroom-based projects across the semester.

The add-on model, although fraught with logistical challenges, can improve on the virtual experience by asking students to plan and enact instruction and reflect on firsthand evidence of student learning. The add-on field experience that I created evolved over time into a longer and more integrated experience with partner schools in my university's professional development school (PDS) program, and eventually resulted in a separate field experience block, models that are described in the following sections.

Model 3: The Partners Model

The add-on model can be strengthened, and logistical challenges reduced, by developing long-term relationships with local teachers and schools. I have experienced two different, but reinforcing, ways of building such relationships. The first method involves working with local teachers in graduate courses and teacher enhancement projects. In these settings, teachers start to think differently about science teaching and learning, and begin to experiment with their own teaching. These teachers are ideal mentors for future teachers, because they are aware of current trends in the field and are open to new ways of teaching science. Over time it is possible to build a cadre of science enthusiast teachers (Abell & Roth, 1992) who enjoy regular involvement with science methods field experiences.

The second method requires an institutional commitment to building more formal partner schools arrangements, often through efforts such as PDS. When my institution initiated its PDS program, I found myself at the helm, steering us on our partnership journey. Collaborating with schools in terms of teacher preparation, professional development, and research made sense given the situated framework we were trying to build around elementary teacher education. We established four official elementary partner schools, which was a boon for the science methods field experience. We found enough interest among teachers in each school (many of whom had already served as science methods mentors), that we could assign each entire section of science methods to one of the partner schools. A teacher from each site agreed to be the science methods liaison. This greatly facilitated consistent communications concerning expectations and schedules. As a class, we visited the partner school regularly throughout the semester; teaching teams spent time in classrooms, and we also found time to debrief as a whole class right at the school. The immediacy of the reflection on experience became another advantage of the partner school model.

Education faculty across the country have developed school–university partnerships (Book, 1996; Darling-Hammond, 1994; Shroyer et al., 1996; Teitel, 2000). Such partnerships are most successful when they exhibit common goals, mutual trust and respect (Abell, 2000), shared decision making, a commitment from teachers and administrators, and a manageable agenda with a clear focus (Robinson & Darling-Hammond, 1994). Although the work of building partnerships takes time and concerted effort, the potential outcomes are well worth it. The partners model, beyond relieving logistical and supervision challenges, builds a setting where teacher learning through apprenticeship with master teachers becomes the standard.

Model 4: The Field Experience Block

Because the future elementary teacher is often also enrolled in subject matter pedagogy courses in mathematics, literacy, and social studies, it is necessary to explore ways to collaborate with colleagues in these courses in creating field experiences. The motivation is both logistical and philosophical. Logistically, it can be more efficient to schedule students into a field experience block associated with one partner school, rather than have students enrolled in separate field experiences for each methods course. Pedagogically, students can benefit from observing students and their teacher in more than one subject matter setting. At my institution, elementary education majors enrolled in science methods the same semester as mathematics and advanced literacy methods. It became a challenge for methods course instructors to design ways for students to achieve learning goals for all three courses during the semester. What evolved was a sophisticated version of the Knowles et al. (1994) model, with students collecting data, planning and enacting instruction, and assessing the instruction via evidence of student learning in cycles across the three courses. Because this model was only recently put into place, it is too soon to tell what benefits will accrue.

Summary: Design Principles for Field Experiences

Given the situated learning framework introduced in this chapter, as well as the challenges and opportunities that field experiences associated with elementary science methods present, I offer the following guiding principles for their design:

1. Start with the learning goals. What kinds of learning are expected through the field experience? Can those goals be framed in terms of situated versus rule-bound learning? How will educators facilitate and assess their achievement?
2. Provide environments conducive to teacher learning. This includes classrooms where teachers are open to teaching in reform-minded ways, where teachers are prepared to coach newcomers into the field, and where the learning goals are agreed on by all stakeholders.
3. Be collaborative. Successful teacher education is not owned by the university, but shared between the schools and the university. When school-based colleagues are regarded as team members in the field experience, many problems are prevented from ever arising.
4. Ask students to perform and think about their performance in this environment. Students need guidance in learning from experience.

Ask students to think about their teaching situationally, rather than asking them to apply abstract rules to generic situations.

5. Make the action settings social. Teams of preservice and veteran teachers must work together to understand problems of practice. Build discourse communities (see Putnam & Borko, 2000) and make their functioning explicit for all participants.

FUTURE DIRECTIONS

The situated perspective on teacher learning, as well as these guiding principles, suggest a research program built around the science methods field experience. First of all, the notions that action is grounded in context and that a teacher's knowledge is situation specific suggest the need for thick descriptions—case studies and collective case studies of preservice teachers learning from experience. Such research does not take as its purpose the derivation of rules of practice, but rather seeks to interpret teacher learning in field settings. Second, if the key to learning is apprenticeship, then there must be a better understanding of the apprenticeship experience. Ethnographies and phenomenological studies of teacher apprenticeships could help to examine what happens as students of science teaching become teachers of science, and what facilitates and constrains their learning. Also, there is a need to learn more about what happens to elementary teachers as they are apprenticed into the role of teacher educator. Finally, because performance is social, it will be necessary to study the social interactions that occur in field experiences and how they affect performance. Discourse analysis and semiotics, for example, are tools that could be useful to find out how discourse communities within apprenticeships (both among apprentices and between apprentices and veterans) evolve and function to support and constrain learning. All of this research demands collaboration with teachers in the field.

This calls for a research agenda that will help in understanding the essence of the science methods field experience through the concept of apprenticeship, not one that focuses on external forces (e.g., the variables of McIntyre et al., 1996). This agenda would "provide detailed looks at the experiences of individual teachers [and their contexts and colleagues] as opposed to central tendencies in populations of teachers" (C. A. Brown & Borko, 1992). Generating this kind of knowledge about science teaching apprenticeships will enable science teacher educators to improve teacher education programs, lobby for institutionalization of these efforts, and inform other science teacher educators about how to help future teachers learn in context. Ultimately, the goal is to generate a dialogic practice of science teacher education—informed by research that in turn has been informed by practice—within a situated learning framework.

ACKNOWLEDGMENT

I would like to acknowledge and thank Fran Arbaugh for her thoughtful comments on an earlier draft of this chapter.

REFERENCES

Abell, S. K. (2000). From professor to colleague: Creating a professional identity as collaborator in elementary science. *Journal of Research in Science Teaching, 37*, 548–562.

Abell, S. K., Bryan, L. A., & Anderson, M. A. (1998). Investigating preservice elementary science teacher reflective thinking using integrated media case-based instruction in elementary science teacher preparation. *Science Education, 82*, 491–510.

Abell, S. K., Cennamo, K. S., Anderson, M. A., Bryan, L. A., Campbell, L. M., & Hug, J. W. (1996). Integrated media classroom cases in elementary science teacher education. *Journal of Computers in Mathematics and Science Teaching, 15*(1/2), 137–151.

Abell, S. K., Cennamo, K. S., & Campbell, L. M. (1996). Interactive video cases developed for elementary science methods courses. *Tech Trends, 41*(3), 20–23.

Abell, S. K., & Julyan, C. (1994). Educational journeys in college classrooms. In C. L. Julyan & M. S. Wiske (Eds.), *Learning along electronic paths: Journeys with the NGS Kids Network* (pp. 236–243). Cambridge, MA: TERC.

Abell, S. K., & Roth, M. (1992). Constraints to teaching elementary science: A case study of a science enthusiast student teacher. *Science Education, 76*, 581–595.

Andersen, H. O., & Gabel, D. (1981). Science preparation of the elementary teacher at Indiana University. *School Science & Mathematics, 81*, 61–69.

Book, C. L. (1996). Professional development schools. In J. Sikula (Ed.), *Handbook of research on teacher education* (2nd ed., pp. 194–210). New York: Macmillan.

Boone, W. J. (2001). A qualitative evaluation of utilizing quad screen technology for elementary and middle school teacher preparation. *Journal of Technology and Teacher Education, 9*(1), 129–146.

Brown, C. A., & Borko, H. B. (1992). Becoming a mathematics teacher. In D. A. Grouws (Ed.), *Handbook of research on mathematics teaching and learning* (pp. 209–239). New York: Macmillan.

Brown, J. S., Collins, A., & Duguid, P. (1989). Situated cognition and the culture of learning. *Educational Researcher, 18*(1), 32–42.

Byrd, D. M., & McIntyre, D. J. (Eds.). (1999). *Research on professional development schools (Teacher education yearbook VII)*. Thousand Oaks, CA: Corwin Press.

Carnegie Forum on Education and the Economy. (1986). *A nation prepared: Teachers for the 21st century*. New York: Author.

Cennamo, K. S., Abell, S. K., George, E. J., & Chung, M. (1996). The development of integrated media cases for use in elementary science teacher education. *Journal of Technology and Teacher Education, 4*(1), 19–36.

Chin, P., & Russell, T. (1996). Reforming teacher education: Making sense of our past to inform our future. *Teacher Education Quarterly, 23*(3), 55–68.

Clandinin, D. J. (1992). Narrative and story in teacher education. In T. Russell & H. Munby (Eds.), *Teachers and teaching: From classroom to reflection* (pp. 124–137). Bristol, PA: Falmer.

Clandinin, D. J., & Connelly, F. M. (1996). Teachers' professional knowledge landscapes: Teachers stories–stories of teachers–school stories–stories of schools. *Educational Researcher, 25*(3), 24–30.

Cochran-Smith, M., & Lytle, S. L. (1993). *Inside/outside: Teacher research and knowledge.* New York: Teachers College Press.

Cochran-Smith, M., & Lytle, S. L. (1999a). Relationships of knowledge and practice: Teacher learning in communities. In A. Iran-Nejad & P. D. Pearson (Eds.), *Review of research in education* (Vol. 24, pp. 249–305). Washington, DC: American Educational Research Association.

Cochran-Smith, M., & Lytle, S. L. (1999b). The teacher research movement: A decade later. *Educational Researcher, 28*(7), 15–25.

Committee on Science and Mathematics Teacher Preparation. (2001). *Educating teachers of science, mathematics, and technology: New practices for the new millennium.* Washington, DC: National Academy Press.

Darling-Hammond, L. (Ed.). (1994). *Professional development schools: Schools for developing a profession.* New York: Teachers College Press.

Druckman, D., & Bjork, R. A. (1994). *Learning, remembering, believing: Enhancing human performance.* Washington, DC: National Academy Press.

Fenstermacher, G. D. (1994). The knower and the known: The nature of knowledge in research on teaching. In L. Darling-Hammond (Ed.), *Review of research in education* (Vol. 20, pp. 3–56). Washington, DC: American Educational Research Association.

Goodlad, J. I. (1991). *Educational renewal: Better teachers, better schools.* San Francisco: Jossey-Bass.

Gore, J. M., & Zeichner, K. M. (1991). Action research and reflective teaching in preservice teacher education: A case study from the United States. *Teaching and Teacher Education, 7,* 119–136.

Grimmett, P. P., & Erickson, G. L. (Eds.). (1988). *Reflection in teacher education.* New York: Teachers College Press.

Grossman, P. L. (1990). *The making of a teacher: Teacher knowledge and teacher education.* New York: Teachers College Press.

Harty, H., Anderson, H. O., & Enochs, L. G. (1984). Science teaching attitudes and class control ideologies of preservice elementary teachers with and without early field experiences. *Science Education, 68,* 53–59.

Holmes Group. (1990). *Tomorrow's schools: Principles for the design of professional development schools.* East Lansing, MI: The Holmes Group.

Huinker, D., & Madison, S. K. (1997). Preparing efficacious elementary teachers of science and mathematics: The influence of methods courses. *Journal of Science Teacher Education, 8,* 107–126.

Knowles, J. G., & Cole, A. L. (with Presswood, C. S.). (1994). *Through preservice teachers' eyes: Exploring field experiences through narrative and inquiry.* New York: Merrill.

LaBoskey, V. K. (1993). A conceptual framework for reflection in preservice teacher education. In J. Calderhead & P. Gates (Eds.), *Conceptualizing reflection in teacher development* (pp. 23–38). Washington, DC: Falmer.

LaBoskey, V. K. (1994). *Development of reflective practice: A study of preservice teachers.* New York: Teachers College Press.

Lave, J. (1900). *Cognition in practice. Mind, mathematics and culture in everyday life.* Cambridge, England: Cambridge University Press.

Lave, J., & Wenger, E. (1991). *Situated learning: Legitimate peripheral participation.* Cambridge, England: Cambridge University Press.

MacKinnon, A. M. (1987). Detecting reflection-in-action among preservice elementary science teachers. *Teaching and Teacher Education, 3,* 135–145.

McIntyre, D. (1993). Theory, theorizing and reflection in initial teacher education. In J. Calderhead & P. Gates (Eds.), *Conceptualizing reflection in teacher development* (pp. 39–52). Washington, DC: Falmer.

McIntyre, D. J., Byrd, D. M., & Foxx, S. M. (1996). Field and laboratory experiences. In J. Sikula (Ed.), *Handbook of research on teacher education* (2nd ed., pp. 171–193). New York: Macmillan.

Missouri Department of Elementary and Secondary Education. (2002). *Temporary authorization certificate.* Retrieved March 29, 2002, from http://www.dese.state.mo.us/divteachqual/teachcert/tempauth.htm

Munby, H., & Russell, T. (1994). The authority of experience in learning to teach: Messages from a physics methods class. *Journal of Teacher Education, 45,* 86–95.

National Commission for Excellence in Teacher Education. (1985). *A call for change in teacher education.* Washington, DC: American Association of Colleges for Teacher Education.

Nichols, S. E., & Tobin, K. (2000). Discursive practice among teachers co-learning during field-based elementary science teacher preparation. *Action in Teacher Education, 22*(2A), 45–54.

Putnam, R., & Borko, H. (2000). What do new views of knowledge and thinking have to say about research on teacher learning? *Educational Researcher, 29*(1), 4–15.

Richert, A. E. (1992). The content of student teachers' reflections within different structures for facilitating the reflective process. In T. Russell & H. Munby (Eds.), *Teachers and teaching: From classroom to reflection* (pp. 171–191). New York: Falmer.

Robinson, S. R., & Darling-Hammond, L. (1994). Change for collaboration and collaboration for change: Transforming teaching through school–university partnerships. In L. Darling-Hammond (Ed.), *Professional development schools: Schools for developing a profession* (pp. 203–219). New York: Teachers College Press.

Rogoff, B. (1990). *Apprenticeship in thinking: Cognitive development in social context.* Oxford, England: Oxford University Press.

Russell, T. (1993). Learning to teach science: Constructivism, reflection, and learning from experience. In K. Tobin (Ed.), *The practice of constructivism in science education* (pp. 247–258). Hillsdale, NJ: Lawrence Erlbaum Associates.

Russell, T., & Munby, H. (Eds.). (1992). *Teachers and teaching: From classroom to reflection* (pp. 90–108). New York: Falmer.

Schön, D. A. (1983). *The reflective practitioner.* New York: Basic Books.

Schön, D. A. (1987). *Educating the reflective practitioner: Toward a new design for teaching and learning in the professions.* San Francisco: Jossey-Bass.

Schön, D. A. (Ed.). (1991). *The reflective turn.* New York: Teachers College Press.

Shroyer, M. G., Wright, E. L., & Ramey-Gassert, L. (1996). An innovative model for collaborative reform in elementary school science teaching. *Journal of Science Teacher Education, 7,* 151–168.

Shulman, L. S. (1986). Those who understand: Knowledge growth in teaching. *Educational Researcher, 15*(2), 4–14.

Smith, D. C. (2000). Content and pedagogical content knowledge for elementary science teacher educators: Knowing our students. *Journal of Science Teacher Education, 11,* 27–46.

Strawitz, B. M., & Malone, M. R. (1986). The influence of field experiences on Stages of Concern and attitudes of preservice teachers toward science and science teaching. *Journal of Research in Science Teaching, 23,* 311–320.

Sunal, D. W. (1980). Effect of field experience during elementary methods courses on preservice teacher behavior. *Journal of Research in Science Teaching, 17,* 17–23.

Teitel, L. (2001). *How professional development schools make a difference: A review of the research.* Washington, DC: National Council for Accreditation of Teacher Education.

Teitel, L. (with Abdal-Haqq, I.). (2000). *Assessing the impacts of professional development schools.* New York: American Association of Colleges for Teacher Education.

Tippins, D. J., Koballa, T. R., Jr., & Payne, B. D. (Eds.). (2002). *Learning from cases: Unraveling the complexities of elementary science teaching.* Boston: Allyn & Bacon.

Tippins, D., Nichols, S., & Tobin, K. (1993). Reconstructing science teacher education within communities of learners. *Journal of Science Teacher Education, 4,* 65–72.

Valli, L. (Ed.). (1992). *Reflective teacher education: Cases and critiques.* New York: SUNY Press.

van Zee, E., Lay, D., & Roberts, D. (2000, April). *Fostering collaborative research by prospective and practicing elementary and middle school teachers.* Paper presented at the annual meeting of the American Education Research Association, New Orleans, LA.

Wildman, T. M., Niles, J. A., Magliaro, S. G., & McLaughlin, R. A. (1990). Promoting reflective practice among beginning and experienced teachers. In R. T. Clift, W. R. Houston, & M. C. Pugach (Eds.), *Encouraging reflective practice in education* (pp. 139–162). New York: Teachers College Press.

Wilson, S., Shulman, L., & Richert, A. (1987). "150 different ways" of knowing: Representations of knowledge in teaching. In J. Calderhead (Ed.), *Exploring teachers' thinking* (pp. 104–124). London: Cassell.

Yeany, R. H. (1977). The effects of model viewing with systemic strategy analysis on the science teaching styles of preservice teachers. *Journal of Research in Science Teaching, 14,* 209–222.

Yeany, R. H. (1978). Effects of microteaching with videotaping and strategy analysis on the teaching strategies of preservice science teachers. *Science Education, 62,* 203–207.

6

Relating Learning Theories to Pedagogy for Preservice Elementary Science Education

Janice Koch
Hofstra University

In this ever-changing climate of standards-driven curriculum, high-stakes assessments, and conflicts about what elementary school teachers should know and be able to do in science, teacher educators must provide the opportunity for exploring, with their students, the implications of learning theories for elementary science pedagogy. This chapter examines aspects of major learning theories and ways of thinking about learning theories that relate to elementary science instructional practices. Connecting theory to practice is a complex undertaking, requiring thoughtful consideration of the ways individuals learn best and what pedagogies foster those processes.

Learning theories address the importance of the existing knowledge of the learner and the active engagement of the learner, not only with the materials and the phenomena, but also with the ideas that form the basis of the conceptual understanding. Current research emphasizes the importance of understanding the family of theories about knowledge and learning referred to as *constructivism*. This is a major influence on elementary science instructional strategies. It has roots in other related learning theories, an understanding of which informs contemporary elementary science pedagogical content knowledge. The nature of elementary science pedagogical content knowledge (see chap. 3) is addressed in many discussions and has several definitions (Gess-Newsome, 1999; McEwan & Bull, 1991; Shulman, 1986; Smith, 1999). For purposes of exploring learning theories and pedagogy, this chapter uses a transformative model of pedagogical content knowledge, where elements of knowledge from subject matter, pedagogical, and context domains are inextricably combined into a new form of knowledge (Gess-Newsome, 1999, p. 11).

LEARNING THEORIES

In *How People Learn,* Bransford, Brown, and Cocking (2000) remind the reader that asking what teaching technique is best to use is a matter of asking the wrong question. For example, "Are some teaching techniques better than others? Is lecturing a poor way to teach, is cooperative learning effective?" Asking these questions, we are told, is like "asking which tool is best, a hammer, a screwdriver, a knife or pliers" (p. 22). The selection depends on the context, that is, the task at hand. That is, there is no universal best teaching practice. However, understanding a core set of learning principles by focusing on the way people learn helps teachers to make the selection of teaching strategies less daunting.

Learning With Understanding

There are many who feel that children have frequently not been learning science in ways that foster useful conceptual understanding. Newer pedagogies in elementary science education emphasize "hands-on activities" as being only a part of the process of helping children to develop conceptual understandings in science. Conceptual understanding in elementary science differs from the transmission and acquisition of ideas because it involves interpreting, synthesizing, and applying this understanding to new situations (Yager, 1995). It is an active process of meaning making in which the learner engages.

The involvement of students in some interactions with materials is a significant element for science learning to take place. Referred to as "hands-on science activities" the shift from text materials to activities, inventions, and even project-based Olympics represented a revolution in the way educators thought about science instruction and resulted in the development of kits of materials and the proliferation of hands-on science activity modules for the elementary school. Although the use of the term hands-on science came to refer to a general approach to instruction, it also evolved into a guiding philosophy of science instruction (Flick, 1993). Exploring the nature of the engagement that students have with materials helps with understanding that the mere physical manipulation of materials in itself is sometimes presumed to mean that students are learning science. Students do not, however, automatically understand the concepts targeted by an activity just because they are actively engaged with materials. The term hands-on science became expanded to "hands-on, minds-on" science (Duckworth, Easley, Hawkins, & Henriques, 1990), implying a need for students to do more than just manipulate materials and recite terms. Yet, many elementary school teachers believed that if students work together, and complete prescribed tasks with materials, they would somehow learn the science concepts around which the activity is developed.

According to Perkins, "Understanding goes beyond knowing. . . . Understanding a topic is a matter of being able to perform in a variety of thought-demanding ways with the topic, for instance to: explain, muster evidence, find examples, generalize, apply concepts, analogize, represent in a new way, and so on" (1993, pp. 28–29).

Designing pedagogical strategies to foster understanding in elementary science requires examination of the processes by which understanding is built. Hence, helping preservice elementary school teachers examine their own beliefs about the ways people learn and immersing them in an exploration of learning theories is primary.

Piaget and Constructivism

An emphasis on understanding helps with the exploration of what it means to come to know something. Swiss psychologist Jean Piaget contributed to the understanding of this process by exposing the importance of existing knowledge for constructing new knowledge. Piaget (1977/1978) explored the processes by which the learner approaches new experiences with a set of preestablished beliefs and naïve theories. Learners change those beliefs and theories only when they cannot reconcile new data with presently held conceptions. He described these cognitive structures as "schemas," which are used to interpret and make sense of new experiences. Piaget concluded that the growth of knowledge is the result of individual constructions made by the learner. In other words, Piaget determined that knowledge is not passively received, but is actively built up by the learner—a process of invention or creation, not reception. As he recognized, this gives tremendous responsibility to the learner. Piaget's description of cognitive structures, the processes of assimilation and accommodation, and the match or mismatch between new knowledge and existing ideas is the basis for cognitive constructivism.

Constructivism refers to a philosophy of knowledge and a set of learning theories as well as a set of instructional practices. It is used widely in differing ways. As a family of learning theories, it challenges elementary science educators to examine their practices in light of how students learn.

A common theme in constructivist theories is that the learner approaches new experiences with a set of preestablished beliefs and naïve theories and that learners change those beliefs and theories only when they cannot reconcile new data with presently held conceptions. Careful examination of student thinking and theories of constructivism has shown that learners generate meaning through iterative mental formulation and reformulation of theories that satisfy the search for understanding (Brooks, 2000). The classroom is one social context in which students and teachers co-construct knowledge and reflect on experiences with objects, phenomena, and people.

Vygotsky and Social Constructivism

Whereas personal knowledge construction is a key element of the constructivist paradigm, constructing an understanding of a concept does not happen in a social or cultural vacuum. Social contexts influence the ideas that individuals construct as they communicate with each other. Vygotsky (1962) described several social mediators of learning, asserting that the most powerful mediator of learning is language. The teachers and children in a classroom use language that is socially and culturally accepted in that particular environment. From the perspective of Vygotsky, the sociocultural nature of learning suggests that work with other individuals is a critical component of the process of knowledge construction. This theory of constructivism is referred to as "social constructivism" and gives rise to a community-of-learners model within educational settings, a model in which learners in search of understanding communicate current thinking with others by formulating and reformulating their thoughts based on peer and expert feedback and by reflecting on that feedback. Over the past 10 years, the issue of whether learning ought to be conceptualized as primarily an individual activity or as a product of more complex sociocultural processes has given way to a move toward social constructivism (Erickson, 2000, p. 279). This acknowledges the importance of context and community in the process of meaning making.

Refuting Developmental Stages and Bruner

Piaget was a developmental psychologist who also described stages of intellectual development, which are widely taught in educational psychology classes. Beginning in the 1920s, Piaget conducted countless interviews and research studies with children of varying ages. He was able to describe *stages of cognitive development* that the children passed through—that is, specific times in the children's development when different mental structures began to emerge. Although he recognized that children vary, he pegged his stages of development to general age ranges. In describing the stages of cognitive development, Piaget demonstrated that at each stage of maturation, the child is ready for a different type of learning. That is, the types of ideas that learners construct vary with each maturational stage. Some critics of Piaget's stages assert that children can be in several stages at the same time and that the stages cannot be neatly defined by specific ages.

Bruner, a leading supporter of Piaget's work, suggested that at any given stage of cognitive development, teaching should proceed in a way that allows children to discover ideas for themselves. He differed from Piaget in believing that children are always ready to learn a concept at some level. Realizing this, Bruner emphasized the importance of returning to science topics at various ages, revisiting them at different stages of the child's development (Bruner, 1960). Bruner suggested that even very young children could be introduced to scientifically appropriate important concepts in a developmentally appropriate fashion. "Our schools may be wasting precious years by postponing the teaching of many important subjects on the grounds that they are too difficult" (p. 12). Bruner's work stressed that understanding the structure of a subject is understanding it in a way that permits many other things to be related to it meaningfully. Hence, to learn structure is to learn how things are related (p. 7).

Ausubel and Meaningful Learning

In *Educational Psychology: A Cognitive View*, Ausubel (1968) offered this thought: "The single most important factor influencing learning is what the learner already knows. Ascertain this and teach him accordingly" (epigraph). Ausubel was influenced by Piaget and he developed his instructional models based on cognitive structures. He distinguished between "rote learning" and "meaningful learning" and he indicated that for meaningful learning to occur, the learner must have relevant concepts in his existing mental structures, the material itself has to have potential meaning and not just be a collection of nonsense syllables, and the learner must incorporate the new knowledge in a purposeful way. A primary process in learning is subsumption in which new material is related to relevant ideas in the existing cognitive structures. A major instructional mode proposed by Ausubel is the use of advance organizers. He emphasized that advance organizers are different from overviews and summaries that simply emphasize key ideas and details in an arbitrary manner. Organizers act as a "subsuming bridge" (Ausubel, 1963) between new learning material and existing related ideas.

Gagne and a Hierarchy of Learning Levels

Gagne's (1965) work contributed to elementary school science programs by identifying simple and advanced science process skills for primary, intermediate, and upper grades. His learning theory had, at its core, the concept that the learning of complex skills and behaviors depends on a hierarchy of successively more complex skills. Gagne, however, failed to introduce frameworks of interrelated science concepts, hence students were just progressing from simple to more complex skills tasks without the requisite context for understanding. His later work (Gagne, Briggs, & Wager, 1988) involved describing instructional events and corresponding instructional processes. It is interesting for preservice elementary school teachers to explore these learning theorists' work and examine the ways in which learning theories are built, their usefulness and their staying power.

Personal Knowledge Construction

Personal knowledge construction is a key element of the constructivist paradigm. The 18th-century philosopher Vico is one of the early writers to put forth the notion that human beings can only know that which their cognitive structures allow them to know (Brooks, 2000). How far a learner can progress is a function of what the learner currently understands and what the learner currently understands is a function of the learner's present mental structures. An understanding of prior knowledge, the children's own ideas, is an essential component for understanding how to translate learning theories into instructional practices. An emphasis on active learning, the importance of individuals taking control of their own meaning making, leads the individual learners to recognize when they understand something and when they need more information. This "metacognition" refers to people's abilities to monitor their own levels of understanding (Bransford et al., 2000).

The teacher plays a role in facilitating the learner's development of new knowledge and new structures through the creation of settings in which the learner may detect discrepancies and the teacher can scaffold the learner's ideas. The goals to which the learner aspires may differ, but the constructivist teacher and researcher are focused on better understanding the learner's conceptual changes over time, the nature of the changes, and the contributing variables (Brooks, 2000).

A Common Misconception

One common misconception regarding "constructivist theories" of knowledge building is that teachers should never "tell" students anything directly, but should always allow them to construct meaning for themselves. "This perspective confuses a theory of pedagogy with a theory of knowing" (Bransford et al., 2000, p. 11). In fact, there are times, after students have struggled with the ideas and have engaged in dialog about them, that teaching by telling is necessary and effective.

PEDAGOGICAL IMPLICATIONS

There are a number of pedagogical implications that arise from consideration of these views about learning. In all cases, the Ausubelian notion of "deep," or "meaningful," learning is adopted as the preferred type of learning for science education and science teacher education.

Complexity and Meaning Making

Learning, as an active process of meaning making, is complex and recursive. The learner needs to reflect on experiences, objects, events, phenomena, people, and ideas. The learner will challenge preexisting concepts when compelled to do so. This has important pedagogical implications. How do teachers create learning experiences for students that help challenge and expand on their preexisting ideas (Koch, 2002)?

Because of its complexity, active meaning making cannot be facilitated by simple hands-on tasks requiring that students follow directions and make language associations with the task. Educators must understand the children's own ideas and then scaffold the learner's reflections in ways that invite them to analyze their own thinking. This involves engaging the children in conversation about their ideas and helping them reflect on and consider other ways of thinking about these ideas. It is the learner's responsibility to construct meaning; it is the teacher's responsibility to scaffold learner reflection in a manner that may generate learner analysis, synthesis, and insight.

Consequently, the process of elementary science teaching must allow for and require the thoughtful considerations of the learners. An understanding of constructivist pedagogy asks teachers to view themselves as listeners, interpreters of ideas, and, often, colearners, exploring the children's thoughts as they are trying to make sense of a concept. In this way, the teacher is like a researcher, exploring how children build their understanding (Duckworth, 1986). Duckworth (1986) respected that knowledge must be constructed by each individual, and described two aspects to teaching:

> The first is to put students into contact with phenomena related to the area to be studied—the real thing—not books or lectures about it—and to help

them notice what is interesting; to engage them so they will be able to think and wonder about it. The second is to have students try to explain the sense they are making, and instead of explaining things to students, to try to understand their sense. (p. 123)

In contrast to exploring students' sense making, elementary school teachers have often newly learned the science background for themselves and are eager to impart it to the students. Often, because they have not had personal experience with the exploration of their own ideas about science, elementary school teachers do not recognize the value and importance of theirs and their students' preexisting ideas or invented theories for developing conceptual understanding.

The implications of the current research on the ways people learn are important, therefore, for elementary science teacher educators as well as elementary school teachers. Elementary science teacher education must be structured in ways that invite teachers to explore nature, interrogate their own ideas, and construct their own stories and conceptual understandings (Abell, George, & Martini, 2002; Koch, 2002).

Thinking Deeply and Making Decisions

When students are asked to design experiments, invent theories, or create artifacts representing what they have learned, they are being asked to think deeply about an event, experience, or phenomena and then to share ideas and make decisions about how and where to proceed. These activities emphasize meaning over memorizing, understanding over awareness, and quality over quantity (Mintzes, Wandersee, & Novak, 1998). Discussing possible theories for phenomena often leads to the struggle to make sense of the phenomena. This struggle, or cognitive dissonance, encourages designing experiments, repeating demonstrations, and reorganizing existing ideas (as noted in the story in the next section).

Meaning Making in the Elementary Science Methods Class

A science methods class is exploring properties of air (Koch, 2005). The science teacher educator does the "egg in the bottle" demonstration where a hard-boiled egg (shell peeled) sits atop a glass milk bottle and after lighting a piece of paper in the bottle, the egg jumps about and finally, slowly, slithers, whole, into the bottle. The preservice teachers (students) are challenged to write about their observations and come up with their own ideas. They are asked to invent theories about why the egg went into the bottle. Prior to this demonstration, the students had observed a candle being extinguished as it sat in a pan of shallow water and a jar was inverted over it. The students observed the water level in the bottle rise as the flame went out. They explored the relation between burning and oxygen consumption and between air pressure both outside and inside the bottle.

When the students revealed their egg in the bottle theories, some suggested that some mysterious force was "sucking" the egg into the bottle:

> I think that the flame in the bottle "ate" up the oxygen, making a vacuum in the bottle, which is how the egg got sucked in. (Student 1)

> I thought that the experiment with the egg and the bottle was remarkable, and interesting. At first I was not sure why or how the egg went into the bottle. I think that what might have happened was, when the fire was put out by the lack of oxygen, it sucked up the air that was in the bottle, and the egg got sucked in. I think I have an idea of how it happened but it is hard for me to say in words. (Student 2)

Other students thought the egg was pushed into the bottle due to a difference in air pressure:

> I am not really too sure why this happened, but I think the egg did this because burning the paper inside the bottle causes a difference in the air pressure between the inside and the outside of the bottle causing the egg to be pushed into the bottle. (Student 3)

> The fire in the bottle used up the oxygen in the bottle creating less air pressure inside the bottle than outside the bottle. The air pressure outside the bottle caused the egg to be pushed into the bottle. (Student 4)

The students were invested in understanding the concepts behind the egg in the bottle demonstration. They also wondered what would happen with a raw egg. Cognitive dissonance was created as students gained access to information from their peers. In this community of learners, the context and the social interaction of the discourse was a necessary prerequisite to meaning making. The students asked the teacher to repeat the demonstration two more times. The teacher did and the students became engaged in talking about the "dance" the egg did as the paper was burning slowly in the bottom of the bottle. The students wanted to test their ideas about the egg in the bottle, using a raw egg and a hard-boiled egg with the shell still on. They were designing new experiments, reorganizing existing ideas about eggs, eggshells, and air pressure. They were sorting out their own ideas and those of their peers and trying to more fully understand what happened by designing new experiments. In this way, the students were at the center of the learning experience. Traditional science teaching placed the teacher and the teacher's thoughts and ideas at the center of the science learning experience (Yager, 1995).

The teacher educator, in turn, was interested in the learners' development of new knowledge; she facilitated the creation of a setting where the learners detected discrepancies between their beliefs and others, and between their beliefs and what actually occurred. Fostering discussion and resolution of these discrepancies gave rise to more intense exploration of the ideas. It involved repeating demonstrations, discourse about the warmed air escaping upward, allowing the egg to "dance," and ultimately discourse about whether the egg was sucked in or pushed in. The students perceived a conflict, experienced frustration, and resolved the frustration by trying new experiments, listening to everyone's theories, conversing with the teacher, and finally, exploring the effect of removing oxygen from the bottle on the air pressure inside the bottle. The conceptual understanding that air exerts pressure and that air pressure varies depending on the temperature and volume of air began with an intense conversation about why the egg went into the bottle. The teacher educator helped the preservice teachers to be metacognitive about their thinking, relating the stages they went through to their preexisting mental structures and to the cognitive dissonance they experienced as well as the resolution as described in constructivist learning theories.

Developing Conceptual Understanding Takes Time

In elementary classrooms, scientific knowledge is often bundled into packets of information, often focused on a manipulative activity or experience. Under time constraints to meet local state requirements for standardized testing, teachers often experience pressure to "cover" curriculum and are forced to ignore the processes required for the thoughtful consideration of a topic.

Everything that is known about how learners construct meaning flies in the face of rushing through activities and their related concepts. Raizen and Michelsohn (1994) stressed a number of qualities associated with effective science teachers. Among them are an awareness of children's informal ideas, prior knowledge, and experience, especially as it relates to science concepts with which they are likely to experience difficulty. Other attributes include an appreciation of scientific reasoning skills (posing questions, designing investigations), as well as an understanding of how to foster them among students; a specialized knowledge of appropriate ways to represent science to children; and the ability to orchestrate science learning through the use of various organization structures, including cooperative groups. These attributes of elementary science teachers can define the elementary science pedagogical content knowledge that is represented by a transformative model mentioned earlier. Melded together is an understanding of the importance of how children learn, the nature of scientific knowledge, and the strategies best suited for a specific teaching context.

The traditional practice of passive delivery of information often prevents elementary school teachers themselves from thinking deeply about their own scientific understandings. Realizing this, teacher educators need to engage preservice and inservice teachers in their own thinking. Teacher educators need to model everything they want others to do. Without appropriate modeling, most future teachers will enter their classrooms with good ideas but no visual templates for action. If teachers are asked to engage children in iterative experiences that demand increasingly more complex reasoning and information, then science teacher education and staff development must create those same kinds of environments (Abell et al., 2002).

Framing Constructivist Pedagogy

An understanding of constructivist epistemology alters the fundamental ways of viewing the teaching–learning process (Jones, 1997). The responsibility for the teacher is much broader than previously conceptualized. The teacher must select a meaningful activity that engages students in relevant experiences that cause them to question their ideas about a natural phenomenon. In engineering the learning environment (Jones, 1997), the teacher is a facilitator who seeks to assist students in developing understandings of these phenomena through interactions with their peers and with the teacher. These interactions are key to creating connections between prior knowledge, new experiences, and the construction of new ideas. Hence, constructivist learning theories are related to a specific set of pedagogical skills. These include, but are not limited to, the following:

- Seeking out and using student questions and ideas to guide lessons and whole instructional units.
- Accepting and encouraging student initiation of ideas.
- Encouraging students to elaborate on their questions and their responses.
- Encouraging students to test their own ideas.
- Encouraging students to challenge each other's conceptualizations.
- Encouraging self-analysis, collection of real evidence to support ideas, and reformulation of ideas in light of new experiences and evidence. (Yager, 1995, pp. 54–55)

CONSTRUCTIVIST PEDAGOGY AND FUTURE
ELEMENTARY SCHOOL TEACHERS

What conditions, then, account for the types of shifts in thinking that future teachers make? The answer lies in understanding the intellectual wiggle room that exists between individuals' present thinking and the concepts, suggestions, or issues they are confronting. Some preservice teachers, for example, perceive a contradiction between what they do in their science methods classes and what they see in their internships. Those preservice teachers who perceive a contradiction and grow to embrace constructivist pedagogies transform their previous ideas about teaching into new ones. The authenticity of these preservice teachers' personal belief systems is visible as they are observed engaging in their science work with children. Other preservice teachers perceive a conflict between their present beliefs or actions and some of those proposed in these pedagogies and also experience frustration, but these preservice teachers resolve it in a different manner. They do not change their internal mental schemes; instead, they adopt the pieces of constructivist pedagogies that do, indeed, make sense to them. Often, this translates into the engagement of children in a hands-on science activity. However, the activity itself does not produce new conceptual knowledge. Further, the activity selection must engage children in exploration of what Ausubel called a meaningful learning task (1963), or a meaningful one that is relevant to the learner's cognitive structure, and one to which the learner can bring prior knowledge.

Meaningful science experiences for the elementary classroom:

- Relate to the students' lived experiences;
- Engage students in explorations;
- Stimulate the students to reflect on their explorations and develop new ideas;
- Reveal the students' thinking;
- Lead to new understandings. (Koch, 2002, p. 14)

Still other preservice teachers, unfortunately, develop dual knowledge structures with little overlap—university "theory," including constructivist ideas, and school-based "practice." They use the former for academic accreditation, and the latter for classroom actions, neither informing the other. Depending on their classroom experience, science may be ignored or taught using non-hands-on strategies. A key challenge for preservice elementary science teacher educators is to maximize the number of graduates who have transformed their science teaching practices.

PEDAGOGICAL DILEMMAS AND ELEMENTARY
SCIENCE TEACHING

How can a successful science learning experience be recognized? Look at an example. One sixth-grade class last year worked on designing cars for speed and stability. Two girls designed car shapes for smoothness to cut down on wind resistance and added a base bigger than the top to ensure stability. They then glued the wheels to the axles and the axles to the chassis, not initially realizing that the wheels would not turn. Quickly, after the hardening of the glue, they learned that they had a major design flaw. In discussion with their teacher, they came to realize that "something" needed to be able to rotate. They had not yet decided if it should be the axles or the wheels. They examined the chassis–wheel–axle relations closely. The girls, through discussion, recognized their errors, but did not have the opportunity to make a new project.

Did students' producing an immovable car, but an accurate analysis of why it did not move, constitute a successful learning experience? Or, were students with an immovable car, regardless of the depth of the accompanying analysis, a failure? Was the car itself the success or failure? Should the students earn passing or failing grades? In a recent grant-funded project, teacher educators differed on their answers to each of these questions. Some were disappointed that the teacher in the previous example did not provide specific steps and procedures that would have resulted in wheels that turn. Some approved of the openness that made an immovable car a possibility. The different responses to this classroom scenario center on the role of student error in learning. Is it better to head off students' mistakes before they occur or is it better to allow students to make their mistakes and then correct them? And what are the parameters on these questions? Clearly, differences in pedagogy arise from teachers' internal belief systems, and these differences greatly influence how quality is assessed.

At the heart of the beliefs informing teaching for understanding and constructivist pedagogies is that well-prepared teachers can actively intervene in classroom conversations to help create shared meanings and lead students to develop scientific concepts. "Human beings are meaning-makers; the goal of education is to construct shared meanings and this goal may be facilitated through the active intervention of well-prepared teachers" (Mintzes et al., 1998, p. xviii).

Perhaps of greatest importance is the relevance of highlighting the complexity of the constructivist teaching experience for both elementary science teachers and elementary science teacher educators. This complexity is found in the multiple approaches to engaging children in meaning making: inclusivity, authentic experiences, and engagement.

Inclusivity

Using an inclusive approach to related pedagogies is in the best interests of fostering meaningful science learning. In Haury's "Circle of Inquiry" (1995), everyday language is used to describe experiences that schoolchildren should be engaged in to construct understandings in science. These include wondering, collecting data, studying data, and making connections. As the circle widens, there are many descriptors for what children do when they wonder, collect data, study data, and make connections. Teachers can relate to these action words so that, for example, wondering includes exploring, noting discrepant events, asking questions, identifying problems, planning investigations. It is complex. As Duckworth suggested (1991), it must remain complex, because it is the complexity that provides the "wiggle room" to tease out meanings and construct understandings.

Authentic Experiences

Experiences need to be authentic to both science and the child's experience. The controversy over what happens in authentic science is an ongoing one. Bauer (1992) suggested that the sweeping generalization and common belief held by elementary teachers that all scientific study must be defined by *the* scientific method was probably a fallacy. Meaning-making activities of scientists do not follow one single method, so meaning-making activities for children should not either. Novak (1964) remarked, "Inquiry is the set of behaviors involved in the struggle of human beings for reasonable explanations of phenomena about which they are curious" (p. 26). This relates to the elements necessary for an elementary constructivist learning experience; namely, the engagement with relevant phenomena about which there is meaning making, reorganizing of ideas, redoing of experiments, and analysis of evidence in a systematic search for understanding.

Engagement

We need engagement with activities and ideas, a key tenet of the major pedagogies in elementary science education. The learning cycle is a model of instruction that has been linked to learning theories (Lawson, Abraham, & Renner, 1989). In the three-phase learning cycle of exploration, concept introduction, and concept application, teachers engage students in activities, ideas, connections to prior knowledge, and the exploration of new knowledge. In the exploration phase, students ask questions, develop hypotheses, collect and analyze data, and come up with their own ideas. This is a time of social interaction as well as materials manipulation.

In the second phase, students report their findings, beliefs, and ideas. This is a time of intense sharing and discourse. By the third phase, they link new understandings and make connections to prior knowledge.

THE PARADIGM SHIFT

Whatever the recommended pedagogies for elementary science teaching are named, the important component for science learning is the teacher's orientation toward teaching (see chap. 3). The centrality of the individual teacher and the teacher's place at the heart of science educational reform is the challenge for elementary science teacher preparation.

This is especially true in elementary science, where many elementary school teachers lack the confidence to deeply interrogate a concept with their students. Engaging preservice and inservice teachers in explorations of science shows them what learning can be like. Only when the teachers get "hooked" on the excitement of learning can they understand the joy of "keeping it complex" (Duckworth, 1991). Duckworth often denounces the price paid for the oversimplification of science teaching, and hence the oversimplified view of nature. This is analogous to the oversimplified view of teaching that is the psychic "baggage" of many future elementary school teachers.

Science teacher educators must therefore empower preservice and inservice teachers to see themselves as "learners," struggling for their own meaning of both science and science teaching. The struggle must be sanctioned and honored as they make their way to new understandings. As stated by Driver et al. (1994), the role of the science educator is to mediate scientific knowledge for learners, to help them to make personal sense of the ways in which knowledge claims are generated and validated (p. 6). Involving preservice and inservice teachers in constructing their own science understandings, and acknowledging the value and importance of this complex process, will hopefully promote a paradigm shift in the elementary school teaching and learning of science.

REFERENCES

Abell, S., George, M., & Martini, M. (2002). The moon investigation: Instructional strategies for elementary science methods. *Journal of Science Teacher Education, 13*(2), 85–100.

Ausubel, D. (1963). *The psychology of meaningful verbal learning.* New York: Grune & Stratton.

Ausubel, D. (1968). *Educational psychology: A cognitive view.* New York: Holt, Rinehart & Winston.

Bauer, H. (1992). *Scientific literacy and the myth of the scientific method.* Urbana, IL: University of Illinois Press.

Bransford, J., Brown, A., & Cocking, R. (Eds.). (2000). *How people learn: Brain, mind, experience, and school.* National Research Council Committee on Learning Research. Washington, DC: National Academy Press.

106

KOCH

Brooks, J. G. (2000). Constructivism. *Encyclopedia of cognitive science, Article 503*. New York: Macmillan.

Bruner, J. (1960). *The process of education*. Cambridge, MA: Harvard University Press.

Duckworth, E. (1986). Teaching as research. In E. Duckworth (Ed.), *The having of wonderful ideas and other essays on teaching and learning* (pp. 122–145). New York: Teachers College Press.

Duckworth, E. (1991). Twenty-four, forty-two, and I love you: Keeping it complex. *Harvard Educational Review, 61*(1), 1–24.

Duckworth, E., Easley, J., Hawkins, D., & Henriques, A. (1990). *Science education: A minds-on approach for the elementary years*. Hillsdale, NJ: Lawrence Erlbaum Associates.

Erickson, G. (2000). Research programmes and the student science learning literature. In R. Millar, J. Leach, & J. Osborne (Eds.), *Improving science education: The contribution of research* (pp. 271–292). Philadelphia: Open University Press.

Flick, L. (1993). The meanings of hands-on science. *Journal of Science Teacher Education, 1*(4), 1–8.

Gagne, R. (1965). *The conditions of learning*. New York: Holt, Rinehart & Winston.

Gagne, R., Briggs, L., & Wager, W. (1988). *Principles of instructional design* (3rd ed.). New York: Holt, Rinehart & Winston.

Gess-Newsome, J. (1999). Pedagogical content knowledge: An introduction and orientation. In J. Gess-Newsome & N. Lederman (Eds.), *Examining pedagogical content knowledge* (pp. 3–17). Dordrecht, The Netherlands: Kluwer Academic.

Haury, D. (1995, April). *Study of a field-developed model of scientific inquiry*. Paper presented at the annual meeting of the National Association for Research in Science Teaching, San Francisco, CA. (ERIC Document Reproduction Service No. ED 381 402)

Jones, M. G. (1997). The constructivist leader. In J. Rhoton & P. Bowers (Eds.), *Issues in science education* (pp. 140–149). Arlington, VA: National Science Teachers Association.

Koch, J. (2002). *Science stories: A science methods book for elementary school teachers* (2nd ed.). Boston: Houghton Mifflin.

Koch, J. (2005). *Science stories: A science methods book for elementary and middle school teachers* (3rd ed.). Boston: Houghton Mifflin.

Lawson, A., Abraham, M., & Renner, J. (1989). *A theory of instruction: Using the learning cycle to teach science concepts and thinking skills*. NARST Monograph No. 1. Manhattan, KS: National Association for Research in Science Teaching.

McEwan, H., & Bull, B. (1991). The pedagogic nature of subject matter knowledge. *American Educational Research Journal, 28*(2), 316–334.

Mintzes, J., Wandersee, J., & Novak, J. (Eds.). (1998). *Teaching science for understanding: A human constructivist view*. San Diego, CA: Academic Press.

Novak, J. (1964). Scientific inquiry. *Bioscience, 14,* 25–28.

Perkins, D. (1993). Teaching for understanding. *American Educator, 17*(3), 28–35.

Piaget, J. (1978). *The development of thought: Equilibration of cognitive structures*. (A. Rosin, Trans.). Oxford, UK: Basil Blackwell. (Original work published 1977)

Raizen, S., & Michelsohn, A. (1994). *The future of science in elementary schools* San Francisco: Jossey-Bass.

Shulman, L. S. (1986). Those who understand: Knowledge growth in teaching. *Educational Researcher, 15*(2), 4–14.

Smith, D. (1999). Changing our teaching: The role of pedagogical content knowledge in elementary science. In J. Gess-Newsome & N. Lederman (Eds.), *Examining pedagogical content knowledge* (pp. 163–197). Dordrecht, The Netherlands: Kluwer Academic.

Vygotsky, L. (1962). *Thought and language*. Cambridge, MA: MIT Press.

Yager, R. (1995). Constructivism and the learning of science. In S. Glynn & R. Duit (Eds.), *Learning science in the schools* (pp. 35–58). Hillsdale, NJ: Lawrence Erlbaum Associates.

7

"Meaning-Making Science": Exploring the Sociocultural Dimensions of Early Childhood Teacher Education

Marilyn Fleer
Monash University

> *How do people participate in sociocultural activity and how does their participation change from being relatively peripheral participants (Lave and Wenger, 1991), observing and carrying out secondary roles, to assuming various responsible roles in the management or transformation of such activities?*
>
> —Rogoff (1998, p. 695)

How do early childhood preservice teachers move from being peripheral participants to becoming full members of a science education community? In drawing on the work of Lave and Wenger (1991) and Wenger (1998), this chapter examines the community of practice known as "science education" and the community of practice known as "early childhood education" and discusses the disjunction between these two communities, noting the following challenges for preservice teachers:

1. The limited research base for supporting science education in the early years
2. Methodological difficulties faced by researchers when studying children under 8 years
3. The perception of "science" as unobtainable and by implication "science education" as difficult
4. Approaches to science teaching based on research and practices suitable for children aged 8 years and older

The work of Bourdieu (see Grenfell, James, Hodkinson, Reay, & Robbins, 1998) is used to examine critically the habitus of science school learning and the difficulties faced by preservice teachers as they enter mainstream science education. Aikenhead's (2002) research on border crossing for different cultural groups is used for discussing preservice teachers' own border crossing during science learning. In drawing on sociocultural research (Rogoff, 1998) and tertiary practices in early childhood science, the chapter documents the building of a community of practice for preservice early childhood teachers known as *meaning-making* science education. In this chapter, the international definition of early childhood education is used—children from birth to 8 years.

COMMUNITIES OF PRACTICE: LEARNING, MEANING, AND IDENTITY IN EARLY CHILDHOOD SCIENCE EDUCATION

The act of teaching occurring in learning environments such as schools, colleges, early childhood centers or classrooms, and universities has been described as separate from the rest of human activities (Wenger, 1998). These specialized learning environments develop their own rituals, routines, practices, artefacts, symbols, conventions, stories, and histories. Wenger argued that centers and schools develop their own community of practice. That is, they evolve a specialized discourse to name and explain their practices. These practices and the terms used to describe the practices become familiar to all the participants of that community; they hold meaning and suggest particular behavior.

In the early childhood community of practice, teachers have built ideologies and specialist language that center all activities and thinking on the individual child. Learning contexts are said to be built from the individual child—their interest, their world, and their experience. The community of practice is not only recognized by its discourse, but also its specialist equipment, furniture, and often, its purpose-built buildings.

Science education has also evolved its own discourse, equipment, furniture, and in many instances its own identifiable spaces (e.g., science laboratories and elementary science rooms). Science education has also constructed a community of practice, with its own artefacts, rituals, symbols, conventions, stories, and histories—in many cases, mirroring community views on the nature of science—that are specialized and different from other human activities (see Barnes, 1989).

Popular images of scientists (e.g., draw a scientist) gained from a range of sources (see Gough, 2001) have assisted with the partitioning of science from the real world. Similarly, symbols such as the science laboratory and white laboratory coat are depicted in popular media images of science, and are features of most secondary science classrooms. As Gough (1993, 2001) noted, school laboratories are an 18th-century view of how scientists work. Yet, these icons of science are represented in the minds of the general public and are held by many children (D. A. Newton & D. P. Newton, 1998) and university students (Jane & Kelly, 2001). These artefacts of history carry with them messages. In the case of science, the laboratory and white coat signal a specialized community of practice, a community of practice not necessarily accessible to all. A significant amount has been written in this area over the years to highlight the social nature of science knowledge construction (from amateur to specialist) and the gradual changes in public perception about the nature of science (e.g., Barnes, 1989; Jacobs, 1991; Woolgar, 1988).

The social construction of science (see Barnes, 1989) and therefore science education, has also been noted by many scholars in science education, revealing concerns for the disembedded teaching of science (science not connected to children's lives, experiences, or even the real world), resulting in children's alienation, disinterest, and inability to access science (see Cross & Fensham, 2000; Gough, 2001). This alienation has been strongly noted by early childhood preservice teachers; for example:

> While I was at school I found it difficult to understand many of the scientific concepts that were presented to me. I learnt the jargon and was able to satisfy the examination requirements by learning, but not always understanding, those things that were deemed to be important. I enjoyed the "doing" stuff in science laboratories, the "hands-on" work but found the rote learning of scientific facts irrelevant and meaningless. (second-year early childhood preservice teacher, 2001)

Wenger's (1998) work on communities of practice adds to calls for "meaning making" by focusing attention on issues of power and access: "Unequal relations of power must be included more systematically in our analysis. Hegemony over resources for learning and alienation from full participation are inherent in the shaping of legitimacy and peripherality of participation in its historical realizations" (Lave & Wenger, 1991, p. 42).

In science education communities where newly developed science curricula are used or under development, the discourse of dominance, and hence power, is that of a Western construction of science (see Fleer, 1996, 1999; Jegede & Kyle, 1999; McKinley & Waiti, 1995)—thus alienating some groups and preventing them from acting as full participants of the community of practice. The question of whose science is being foregrounded in curriculum is not asked. For instance, Hill, Comber, Louden, Rivalland, and Reid (1998), citing Delpit (1988), argued that "the rules of the culture of power are a reflection of the rules of the culture of those that have power"

(p. 29). In Australia and other Western countries, Western science, with its legitimation through Western research, is the prevailing view of learning in science education curriculum. Western research with its enactment through curriculum development, positions children from Western cultures as the dominant and privileged group. Unfortunately, "those with power are frequently least aware of or least willing to acknowledge its existence, while those with less power are often most aware of its existence" (Hill et al., 1998, p. 29). As such, the "taken-for-granted" practices become the accepted community of practice (Wenger, 1998); the predominant way things are done becomes habitual. Bourdieu (see Grenfell et al., 1998) suggested that the dominant habitus becomes a form of cultural capital that *science education professionals take for granted*. A critical analysis of the taken-for-granted practices from the perspective of those who are not part of the culture with power is necessary if early childhood education is to be inclusive of science education—as noted by preservice students:

> Science had always meant to me that you found out the answer to some externally asked questions and it didn't really matter whether that answer was meaningful or even understandable. It just had to be right. In fact, some of the questions that we were supposedly seeking answers for seemed to me so very abstract in nature that I was stumped by the question even before attempting to find the answer. (second-year early childhood preservice teacher)

As argued by Wenger (1998), simply living in a world does not always afford individuals access to all the meaning underpinning a particular community of practice. People must negotiate meaning. Negotiating meaning among people in a science learning community takes into account past practices, commonly understood routines, and negotiated pathways. In order to access the taken-for-granted practices of the science education community, important communication tools are needed if early childhood preservice teachers are to feel they can move from being peripheral participants to full participants.

The term *reification*, put forward by Wenger (1998), is a helpful construct for realizing how negotiated meaning takes place in a community of practice across science education and early childhood education. Reification means "making into a thing"—that is, "aspects of human experience and practice are congealed into fixed forms and given the status of objects" (Wenger, 1998, p. 59). Science has undertaken this transformation. The terms *science* and *science education* are talked about as though they have form and substance. They are reified terms and allow people to communicate complex ideas quickly and succinctly. In early childhood education, terms such as *child-centeredness, planning for individuals,* or even *play* are all reified terms used within that community of practice. Wenger argued that reification shapes individuals' experience, because it provides them with a concrete tool to perform an activity and therefore it changes the nature of that activity. For example, in science education, "reifying the concept of gravity may not change its effect on our bodies, but it does change our experience of the world by focussing our attention in a particular way and enabling new kinds of understanding" (Wenger, 1998, pp. 59–60).

Wenger (1998) argued that although reification is a powerful tool, it is also a "double edge sword." He suggested that in the process of concretizing, words become more succinct, and deep and complex ideas become simplified. Terms such as *quality* no longer hold the depth of meaning once afforded them. In fact, the term can be used in many different ways and hold a range of meanings for those using this term in their discussions. In an early childhood context, knowing about and analyzing the use and abuse of reified terms is important, particularly if there is a perception that science, and by implication science teaching, has been reified to be exclusive, elitist, and unobtainable for many early childhood teachers. For instance, Fensham (1991), in citing data from the Australian Disciplinary Review of Teacher Education in Mathematics and Science (Department of Employment, Education, and Training, 1989), indicated that although 74% of early childhood student teachers graduating at that time had studied at least one science subject in grades 11 or 12, most perceived that they had very little science knowledge and held negative attitudes toward science. Fensham argued that early childhood preservice teachers were focused mostly on the science content they had *not* studied (i.e., physics, chemistry) rather than what they knew (i.e., biology, human biology).

Watters, Diezmann, Grieshaber, and Davis (2001) evaluated the outcomes of early childhood inservice programs in science, and identified teacher knowledge of science and teacher knowledge of early childhood science education pedagogy as the areas of most concern for early childhood professionals: "Teachers also raised issues related to teaching science in the early childhood years . . . how to explain scientific concepts to young children; and ideas for observation and evaluation of young children's scientific knowledge" (p. 5).

Preservice teachers have also echoed these concerns. For example Carmel, a second-year early childhood preservice teacher, expressed a feeling common among early childhood student teachers: "The thought of teaching science was pretty overwhelming because I had never thought of myself as having much scientific knowledge."

Fensham (1991) argued that during the review of science education in Australia in the early 1990s, early childhood undergraduates were the group least likely to have prior formal science learning. Access to the discourse and knowledge of science by early childhood professionals is clearly limited. Early childhood preservice teachers must rely more on science education and science education research to inform their general understandings and to build their confidence to teach science to young children.

MEANING IS MORE THAN AN INDIVIDUAL CONSTRUCTION

Over the past 20 years, research in science education has amassed data and findings on individual children's thinking of earth and beyond, natural and processed materials, energy and change, and life and living (see Curriculum Corporation, 1994; Gabel, 1994; Leeds, 1992; Northfield & Symington, 1991). Although these findings are invaluable for teachers, early childhood teachers have generally not accessed this body of knowledge housed in the science education community.

Closer examination of the studies undertaken in Australia and elsewhere (Fleer & Hardy, 1993; Fleer & Robbins, 2001; Goodrum, 1993; Goodrum, Hackling, & Rennie, 2000; Raizen & Michelsohn, 1994) demonstrate that the majority of research has concentrated on children aged 8 years and older. Very little research has been undertaken with children under age 8 years. As a result, early childhood professionals and preservice teachers are poorly served by research to inform their practice and understandings.

In addition, in science education, the focus has been on finding out what the individual thinks. Although acknowledgment of peer influence is granted as an important factor, the construction of most research, and subsequent pedagogy (e.g., discovery learning; interactive teaching; see also Kamii & DeVries, 1993), has been on the individual and their thinking (Segal & Cosgrove, 1992). Fleer and Robbins (2001) argued that in early childhood education contexts, gathering data on individuals using accepted practices such as the "interview about incidents" (Osborne & Freyberg, 1985) has made it harder, rather than easier, to gain understandings of children aged 8 years or younger.

Fleer and Robbins (2001) demonstrated that gathering data on what individual children think using such standard practices has generally not worked well in ascertaining very young children's thinking. Mostly, very young children require a significant amount of time to discuss their ideas, and draw on nonconventional areas such as singing, artwork, or storytelling in the sand to assist them with their thinking. More importantly, it has been found that young children give what appear to be contradictory statements within a few minutes after a one-off interview (see also Fleer, 1992; Robbins, 2000a, 2000b, 2001a, 2001b, 2002). However, when interviewing occurs informally over an extended period with the assistance of conceptual tools such as singing and drawing, a fuller understanding of children's thinking can be gained (see Robbins, 2000b). This approach contrasts with that of asking children specific questions about incidents (see Biddulph & Osborne, 1985) or documenting their ideas through various forms of concept mapping (White & Gunstone, 1992).

In Western communities, the emphasis of research and theory development in science education has tended to focus on individuals and their individual experiences. Yet, as Wenger (1998) suggested, learning to be a part of the community of practice is a distributed process, held by all the participants (newcomers and old-timers), rather than being held in the mind of individuals or as a "one person act." This moves the focus from the individual to the broader community: the community of practice. That is, "to understand how individuals learn and develop through participation in the social world, it is necessary to grant that meaning is more than a construction by individuals" (Rogoff, 1998, p. 686).

These sociocultural perspectives foreground the notion that understanding is more than an individual construction: Meaning occurs in the context of participation in the real world. Ideas are socially mediated and reside not in individuals but are constituted in collectives, such as a particular community of practice (Wenger, 1998). Because meaning and therefore understanding are enacted in social contexts, research into young children's understandings must be viewed as a transient and fluid organism (see Wenger, 1998). Consequently, research that follows a sociocultural perspective must be framed to map the transformation of understanding and not some end point. According to Rogoff (1998, p. 691), "What is key is transformation in the process of participation in community activities, not acquisition of competences defined independently of the sociocultural activities in which people participate."

When these issues are taken together, the outcomes of such research in science education hold limited immediate relevance for early childhood preservice teachers. Not only are the findings of this research mostly concentrated on children aged 8 years, but studies with young children have identified flaws within the current science education research paradigm for early childhood education; and by implication, lead to a questioning of the theories of teaching science that have been built from this research. For example, Watters et al. (2001, p. 5) found that practicing early childhood teachers "reported a need for activities and investigations that integrate science with other curriculum areas and *are appropriate for preschool children* [that is, the year prior to commencement of formal schooling]" (emphasis added).

Although approaches to the teaching of science outlined in the literature have been demonstrated to be useful for children aged 8 years and
older, there is little research evidence to support their use with children
under that age (Fleer, 1992). Consequently, it can be argued that early
childhood teachers crossing the borders from early childhood community
of practice to the science education community of practice may find little
of immediate value for informing their teaching of very young children.
As Watters et al. (2001) suggested, "Successful professional development
[for practicing early childhood teachers] . . . requires a 'meaning-oriented'
approach" (p. 6). It can be argued that preservice teachers, as the next generation of teachers, not only require a meaning-oriented approach, but
also need to see the immediate relevance of their preservice science experience—that is, it should include research and approaches to teaching that
focus on children birth to 8 years of age. Otherwise, early childhood student teachers must move from a community of practice known as early
childhood education, to a community of practice known as science education—each with its own clearly defined borders.

THE CONSTRUCTION OF BORDERS: BORDER
CROSSING OR BORDER CREATING?

Aikenhead (1998, 2002) provided a rich and conceptually useful framework for thinking about the science borders that children and adults cross
when moving from home to school. He suggested that border crossing occurs when the everyday lived science experiences or worldviews of students are different from those experienced or expected in school science.
Jegede and Aikenhead (1999) suggested that when one worldview is different from another, individuals must cross borders between the different
worldviews. They suggested that border crossings can be harmonious
when the worldviews or lived experiences match. However, some border
crossings may be extremely difficult when the disparity is too great.
Jegede and Aikenhead (1999, p. 5) argued that border crossing from a cultural worldview to a school science worldview can be categorized as:

1. *Potential scientist,* whose transitions are *smooth* because the cultures
   of family and science are congruent;
2. *Other smart kids,* whose transitions are *manageable* because the two
   cultures are somewhat different;
3. *"I don't know"* students, whose transitions tend to be *hazardous* when
   the two cultures are diverse; and
4. *Outsiders,* whose transitions are virtually *impossible* because the cultures are highly discordant.

As border-crossing experiences will influence success or failure in school science, it is important to understand how people navigate border crossing.

In early childhood preservice teaching contexts, crossing conceptual borders is also applicable. For some students, Fatima's rule (giving the teacher the answer they want) is applied in order to pass courses, as noted by Carmel: "I learnt the jargon and was able to satisfy the examination requirements by learning, but not always understanding" (second-year early childhood preservice teacher, 2001).

As Aikenhead (2002) suggested, for other students who attempt border crossing, negotiating a pathway is treacherous and can result in a complete displacement of prior knowledge, a complete rejection, or the creation of dual understandings suitable for different contextual environments.

Added to the challenges facing early childhood preservice teachers is the notion that not all the research and learning theories espoused in the general literature and in tertiary textbooks necessarily match their experiences of working with very young children (i.e., children do not necessarily draw concept maps and answer questions), or help them know what children younger than 8 years might actually think about a given area. Consequently, there is also a teaching–learning theory border that these students must negotiate. Their worldview of working with, and learning about, very young children's thinking is a different worldview from that of the general science education community.

Early childhood preservice teachers must cross the borders from early childhood pedagogy (theory and practice) to that of science education pedagogy. Often early childhood preservice teachers are asked to take back to their teaching context research and teaching approaches for children 8 years and older and apply what they can, or construct this material as their own, in order to develop a range of ways of working with young children to support science learning. However, rather than crossing the borders from early childhood ideology to a science education paradigm built on research based on much older children, student teachers frequently play Fatima's rule or make rejection statements such as the following: "Young children don't ask scientific questions—so why would I use an Interactive Approach[1] to teaching science?" (third-year early childhood preservice teacher, 2001).

Taking into account all these factors, it is little wonder that Fensham (1991) made the following statement: "The Review [of science teacher ed-

---

[1]The Interactive Approach (Biddulph & Osborne, 1984) suggests that science activities be devised to help children answer questions about science topics that they have raised themselves.

ucation] and the discussions in this paper make depressing reading about the prospects for science in early childhood education" (p. 7).

BUILDING A COMMUNITY OF EARLY CHILDHOOD SCIENCE PRACTICE: MEANING-MAKING SCIENCE

It has been argued in previous sections that the science education research paradigm and the community of practice that has built from this research cannot be easily accessed by early childhood inservice and preservice teachers due to the limited science education research base for children under the age of 8 years; approaches to teaching science being built on research featuring children older than 8 years; and research approaches using particular methods and situated within a particular paradigm that do not easily translate to an early childhood context or group of children.

An alternative is that preservice education can be designed to help early childhood preservice teachers build their own community of early childhood science practice. As researchers of young children's thinking in science, they can negotiate a pathway into the community of practice called *early childhood education*. As newcomers, they have agency and can contribute to the discourse of early childhood education, influencing the old-timers and therefore the community of practice known as early childhood education. In building a community of practice that affords science teaching, preservice teachers need to negotiate meaning—as such, preservice science teaching courses need to recognize and therefore make explicit the limited research base and underdeveloped theories for teaching science pertinent to children under 8 years. Preservice teachers need to be positioned as researchers of young children capable of building a community of practice for science teaching; and preservice teachers need to be supported as they cross the border from the university science education community to the early childhood education community. If this journey is seen as a community-building process, where understandings need to be negotiated, then meaning making in early childhood science education is possible for student teachers.

The building of an early childhood science community of practice needs to take into account the "entry," the border crossing, and the agency afforded to the preservice teacher. For example, a second-year preservice teacher reported on her discussion with child-care center staff about how she was considering tackling science teaching on her visit:

> When I first began this experience [science unit on bees] I took it to all the staff and told them I was interested in doing this unit on bees. I took all the resources that I had, plans I had written, and I gave them to the staff and I

told them about my ideas [about teaching science]. One staff member looked at me and had a funny look on her face and I said, "You don't look very confident." She said, "Don't you think that is a bit advanced?" "Yeah, now that I am looking at it myself, I do. They are not going to have any idea are they?" ... And we all started to get a bit sceptical. Should I just pull the plug on the whole thing?

Entry into the early childhood setting is an enormous border-crossing experience for students. They live in a tertiary science community of practice, drawing on theories and research focusing mostly on children aged 8 years and above, and move to an early childhood teaching setting where they must apply their newfound knowledge. As already discussed, the early childhood community of practice contrasts with that of the science community of practice. As such, preservice students experience a "sceptical" reception when they enter their new community of practice and discuss the teaching of science to young children.

Perceptions about science and the teaching of science to young children will also influence the ease of border crossing for early childhood preservice teachers. For example, when teachers perceive that young children should experience science through the manipulation of objects and conversations around these objects, but have no particular framework for finding out what sorts of ideas or questions children have, then border crossing for preservice teachers is difficult. As one second-year early childhood preservice teacher put it, early childhood children "are used to exploring things, but they are not really used to formulating questions [as suggested in the Interactive Approach]. All children ask questions but not on a set topic."

Organizing learning experiences based on questions asked by the children has been shown to be a conceptually very rich teaching approach with older children. However, within the early childhood community of practice, this approach to teaching science is outside of many early childhood teachers' experiences. Consequently, it is difficult to imagine how this might occur with young children. As such, it is not surprising when practicing early childhood teachers say, "Don't you think that is a bit advanced?"

This contrasts significantly with the experiences of those preservice teachers who were placed in early childhood centers where staff had been involved in inservice education following an Interactive Approach to Teaching Science, or who were placed in the University Childcare Centre where extensive science education practice and research were ongoing features. Elizabeth, a third-year early childhood preservice teacher working in the University Childcare Centre, outlined the community of practice in which she was involved:

At the beginning of the preschool year I became engaged in conversation with one of the children during outside play.

Bronte: Look Elizabeth my shadow doesn't want to go on the other side even if I turn around. It just stays on this side.

Elizabeth: That's very interesting Bronte—why do you think you have a shadow?

Bronte: I think it's because of the sun.

Elizabeth: You're absolutely right Bronte, it is because of the sun.

Bronte: But I also have a shadow at night even when the sun's gone to sleep, cause I see my shadow in my room.

Elizabeth: What do you think might give you the shadow in your room Bronte, if there's no sun in your room? What is in your room that is like the sun? Is there something bright with lots of light?

Bronte: My light is bright.

Elizabeth: That's right Bronte, but why do you think a light helps to make a shadow? Isn't a shadow dark?

Bronte: I don't know how it makes it.

Elizabeth: Look what happens if I stand side ways. What happens to my shadow?

Bronte: It gets skinny.

Elizabeth: What happens if I turn with my whole body to face the sun? What happens to my shadow then?

Bronte: It gets all fat again.

Elizabeth: Can you see how my body is getting in the way of the sunlight? The light can't go through my body so I make a big dark spot just like night time. I can make day time and night time with my body. You can do the same with the light in your room because your body is blocking the light.

This conversation gave me an idea for a day and night program that would challenge many common ideas that preschoolers have about the occurrence of day and night and where the sun goes during the night time. I embarked on a two week program that saw children first sharing their preconceptions about the sun and day and night and then interacting with a working model of the sun and the earth and discussing their findings.

A graduate working in the same center provided a further example:

In one of the introductory sessions I introduced a group of children to the globe and told them that this was our "home" and it was called Earth. One of the four year olds put her hand up and said, "But I live in Florey [suburb] and it's flat because the floor is flat and the road is flat." I had assumed that

the model [globe] would be meaningful for the children and hadn't anticipated that it would be such a difficult concept for them to adopt. It took
many other experiences of looking at books and videos before we could really start to use the globe for our discussions about night and day. Even so, I
wonder how many of those children still clung to the belief that the world
was really flat? (first-year graduate)

Student teachers who worked in the center each week (as part of a
mentoring and employment scheme for early childhood undergraduates)
contributed in tutorials in ways that demonstrated to other students they
were "researchers-in-action." As researchers of young children's thinking,
meaning making in early childhood education science education was being constructed in the group. Conversations featured what ideas children
held and the range of ways contexts were created for science conversations with children aged 5 years and younger. Early childhood science education was being constructed and negotiated within class: Meaning making was occurring for student teachers, and broader understandings of
early childhood science education were being developed.

When preservice teachers are in less friendly science learning centers,
meaning making in early childhood science is more challenging and requires them to have a strong basis for their pedagogical choices: "I was really nervous about doing it, but then . . . within the first 15 minutes of
[teaching the interactive] approach I new that I had done the right thing.
You just can't underestimate children" (second-year early childhood preservice teacher).

By contrast, when student teachers view themselves as helping build a
community of early childhood science education, they have permission to
develop learning experiences and frameworks to support science learning. They do not have to "get it right" or feel they need to apply a particular approach; rather, they view themselves as researchers developing professional experience about science learning within an early childhood
community. According to one second-year student teacher,

> When I first began, the thing that drew me into teaching a unit on bees was
> that I had a child ask me, "Why do we step on bees when we see them?" This
> struck me as something quite a difficult question for me to answer, because I
> don't think if we see bees we should step on them. . . . A student made a
> book, *Buzzy the Bee*, and read this to all the children. In the midst of all this, a
> very quiet child put her hand up and said, "I have a question." And I said,
> "What's your question?" And she said, "Are there boy bees and girl bees?" I
> thought, "Oh Wow. That was just fantastic. From her own mind . . . and all
> by herself!"

Getting early childhood children to ask questions had been identified
by the early childhood preservice teachers as very difficult. As a result, the

community of early childhood preservice teachers worked together to develop understandings about the contexts that would facilitate question asking by children. The following extended interview highlights the movement in and around an evolving community of practice as student teachers work to make meaning of early childhood science teaching.

Some of the questions were:

What do they eat?

What do they drink?

How many legs do they have? That was one they asked out in the garden.

How do bees move? That's another one they asked out in the garden—once we had started the whole process.

I had a list of questions, but as we started to investigate, they came up with more, so we just added them to the list:

How many bees are in a hive? Although the children can't read, they took the book and said, "There are lots of bees here. How many are there?" They couldn't count them, so we counted them, and then they drew pictures of bees on their pieces of paper. I think the children were there all day drawing pictures of bees on their pieces of paper.

What colour are bees? That was brought forward when they were out in the garden. The child was looking at a bee and saying, "Oh what colour are they? Let's get a bit closer."

How do bees hurt you? This came up in group time. I was really astounded, as this child [who asked the question] doesn't really tend to engage much in group work. Primarily, where topics are given to him, he likes to investigate from his own interest and pathway, but that was really great. We looked up that one as well. That was a hard one.

What sound do bees make? Again that came from the garden, so the experience [of going into the garden] drew a lot more questions out.

How do bees communicate? That was one they asked on the telephone [to a bee keeper].

What do bees collect from flowers? That was an interesting one too. A child who responded to that question wanted to find out all of the answers to another question, "What do bees eat?" [shows name tag with question on it that was worn by the child], "What do bees collect from flowers?"—she wanted to answer all of them, so we ran with that. (second-year early childhood preservice teacher)

When student teachers act as co-constructors of knowledge, meaning-making science evolves. Therefore, student teachers have success with creating science learning experiences that lead to deep and powerful learning, as exemplified in the range of questions that evolved over the duration of the science experience described earlier.

CRITICAL INSIGHTS

No one really knows the best way to teach science to young children. Similarly, little is known about how to support preservice early childhood teachers crossing the borders between the science education community and the early childhood education community. What is known is that the research base in science education has concentrated on children aged 8 years and older. Also, this research has been used to inform the development of teaching approaches in science education. As such, more thought should be given before applying a deficit model to early childhood science education. The victim cannot be blamed. Preservice students cannot be expected to somehow "plug the gap" if the research and theory falls short. Educators need to take responsibility for the quality of teaching in the early childhood profession—they are, after all, the people educating the new generation of early childhood science education teachers! Why should the treacherous task of border crossing lie with our preservice students? Surely, educators are experienced enough in science education to consider making the journey into early childhood education, mapping as they go, so that they can make border crossing for early childhood student teachers smoother and meaningful. Tertiary science teaching is about early childhood preservice teachers being positioned as coresearchers, and being encouraged to evolve their own community of early childhood science practice. According to one teacher:

> After participating in the science [course] in uni [university] I discovered that I didn't have to have all the answers. I'm now happy to go along with the children, learning as we investigate together. I'm more interested in the process of learning and constructing knowledge than just being able to didactically "teach" children scientific information. . . . I still don't think of myself as having a great deal of scientific knowledge, but I'm not afraid of teaching science to children. I find that *we can learn together*. (first-year early childhood teacher; emphasis added)

Preservice science education can also be viewed in this way. Tertiary educators can learn together with their early childhood student teachers, researching children's ideas and thinking, and ways of constructing knowledge in early childhood contexts. Positioning preservice early childhood teachers as coresearchers creates an environment for meaning-making science in an early childhood community of practice. There are no borders for students to cross, but rather a community of learners who construct meaning-making science together, and in the process teach science to very young children.

REFERENCES

Aikenhead, G. (1998, June). *Many students cross cultural borders to learn science: Implications for teaching.* Keynote address given at the annual conference of the Australian Science Teachers' Association (CONASTA), Darwin, Australia.

Aikenhead, G. (2002, March). *Border crossing in science and mathematics education.* Science Seminar Series, Monash University, Melbourne, Australia.

Barnes, B. (1989). *About science.* London: Blackwell.

Biddulph, F., & Osborne, R. (1984). *Making sense of our world.* Hamilton, New Zealand: University of Waikato.

Biddulph, F., & Osborne, R. (1985). *Learning in science.* Working paper of the Learning in Science Project. Hamilton, New Zealand: SERU, University of Waikato.

Cross, R. T., & Fensham, P. J. (Eds.). (2000). *Science and the citizen. For educators and the public.* Melbourne, Australia: Arena Publications.

Curriculum Corporation. (1994). *A statement on science for Australian schools.* Carlton, Victoria, Australia: Author.

Delpit, L. (1988). The silenced dialogue: Power and pedagogy in educating other people's children. *Harvard Educational Review, 58,* 280–298.

Department of Employment, Education, and Training. (1989). *Discipline review of teacher education in mathematics and science* (Vols. 1–3). Canberra, Australia: Australian Government Publishing Service.

Fensham, P. J. (1991). Science education in early childhood education—a diagnosis of a chronic illness. *Australian Journal of Early Childhood, 16*(3), 3–8.

Fleer, M. (1992). The suitability of an interactive approach to teaching science in early childhood. *Australian Journal of Early Childhood, 17*(4), 12–23.

Fleer, M. (1996). Early childhood science education: Acknowledging and valuing differing cultural understandings. *Australian Journal of Early Childhood, 21*(3), 11–15.

Fleer, M. (1999). Children's alternative views: Alternative to what? *International Journal of Science Education, 21*(2), 119–135.

Fleer, M., & Hardy, T. (1993). How can we find out what 3 and 4 year olds think? New approaches to eliciting very young children's understandings in science. *Research in Science Education, 23,* 68–76.

Fleer, M., & Robbins, J. (2001, July). *"Hit and run research" with "hit and miss" results in early childhood science education.* Paper presented at the 32nd conference of the Australasian Science Education Research Association, Sydney, Australia.

Gabel, D. L. (1994). *Handbook on research on science teaching and learning.* New York: Macmillan.

Goodrum, D. (Ed.). (1993). *Science in the early years of schooling: An Australian perspective.* Monograph Number 6. Perth, Australia: National Key Centre for School Science and Mathematics, Curtin University.

Goodrum, D., Hackling, M., & Rennie, L. (2000, July). *Science teaching and learning in Australian schools.* Paper presented at the annual conference of the Australasian Science Education Research Association, Perth, Australia.

Gough, N. (1993). *Laboratories in fiction: Science education and popular media*. Geelong, Victoria, Australia: Deakin University Press.

Gough, N. (2001). Teaching in the (Crash) zone: Manifesting cultural studies in science education. In J. Weaver, M. Morris, & P. Appelbaum (Eds.), *(Post)modern science (education): Frustrations, propositions and alternative paths* (pp. 249–273). New York: Peter Lang.

Grenfell, L., M., James, D., Hodkinson, P., Reay, D., & Robbins, D. (1998). *Bourdieu and education. Acts of practical theory*. London: Falmer.

Hill, S., Comber, B., Louden, W., Rivalland, J., & Reid, J. (1998). *100 children go to school. Connections and disconnections in literacy development in the year prior to school and the first year of school* (Vols. 1–3). Canberra, Australia: Australian Government.

Jacobs, S. (1991). *Theories of science*. Geelong, Victoria, Australia: Deakin University Press.

Jane, B., & Kelly, L. (2001, July). *Transferring experience from science education research to technology education research*. Paper presented at the 32nd conference of the Australasian Science Education Research Association, Sydney, Australia.

Jegede, O. J., & Aikenhead, G. (1999). *Transcending cultural borders: Implications for science teaching*. Retrieved April 3, 2000, from http://www.ouhk.edu.hk/cridal/misc/jegede.htm

Jegede, O., & Kyle, W. C., Jr. (1999, June). *Equitable science and technology education in the postmodern era: Four critical issues*. Paper presented at the ninth International Organization for Science and Technology Education, Durban, South Africa.

Kamii, C., & DeVries, R. (1993). *Physical knowledge in preschool education. Implications of Piaget's theory*. New York: Teachers College Press.

Lave, J., & Wenger, E. (1991). *Situated learning: Legitimate peripheral participation*. Cambridge, England: Cambridge University Press.

Leeds. (1992). *Leeds National Curriculum Science Support Project: Resources for supporting pupils' learning at Key Stage 3*. Leeds, UK: Leeds City Council, Department of Education, the Children's Learning in Science Research Group at the University of Leeds.

McKinley, E., & Waiti, P. (1995). Te Tauaki Marautanga Putaiao: He tauira—The writing of the national science curriculum in Maori. *SAMEpapers 1995*, 75–94.

Newton, D. A., & Newton, D. P. (1998). Primary children's conceptions of science and the scientist: Is the impact of a national curriculum breaking down the stereotype? *International Journal of Science Education, 20*(9), 1137–1149.

Northfield, J., & Symington, D. (Eds.). (1991). *Learning in science viewed as personal construction: An Australasian perspective*. Monograph Number 1, National Key Centre for School Science and Mathematics. Perth, Australia: Curtin University of Technology.

Osborne, R., & Freyberg, P. (1985). *Learning in science: The implications of children's science*. Auckland, New Zealand: Heinemann.

Raizen, S. A., & Michelsohn, A. M. (Eds.). (1994). *The future of science in elementary schools*. San Francisco: Jossey-Bass.

Robbins, J. (2000a, January). *"It isn't raining on Wednesdays!" Young children's explanations of natural phenomena*. Paper presented at the eighth annual Australian Researching Early Childhood Education Conference, Canberra, Australia.

Robbins, J. (2000b, July/November). *What's Mr. Whiskers go to do with the sun? The importance of finding out what children understand. High expectations: Outstanding achievements*. Conference proceedings of the Early Years of Schooling P–4 Conference, Victoria, Australia.

Robbins, J. (2001a, January). *"My mummy can tell me other things, and I'll be able to tell you": Interviewing young children to find out what they understand*. Paper presented at the ninth annual Australian Researching Early Childhood Education Conference, Canberra, Australia.

Robbins, J. (2001b, December). *Shoes and ships and sealing wax—taking a sociocultural approach to interviewing young children*. Paper presented at the 2001 New Zealand Early Childhood Research Symposium, NZARE National Conference, Christchurch, New Zealand.

Robbins, J. (2002, January). *Moving through understanding rather than to understanding: A sociocultural perspective on young children's conceptions of the rain*. Paper presented at the 10th annual Australian Researching Early Childhood Education Conference, Canberra, Australia.

Rogoff, B. (1998). Cognition as a collaborative process. In W. Damon (Chief Ed.) & D. Kuhn & R. S. Siegler (Vol. Eds.), *Handbook of child psychology: Cognition, perceptions and language* (5th ed., pp. 679–744). New York: Wiley.

Segal, G., & Cosgrove, M. (1992). Challenging student teachers' conceptions of science and technology education. *Research in Science Education, 22,* 348–357.

Watters, J. J., Diezmann, C. M., Grieshaber, S. J., & Davis, J. M. (2001). Enhancing science education for young children: A contemporary initiative. *Australian Journal of Early Childhood, 26*(2), 1–7.

Wenger, E. (1998). *Communities of practice. Learning, meaning and identity.* Cambridge, England: Cambridge University Press.

White, R., & Gunstone, R. (1992). *Probing understanding.* London: Falmer.

Woolgar, S. (1988). *Science the very idea.* New York: Ellis Horwood & Tavistock.

II

INTERSECTIONS OF CONTENT, PEDAGOGY, AND PRACTICE

A number of the practical issues affecting elementary science teacher education courses are considered in this section, where matters of content, pedagogy, and practice dominate. In chapter 8, Olson and Appleton explore the nature and composition of methods courses by describing aspects of their own courses, looking at underlying commonalities in philosophy and practice. A key demand on elementary science programs is that they develop scientific literacy in students. What this means, how it relates to language literacy, and the implications for preservice courses are analyzed in chapter 9 by Prain and Hand.

Related to this, but often in contrast to it, is the pressure experienced by teachers arising from an overemphasis on standards and high stakes testing. In education systems where such an emphasis dominates, science teacher educators have sought ways of working within the system to advance elementary science. Wieseman and Moscovici describe what they call a "standards-infused" approach to curriculum that incorporates science in chapter 10. This is an important approach to interdisciplinary teaching that provides one way forward in the current teaching climate.

Jones discusses a very different aspect of the elementary curriculum in chapter 11: the role and place of technology and the development of technology literacy. He outlines the difference between technology and information technology, and argues for a distinct and separate curriculum for technology that is not subsumed by, or part of, the science curriculum. This view holds a number of implications for preservice elementary programs, and in particular, elementary science courses.

In chapter 12, Harrison outlines the recent research findings related to assessment—in particular, assessment for learning. Many of these stand in stark contrast to normal practice in elementary science classrooms, and in elementary science methods courses. She provides a number of important principles for building assessment for learning into the curriculum.

8

Considering Curriculum for Elementary Science Methods Courses

Joanne K. Olson
Iowa State University

Ken Appleton
Central Queensland University

> *It's not what you don't know that hurts you. It's what you do know that ain't so.*

The preceding quotation (attributed to both Mark Twain and Will Rogers) exemplifies one of the fundamental challenges of elementary science teacher preparation. Literature from the nature of science, as well as research on human learning, indicates that new experiences are not viewed in an objective, detached manner, but are understood through the use of an existing mental framework (Freyberg & Osborne, 1985; Kuhn, 1962). Preservice elementary teachers enter teacher education programs with well-formed views of teaching, learning, and the purposes of schooling; unfortunately, these views too often are inconsistent with research on effective teaching and student learning (Kennison, 1990; Windshitl, 2003). Unless preservice teachers become dissatisfied with their naïve views, they will likely do what science students do: take those elements from instructional experiences that fit with their existing frameworks and reject those that do not fit, resulting in a piecemeal learning experience that maintains their prior ideas.

Because learning to teach science effectively requires most students to change prior ideas, teacher educators must consider the demanding nature of learning to teach, including both cognitive and emotional aspects, as summarized by Driver (1997):

We know that changing the way you think about things can often be not just difficult, but emotionally taxing, because it means giving something up. It means letting go of some knowledge that you have used in the past. It means letting go of something while you are still unsure about what it is you are grasping after in terms of a new way of seeing things.

Appleton (1997) proposed a model of learning that takes into account students' thinking when they encounter experiences at odds with their existing ideas. The model recognizes that students seek equilibrium, and will exit from instruction when they perceive that the new idea either fits their previous ideas, works better than previous ideas, or can be safely rejected. Of particular importance is the student's desire to exit from instruction (i.e., reach equilibrium), and that students may prematurely exit from instruction with a perceived congruence between new and existing ideas, when in fact substantial inconsistencies may exist. Structuring teacher education experiences, and methods[1] courses in particular, requires that teacher educators take into account the complexities of conceptual change, providing an experience that: confronts preservice teachers' ideas and beliefs, helps them understand more effective frameworks for science teaching, and increases the pressure to remain engaged in instructional experiences, decreasing the likelihood they will exit from instruction with naïve ideas intact. However, teacher educators must also consider the nature of the learner and provide substantial support for students as they engage in what can be an emotionally taxing experience of rethinking what it means to teach and to learn. These needs raise important issues that must be addressed when preparing prospective elementary teachers to teach science, including what should be part of a preservice science education experience.

Science methods courses are a common feature of most preservice teacher education programs (Anderson & Mitchener, 1994). In some programs, the science methods course may be the only compulsory course in science education; in others it may be part of a suite of courses. Whichever may apply in an institution, a perennial question is, given the vast range of knowledge and skills that a neophyte elementary teacher needs, and the often substantial discrepancies between their existing views of teaching and effective practice, what should be included in the brief time available? And perhaps even more importantly, why? This chapter considers these questions by comparing two methods courses from very different contexts: one from an Australian university in Queensland, and one from a university in Iowa, United States. Despite Kennedy's (1990) assertion

---

[1]As in chapter 5, the traditional use of the term *methods course* has been adopted to make the discussion explicit, recognizing that it may not be the most appropriate term.

that teacher educators "are unable to agree on a requisite knowledge base or an appropriate pedagogy for their profession" (p. 819, in Anderson & Mitchener, 1994, p. 5), a discussion of the chapter brought up how many similarities exist, and how alike our thinking was concerning what a methods course should look like.

Although the term *methods course* seems to be commonly used in teacher education programs, there is a need to clarify what this means. The perspective of a science methods course adopted in this chapter is a course that focuses on how to effectively teach science in the elementary school: Its main aim is to equip prospective teachers to begin teaching science well.

The discussion first outlines some principles that are important in guiding curriculum for methods courses, then describes two case studies—one from Central Queensland University and the other from Iowa State University—and concludes with a discussion of these cases.

GUIDING PRINCIPLES

Whereas a number of guiding principles affect preservice teacher preparation, the focus here is on two: the nature of prospective elementary teachers and conceptual orientations to teacher education.

The Nature of Prospective Elementary Teachers

There are two essential aspects regarding the nature of prospective students.

Attitudes Toward Science.   The majority of prospective elementary teachers dislike science; their choice of a career in elementary education is in part a choice not to do science (e.g., Australian Science and Technology and Engineering Council [ASTEC], 1997; Schoon & Boone, 1998). Many of these prospective teachers have had little or no success in their own study of science, and may have experienced outright failure (Palmer, 2001). Their attitudes toward science are largely based on their personal memories of science experiences. In our courses, few prospective teachers can remember doing any elementary science; some recall an exciting isolated instance, whereas others recall reading and writing. Most recount boring and irrelevant secondary school experiences that were largely text-based and had limited or no practical work. Groups in both countries cite the emphasis on memorization as a main reason for their intense dislike of secondary science (see also Moore & Watson, 1999). Only a small minority of prospective elementary teachers like science, have had success in it at school, and have taken more than one secondary science subject by grade 12. Thus, they have a low level of confidence in both their understanding of and ability to teach science (De Rose, Lockard, & Paldy, 1979; Donnellan, 1982; Schoon & Boone, 1998; Young & Kellog, 1993).

Science Content Knowledge. The overwhelming majority of prospective elementary teachers have limited formal science knowledge (Bethel, 1984; Heikkinen, 1988; Schoon & Boone, 1998). Our students tend to have similar knowledge levels to grade 9 or 10 students, and hold many of the misconceptions documented in the literature. Four main reasons have been cited for this lack of content knowledge, including poor college science experiences (Abell & Smith, 1994; Helgeson, Blosser, & Howe, 1977; Hurd, 1983), concurrent enrollment of education students and science majors in college science classes (National Science Teachers Association [NSTA], 1984), incongruent tertiary and secondary science coursework (Hurd, 1983), and limited science coursework in many elementary education programs (Department of Employment, Education, and Training [DEET], 1989).

Many prospective teachers have reasonable levels of "everyday" knowledge of science, especially in the life sciences and often through hobby and special interest areas. However, everyday knowledge tends to be episodic rather than conceptually organized, so prospective teachers have difficulties applying it to new contexts. Unfortunately, they tend not to recognize this knowledge as science; they frequently view science as limited to knowing lots of scientific terminology and complex ideas, reflecting inadequate understandings of the nature of science (Tamir, 1983). Their views of the nature of science are congruent with secondary student misconceptions noted by Rowell and Cawthron (1982) and Clough (2001).

Conceptual Orientations to Teacher Education

Feiman-Nemser (1990) identified five conceptual orientations for teacher education programs. Each of the five main perspectives is reflected in our methods courses by design. They include *academic*, emphasizing subject matter knowledge, knowledge of the nature of the discipline, and development of pedagogical content knowledge; *practical*, emphasizing experiences in classrooms and reflective practice; *technological*, focusing on developing skills necessary to teach effectively, such as wait time and effective questioning; *personal*, emphasizing the needs, interests, and development of the learner and the role of the teacher in interactions and meaning making; and *critical/social*, focusing on group problem solving, social equity, and democratic values.

Anderson and Mitchener (1994) asserted that no consensus exists regarding a single theoretical perspective that guides teacher education. Rather than attempting to adopt one conceptual orientation at the expense of the others, we assert that each of these orientations has value and is necessary in teacher education. Adopting a single orientation will likely limit a teacher's effectiveness. For example, content knowledge alone is insufficient for effective teaching (Akerson & Abd-El-Khalick, 2002; Shulman, 1986). Likewise, excessive field experiences tend to perpetuate the status quo and result in imitative practices that lack flexibility and conceptual understanding of teaching and learning (Feiman-Nemser & Buchmann, 1985; Ohana, 1999). In addition, Tobin and Garnett (1988) found if teachers lack an understanding of the content they are teaching they cannot ask questions to effectively help students see problems in their thinking. Thus, a more powerful teacher education experience results from addressing all five orientations and their interactions. Based on the context of the specific teacher education program, a science methods course may emphasize one or a few of these orientations more than others, but each orientation needs to be addressed and the importance of their interactions made clear. This interplay of orientations with differing emphases is apparent in the two cases that follow.

DECISIONS REGARDING THE CONTENT OF A SCIENCE METHODS COURSE

Based on the five orientations of teacher preparation and the nature of these prospective teachers, several issues become important to consider when making decisions about the content of an elementary science methods course. The following is not an exhaustive list of the issues:

- Increasing confidence in doing scientific inquiry and teaching science.
- Understanding fundamental ideas in science and the nature of science.
- Developing pedagogical content knowledge for elementary science.
- Observing and reflecting on effective elementary science teaching.
- Practicing science teaching and analyzing the experience.
- Becoming aware of and changing personal beliefs about teaching and learning.
- Developing understanding and proficient use of effective teacher behaviors and strategies understanding their interactions.

- Understanding children's thinking and development, and the arising implications for teaching.
- Effectively designing lessons for elementary children that take into account children's thinking and goals for science education.
- Understanding equity issues in science, including gender issues, adaptations for special-needs students, and differentiating instruction.
- Developing proficient use of social learning strategies such as cooperative learning, class discussions, and laboratory work.
- Understanding how to assess students' science understandings.
- Becoming familiar with resources and curricular materials available for science teaching, including commercially available science programs, Internet resources, national and local standards documents, and school curricula.
- Appropriately modifying curricular materials to better match the nature of the learner and student goals.

Given the extensive nature of these issues, priorities must be assigned. Understanding the context of the science methods course in the larger elementary education program will help determine what issues require more attention than others. However, when making these decisions, revisiting important issues in multiple contexts is necessary for prospective teachers to understand the generalizability of many pedagogical strategies across subject areas. Further, "coverage" of particular content or strategies in one course does not imply understanding. Ideas need to be revisited in order for students to develop an orientation to science teaching more consistent with effective science instruction (Magnusson, Krajcik, & Borko, 1999), and to develop appropriate science PCK (see Fig. 3.1 in this volume).

A second caution is to avoid the common practice of selecting topics and addressing them in a piecemeal fashion—a topic-of-the-day approach. A deep understanding of the complexities of teaching and learning means that topics are understood in the context of their interrelations. Unfortunately, teacher education has been accused of being haphazard (Atwood, 1973), and more recently, unnecessary (U.S. Dept. of Education, 2002). Clough (2003) asserted that problems in teacher education are due, in part, to the lack of a coherent framework that connects multiple elements of effective teaching. Therefore, effective science methods courses by necessity will involve a complicated balancing act between activities, content, viewing effective teaching, assessment, teacher behaviors, practice teaching, and so on, with careful emphasis placed on the synergy between these elements.

Two case studies of methods courses from different universities and very different contexts illustrate how these elements come together.

CASE STUDY 1:
CENTRAL QUEENSLAND UNIVERSITY

The Context

Central Queensland University is located on campuses in five regional areas in the state of Queensland, Australia. This case study focuses on the preservice elementary science methods course offered at the Rockhampton campus as part of the new Bachelor of Learning Management program (replacing the Bachelor of Education). This is a 4-year undergraduate program that most students complete in 3 years by taking summer courses. There are four practicum sessions, one per year, ranging from 10 single-day school visits to a 4-week block, concluding with an internship. The science methods course is situated in the same semester as the second practicum session, and is taught intensively over 10 weeks for 3 hours per week. It is followed immediately by the 3-week practicum in which students are required to teach a science unit of work linked to what they have done in the methods course. Students commencing the course have completed an introductory educational psychology course and courses dealing with general pedagogy.

Aims and Content

The main aim of the course is to prepare teachers who have the attitude, understanding, and skills to successfully begin teaching elementary science. This is addressed through attitudes and key areas of knowledge that, as explained in chapter 3, are related to the development of science pedagogical content knowledge. The development of self-confidence in teaching science is particularly important (Palmer, 2001). Specific areas of teacher knowledge addressed include science content, general pedagogy, curriculum, nature of science, equity and social justice, pedagogy based on constructivist learning theories, and aspects of science PCK related to these.

Course Rationale

A key belief, on which the course is based, is that success and failure can have a major impact on people's attitudes. As success is likely to create attitude improvement, the course is structured so that the preservice students experience success in both science learning and in science teaching. For instance, the preservice teachers take part in science inquiry activities to experience science, and therefore science education, as an enterprise done by people, using investigative hands-on, minds-on science strategies. The instructor addresses only a few topics in order to create confidence; if students have success in one or two science topics, they develop confidence in science more generally (Appleton, 1995).

A dilemma in the course is the instructor's awareness that if students fail at this inquiry, they are probably going to be turned off teaching science. On the other hand, required standards must be met. Consequently, multiple opportunities exist for the prospective teachers to achieve the outcomes of the course, so they are more likely to have success in passing it and achieve the required standards.

Another focus is to always tie science content to pedagogical issues (e.g., when investigating magnetism, students discuss how a magnetic field could be explained to children, and what analogies could be used). The integration of discipline content and pedagogy is important for promoting attitude change in education students (Schoon & Boone, 1998; Skamp & Mueller, 2001).

A third focus is to help preservice teachers understand that if they have not been successful in science before, it was not necessarily a personal failure. They tend to attribute their failure to themselves, so the instructor promotes the idea that it was a failure of the education system rather than a personal failure. If the preservice teachers can see that the way they were taught science in the past was not necessarily the best way to teach the subject, then they may shift their feelings of failure from themselves onto the system.

A fourth focus is that good science teaching should be based on constructivist principles; for instance, in order to learn something children have to link new information to what they already know. Therefore, teachers must consider children's existing ideas, identify specifically the science understandings that they want children to learn, and use pedagogies that will provide a scaffold for children to progress from what they know to the "target" learning. An essential part of providing a scaffold is an awareness that the classroom is a social environment, and that an accepting, low-risk learning environment is necessary (Palmer, 2001). This also plays a major role in working with the preservice teachers, as I (second author) believe that attitude change is dependent on a high level of trust between the students and the teacher. However, it is more difficult to build the sort of relationship that will engender attitude change with recent trends toward increased numbers of preservice teachers and larger classes.

Pedagogy

During the first 6 weeks of the course, the classes follow a similar pattern. Each session includes:

- an area of science content such as Phases of the Moon, Magnets, Electric Circuits, and The Body;
- examples of how the content is related to curriculum requirements and outcomes;
- selected hands-on activities dealing with the science content suitable for the preservice teachers, but adaptable to the elementary school;
- pedagogical scaffolds that the preservice teachers could use to sequence the activities to enhance learning, such as the learning cycle (Lawson, Abraham, & Renner, 1989), or the interactive approach (Biddulph & Osborne, 1984);
- general pedagogy related to the activities and pedagogy, such as small group work, questioning, and equipment management;
- examples of unit plans based on the week's selection of science content and pedagogy; and
- discussions of assigned readings about curriculum, the nature of science, learning theories, and equity.

Each week, one of the areas of knowledge has a more predominate focus. In the first week, the official curriculum documents are examined. In the second, constructivist-derived pedagogies, such as discrepant events and various teaching approaches, are discussed and modeled for the preservice teachers. Videos of science teaching are used as well. The nature of science and its influence on science education is considered in the third week, using a number of accounts of scientific discoveries and exploring how students may "work scientifically" (a term used in the official curriculum to describe a range of science process skills). During the fourth and fifth weeks, different aspects of equity and social justice are considered using case studies.

Two important foci of this section of the course are to ensure that the students experience success in learning science for themselves, and to illustrate how the science experiences that turned them off science were inappropriate—their lack of success in science was not necessarily due to a personal failing, but resulted from problems in their schooling. A constructivist-based pedagogy is employed consistently.

The remaining 4 weeks of class are devoted to planning units of work in science. A unit of work is co-constructed in each of sessions seven and eight, so the preservice teachers can see how and why activities are selected and sequenced to form a scaffold. Working from the official curriculum, learning outcomes and science content are selected and used to initiate a search for possible activities. Selected activities are worked by the preservice teachers, who identify those that would be suitable for initiating a unit of work, that help children achieve understanding of ideas, and that are suitable for summarizing and concluding. They help identify specific understandings that children might achieve, and select suitable assessment strategies. They also view videos of some science teaching, such as the interactive approach, and use such approaches as an aid to working out a pedagogical scaffold for each unit. Although the preservice teachers have previously prepared lessons, this is their first experience of preparing a unit plan. They need specific help in order to move from thinking of individual lessons to a whole unit and how activities might be sequenced to enhance learning.

During the final 2 weeks, the preservice teachers prepare a science unit that they will teach during their forthcoming practicum. They have two prepracticum contact days during which they obtain the science topic that they will teach, and familiarize themselves with the class. They also identify some of the preconceptions about the science topic held by the children. This information is used as an input for their planning. They are also shown electronic aids for planning, such as an online unit plan template as well as electronic and other resources. Much individual conferencing occurs during these sessions. In the final week, the students are required to conference with their professor about their unit plan. A key focus of this component of the course and its link to the practicum is for the preservice teachers to experience success in teaching science.

During the practicum, the preservice teachers are expected to teach their unit of at least three science lessons. The unit that they teach was prepared in consultation with both their teacher-mentor and their professor.

Assessment

The Bachelor of Learning Management is standards based, so relevant program standards have been identified and assessment indicators for the course derived from these. Two items of assessment are used to determine preservice teachers' achievement of the assessment indicators. The first is a show portfolio demonstrating their understanding of the key areas of knowledge addressed during the first 6 weeks. The second is their science unit plan prepared for their practicum. A component of this is a reflective evaluation of the implementation of their unit. In this, they must refer to the children's achievement of learning outcomes and the evidence to support their judgments. Detailed rubrics of criteria are used to grade the assignments.

Evaluation of the Course

The course was evaluated in 2001 using a combination of professor journals and interviews with five preservice teachers conducted by David

Palmer, from the University of Newcastle. The preservice teachers were selected randomly from those in my most recent class.

The journals identified important components of the course for the preservice teachers' learning as modeling of pedagogy and planning, conferencing with them about their unit plans, and the opportunity for them to teach science to children. More importantly, the preservice teachers themselves explained in their interviews about how they had changed from previously negative attitudes about teaching science:[2]

> I've always had science up there with maths—they're both going to be really hard to teach. It [the methods course] just took away all the fear of teaching. Equipped with what [the professor] had shown us, I didn't have any problems going out to teach science in the classroom. It was a really enjoyable experience for all of us, the children as much as ourselves. (Interview 1)

> From [the methods] course, I decided to do a science [discipline] major. You have to have a [discipline] major for primary, so now I'm doing science and maths. (Interview 2)

> When I first saw science as one of our subjects I thought, "Oh, no. I can't do this because it's a science." But when [the professor] got us through it I thought "Maybe I can. I'll have a go." But then at the end I really thought "Well, I can do this now." Then when we went out on prac and they asked us to do a bit of science. I loved it. (Interview 3)

> With watching [our professor], it was just an eye opener. And it was exciting, the way he carried himself, and the way he challenged us with those discrepant events. I think if you can just get your students' attention and make them want to be a part of learning about science, I think it will come easily. I think it does depend on the teacher and how they present themselves. I'm sure that I can do it. (Interview 4)

> It's more relevant to me now. I'm taking a lot more notice now of things around me and just thinking about how things work and listening to things on the radio and TV. There's the [television] science show on Thursday night that I tape now. (Interview 5)

In the interviews, the preservice teachers attributed different aspects of the methods course to its success in equipping them to teach science.

The professor's enthusiasm: "You could tell he really wants to make a change for children in the future and starting with us. This is how he

---

[2]Adapted from *Clever Teachers, Clever Sciences: Preparing Teachers for the Challenge of Teaching Science, Mathematics and Technology in 21st Century Australia,* by J. A. Lawrance and D. H. Palmer, 2001, Unpublished report prepared for the Department of Education, Training and Youth Affairs, Commonwealth of Australia. Adapted with permission.

thinks that he can make a change and he really did model that. He had a real passion for it."

Discrepant events demonstrations (i.e., demonstrations that have a surprising result): "It was a combination of different things, but to me, when [our professor] started each session he would have a discrepant event, and that would suck you in. Before you know it, you are away and the lesson has started."

The professor's modeling of teaching strategies: "[The professor's] modelling of how you should go about teaching science, or interacting with the children in learning about science, was the most important thing. It was his enthusiasm and the fact that everybody's answers are valued."

Valuing the students' answers: "[The professor] kind of made everyone feel clever, just for having an answer even if it was wrong. It was like, you've had a think about it and you've come up with some sort of answer so you've done a good job anyway. So that was really good because you didn't leave feeling that you were a real dodo."

The professor's constructivist approach: "I really think by [the professor's] attitude of a constructivist, that's really changed my attitude. Because I know now how to go about teaching students. In the future I'll be really interested with science."

Making science relevant to real life: "We were just on the go the whole time and we were really interested and the day flew. I started to think that I could learn science, and it wasn't all just formulas and grids and graphs, it was real life relevant."

Having a chance to practice teaching science to children: "The teaching in class at the school—the kids just loved it. The teacher was very supportive of us as well. It made science interesting to teach. . . . The children just enjoyed it because it was hands-on. It made me feel that I had really taught something and the children had taken it in and enjoyed it and I enjoyed teaching them."

Summary

In terms of Feiman-Nemser's (1990) orientations to preservice curriculum, three orientations predominate in this case: academic, practical, and personal. The technological and critical/social orientations receive more limited attention. The strongest emphasis could be considered the academic orientation, in the sense that development of science PCK is an overriding goal—recognizing that only a limited amount of science can be covered in the short time frame of the course. As outlined in chapter 3, confidence to teach science exerts considerable influence on the development of science PCK.

The choice of science content was based on findings that elementary teachers are least confident in the physical sciences (Appleton, 1995). The choice of other aspects to include in the course was influenced by the placement of the course in the program, and prior and concurrent courses.

The evaluation data reflect the emphases of the course; but one further matter is worth noting: the influence of the professor in promoting student learning and encouraging positive attitudes to science and science teaching. Key aspects of this role mentioned in the evaluation data were the professor's passion, his views of learning and of teaching, and his valuing of students and their thinking.

CASE STUDY 2: IOWA STATE UNIVERSITY

The Context

Iowa State University is a large public university located in the rural midwestern United States. This case study focuses on the preservice elementary science methods course offered as part of the elementary education bachelor's degree program. This is a 4-year undergraduate program that includes two practicum sessions, each consisting of a midsemester 4-week teaching experience. The program concludes with a semester of full-time student teaching. The science methods course occurs in the same semester as the second practicum experience, typically the semester prior to student teaching. Students enrolled in the course have completed a course in general pedagogy, an educational psychology course, an educational foundations course, and courses in methods of teaching mathematics, social studies, and primary reading.

Aims and Content

Like the Queensland program, elementary science methods at Iowa State University is designed to prepare prospective elementary teachers to successfully begin teaching science. Considering the orientations described earlier, the emphasis on the practical, technical, and personal orientations is more heavily emphasized than the academic and critical/social, although the latter two are present in the course. This is primarily because content knowledge, confidence, general pedagogy, and knowledge of students are relatively weakly developed.

An important concern in the design of this course is providing students with too many "activities that work." Given the central nature of such activities in the development of PCK, this may seem counterintuitive. However, if students have effective activities, yet little or no general pedagogical knowledge, particularly related to the role of the teacher, then a real danger exists of perpetuating the common perception that by doing an activity correctly, expected learning should occur with little or no further action by the teacher required (Appleton, 2002).

The course is designed to prepare teachers who:

1. Forcefully examine their beliefs about teaching, learning, students, and schooling.
2. Identify student goals consistent with the science education community.
3. Precisely articulate how they will know student goals are being met.
4. Inform classroom decisions with knowledge of how students learn and research on effective practice.
5. Have a robust understanding of the teacher's role in the classroom.
6. Understand how to select content, materials, and activities to promote student goals and diagnose student thinking, while being consistent with research on how students learn.
7. Accurately self-assess their performance and possess a clearly articulated desired state of teaching practice and a practical plan to work toward attainment of that desired state.

Course Design and Rationale

Careful consideration is made in the design of the course to balance both the development of general pedagogical strategies and the application of those strategies within science instruction. Due to the need to develop so many aspects of effective teaching concurrently, the course is designed to both reflect and develop a research-based framework for teaching science (Clough, 2003; Krajick, Penick, & Yager, 1986). A research-based framework (RBF) for teaching science is a holistic framework designed to inform classroom decision making (see also chaps. 13 and 14). Rather than making instructional decisions based solely on past experiences or personal opinion, a research-based framework helps teachers focus on what goals they have for students, the foundational importance of knowledge of how students learn, and how they will promote their goals in the classroom through the deliberate selection and implementation of content, materials, activities, strategies, and specific teacher behaviors. The RBF is designed to help preservice teachers understand the synergy of these critical aspects of teaching, rather than viewing components in isolation. This is a deliberate attempt to prevent prospective teachers from substituting "activities that work" for science PCK (see chap. 3)—a major concern as these prospective teachers have very weak general pedagogical knowledge and equally weak science content understandings.

In the context of this science methods course, students develop a written paper that identifies their goals for students, how they will promote those goals, and how learning theories will inform their practice. The paper is orally defended with the course instructor at the end of the semester. More than an assessment strategy, however, the course itself reflects a research-based framework and students are frequently attended to the behaviors and decisions made by the instructor and are asked to compare them to course readings, previous science experiences, and so on. A model for a research-based framework is provided in Fig. 8.1.

This course is situated in a program that begins with courses on educational foundations and psychology, followed by a series of up to six disconnected methods classes, concluding with a semester of student teaching. Unfortunately, preservice teachers in this program frequently perceive effective activities as the sum of necessary knowledge of teaching rather than part of knowing how to teach. Returning to Driver's (1997) observations, changing these views is emotionally taxing for the learner. Teaching a course such as this requires that the methods instructor be a model of effective teaching and provide sufficient support for students. The course involves no lectures, no declarative knowledge to be memorized, and no tests. It uses science activities, critical incidents, video clips of lessons,

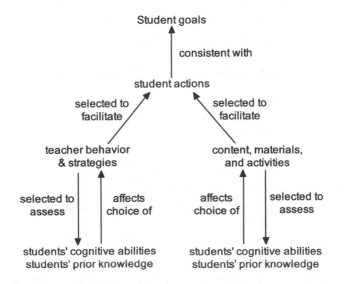

FIG. 8.1. Components of a research-based framework for teaching science. From "Improving Engineering Education: A Research-Based Framework for Teaching," by M. P. Clough and K. J. Kauffman, 1999, *Journal of Engineering Education, 88*, p. 2.

readings, and student questions as springboards for class discussions to make students aware of their ideas, challenge them, and pose new possibilities for conceptualizing science instruction. This case provides a brief overview, however, the course is more thoroughly described in Olson (2003).

Pedagogy

Where Are We? Where Do We Want to Go?   The course begins with small group discussion about prospective teachers' personal science experiences from their earliest memories to the present. Personal experiences are then compared to readings describing the current and persistent problems in science education. The intention of this portion of the course is to help these prospective teachers become aware of their perceptions and develop a need to teach differently than they were taught. Following this, the course addresses student goals. Students are asked, "Where do we want students to be at the end of a K–12 education?" As a class, they generate a long list, and are then challenged to reduce the list to 10–15 goals. Students quickly see that many goals overlap. Typical goals include developing students who will:

1. Exhibit a deep understanding of science content.
2. Understand what science is and how it works.
3. Demonstrate appropriate social skills.
4. Become lifelong learners.
5. Be creative and curious.
6. Be critical thinkers.
7. Effectively identify and solve problems.
8. Respect themselves and others.
9. Be resourceful.
10. Be productive citizens.
11. Use effective communication skills.
12. Apply knowledge of science to their daily lives.

Clough (2003) pointed out that whereas most teachers have these same goals, the vague nature of such goals perhaps prevents teachers from actively promoting them. Therefore, students work in groups to select a goal, generate a list of 10 or more student actions, and send lists to one another via e-mail for feedback. The resulting vision for science teaching sets a context for examining teaching decisions regarding science content, materials, activities, the role of the teacher, and utilizing learning theories to inform instruction.

How Do We Get There? How People Learn and the Role of Content, Materials, and Activities.   Content, materials, and activities are typically the central focus of elementary preservice teachers (Tolman & Campbell, 1989). This reflects their main concern of survival (Fuller, 1969), coupled with the perception that survival is best accomplished by knowing what they need to teach and having enough "good activities" to use with students. This focus on activities is used to set the context for a need to revisit learning theories. Two activities are conducted with students that model the learning cycle (Lawson, Abraham, & Renner, 1989). After these activities, the stages of the lessons are identified by students and the learning cycle is then formally introduced. Consistent with the learning cycle itself, prospective teachers then apply this knowledge by teaching a learning cycle lesson during their practicum teaching experience. One activity is selected, however, that is developmentally inappropriate for elementary students. The university students rarely determine the science content the activity is intended to illustrate. This raises two important issues: the way to determine what content is appropriate for elementary students, and the critical role of the teacher in teaching content—activities do not teach themselves and students often develop alternative understandings from activities alone. The prospective teachers now perceive the need to revisit their understanding of how people learn.

Although the prospective teachers in the course have completed a course in educational psychology, they struggle to apply this knowledge to the classroom. Students review social, developmental, behavioral, and constructivist learning theories, followed by the viewing of videotaped segments of elementary science teaching to illustrate that all four influence the classroom with varying degrees of importance corresponding to specific goals the teacher is promoting. Two class sessions are spent viewing *Minds of Our Own* (Annenberg/CPB, 1997) and discussing the main ideas and implications of these ideas for teaching.

Additional activities are conducted and students are asked to determine how the activity promoted or inhibited their goals, how learning theories influenced the activity, the lesson structure employed, ideas for selection of materials for science, and how to select science content appropriate for elementary students.

How Do We Get There? The Critical Role of the Teacher.   Portions of two videotapes are shown to the class, both depicting learning cycle lessons. Whereas the lesson structure is the same, the teachers' interaction patterns are vastly different. As a result, the impact on student engagement is dramatic. Following discussion of the lessons, a coding sheet modified from Abraham and Schlitt (1973) is introduced to students as one way to determine a teacher's interaction patterns with students. They practice coding a small segment of each lesson and discuss the interaction patterns. Extensive discussions and analyses of teachers' behavioral patterns occur, highlighting their pervasive influence in promoting or inhibiting student goals. Complex interactions between teachers' behavioral patterns and decisions of content, materials, and activities are revisited in increasingly complex ways.

Practicum Teaching Experience.   Students then test and expand their developing understandings in the complex reality of the classroom by teaching 2 days per week in an elementary classroom for 4 weeks. They teach a learning cycle lesson and audiotape themselves to analyze their interaction patterns with children. The course instructor visits as many students as possible during this time.

Formally Making Sense of It All: The RBF.   After their practicum teaching experience, these prospective teachers experience significant cognitive dilemmas. After analyzing their audiotapes, they return to class with the realization, for perhaps the first time, that they do not know what they thought they knew about teaching. Their questioning patterns typically resemble one of the tapes they criticized earlier, and they realize teaching requires more than having activities for students. Instructor support has to be very high to recognize the emotionally taxing nature of this experience, and to help students persevere and move forward.

Course experiences for the rest of the semester focus on revisiting previous ideas in increasing complexity to help these prospective teachers develop a better understanding of how to teach science. Activities are conducted that illustrate how to conduct effective field trips, assess student understanding, teach the nature of science, and modify activities to better promote student goals and teach students with special needs (Cox-Petersen & Olson, 2002; Olson, 2003; Olson, Cox-Petersen, & McComas, 2001).

Class time also involves the presentation of student goals. In pairs, students prepare from 10 to 12 summary statements from literature and research abstracts regarding a goal they chose. They present the goal, how they will promote it, and how their actions in the classroom take into account student learning. This process prepares them for the oral defense (see later), provides them with research they need for their RBF, and makes quite obvious the overlap of strategies used to promote their goals. In the final 3 weeks of the semester, students begin writing their 20- to 30-page RBF paper, articulating their goals for students and how they will teach science in a manner that both promotes their goals and is consistent with how students learn.

Assessment

Students in the course are assessed on four assignments: an analysis of their interaction patterns with students, a case study, the RBF paper, and an oral defense. None of these assignments receives a letter grade. Instead, extensive feedback is written on all papers; students must read the feedback, thoughtfully consider their own performance, compare the feedback and their perceptions to criteria provided in the syllabus, and self-assess their work.

The case study assignment is given to small groups of students early in the course. They receive a videotape of an elementary science lesson and must use the schematic in Fig. 8.1 to analyze the lesson, followed by a rewrite in the form of two lesson plans. The task serves to focus their learning during the first several weeks of the course, helping them develop a framework through which to analyze teaching practice. Having the videotape is important: to help prospective teachers better conceptualize a new way of teaching science; to "perturb student thinking" (Abell, Bryan, & Anderson, 1998, p. 507); to promote self-assessment skills (Olson, 2003); to reduce simplistic good/bad, worked/didn't work thinking; and to provide a common experience for group and personal reflection.

The audiotape analysis engages students in audiotaping their own teaching practice, analyzing their interaction patterns and developing a concrete plan for desired improvements. This process of self-assessment is later expanded in the RBF paper and oral defense.

The RBF paper is intended to engage students in clearly articulating the desired state of their teaching and how they will reach their goals. They must tie together all of the elements of the schematic in Fig. 8.1, making the connections between strategies and behaviors, utilizing ideas gained from other courses where appropriate, and supporting their assertions with education research.

Students may write on paper what they want to accomplish in their classroom, but the understandings they possess in their minds are what they will more likely use when they teach. The oral defense presents students with critical incidents and other questions modified from Clough and Berg (1995). They are asked questions about how they will promote specific goals, determine if they are successful in the classroom, know if students are meeting their goals, respond to parents who question their use of cooperative learning groups, and so on. This conversation between student and instructor can last up to one and a half hours. At the end of the oral defense, students use the course criteria and self-assess their performance in the course. The grade must be based on criteria and defended to the satisfaction of the instructor.

ELEMENTARY SCIENCE METHODS: REFLECTIONS ON CURRICULUM

At first glance, both courses may appear quite different. However, beneath the structural differences, each course sets out to promote positive science teaching self-efficacy, make clear that effective science teaching demands far more than good activities, and address the importance of all five conceptual orientations to teacher education and their complex interactions. Toward these ends, both courses have prospective elementary teachers:

Academic

- Engaging in science activities and addressing issues of content and the nature of science.
- Studying the pedagogy required for those activities, promoting development of PCK.

Practical

- Observing and analyzing effective science teaching via videotape.
- Practicing teaching during a practicum experience with instructor support.
- Reflecting on teaching experiences.

Technological

- Analyzing the role of the teacher in science activities, including questioning, wait time, and nonverbal behaviors.
- Developing more informed decision making and reflection on-action and in-action.

Personal

- Studying student learning via case studies and classroom modeling.
- Assessing student thinking during practicum teaching.
- Considering student thinking when developing unit plans/lessons.
- Developing lessons that require synthesis of science content, consideration of students and their prior ideas, the role of the teacher, and appropriate activities.

Critical/social

- Modifying activities to take into account children's needs and interests.
- Accommodating lessons for diverse learners and children with special needs.
- Engaging in activities that incorporate social learning strategies such as cooperative learning.

DISCUSSION

In all these efforts, the PCK development model in chapter 3 guides decisions regarding the science methods course content, structure, and instruction, as it highlights the different forms of teacher knowledge that are drawn on in constructing science PCK. The variation in emphases between the Queensland and Iowa cases is primarily due to the strengths and weaknesses of the program in which each course is embedded. Queensland students have higher general pedagogical knowledge; thus the course emphasizes developing PCK within the context of activities that work and increasing confidence, while expanding on the general pedagogical knowledge they possess. Because Iowa State preservice elementary students enter their science methods course unaware of the teacher's critical role in effective teaching, more time and effort is directed toward teacher decision making (e.g., choice of content, activities, materials, and teaching strategies) and interaction patterns (e.g., the audiotape analysis of their own questioning, wait time, and responding).

Effective teacher knowledge, including general pedagogy, content knowledge, and PCK, is understanding-in-action (Mellado, Blanco, & Ruiz, 1998; Schön, 1983). Developing this understanding demands that prospective teachers have ample field experiences that are closely linked to their emerging understanding of the five conceptual orientations to teacher education and their interactions, the ubiquitous role of the teacher, and the nontrivial decisions that teachers constantly make (Palmer, 2001). Practicum experiences alone are insufficient for developing this understanding for the same reason that experience alone does not lead to a scientific understanding of the natural world. Hence, experiences such as videotape and audiotape analysis of teaching sessions; lesson planning, development, feedback, implementation, and reflection; critical incidents and case studies; "what if" scenarios; and analyzing the practices of teachers along the continuum of beginner to expert are indispensable. Both instructors in the two case studies accomplish these ends in different ways.

CONCLUSIONS

Whereas this chapter has emphasized the content and structure of effective preservice elementary education methods courses, the role of the methods professor—the decision-making and interaction patterns—has created the enacted curriculum. That is, a teacher-proof curriculum is neither possible nor desirable in science teacher education (or in elementary science teaching either). Understanding prospective elementary teachers' ideas, playing off those ideas, deciding when and how to use content, activities, and materials, and myriad other nontrivial decisions made in the act of teaching determine whether the planned curriculum moves students effectively toward desired ends. Moreover, those engaged in teaching teachers convey messages about teaching by what they espouse as well as what they do. Prospective teachers could easily exit from instruction with their fundamental notions of learning and teaching unchanged if the methods instructor teaches in a manner inconsistent with research-based practices. Hence, modeling effective practices and attending students to both those practices and the decision-making processes that resulted in the exhibited practices are essential. The sharing of PCK naturally fits with this effort of making explicit the complex interactions between goals, research, context, and prior personal experience. In sharing this expertise, students gain trust in their professor and the value of the course. What this means is that highly effective teacher education demands that methods professors model and convey the complexities of effective learning and teaching. To do so requires both science teaching PCK (Abell, Magnusson, Schmidt, & Smith, 1996), as well as science PCK developed through study and extensive effective teaching experiences with children in authentic school settings.

REFERENCES

Abell, S. K., Bryan, L. A., & Anderson, M. A. (1998). Investigating preservice elementary science teacher reflective thinking using integrated media case-based instruction in elementary science teacher preparation. *Science Education, 82,* 491–510.

Abell, S. K., & Smith, D. C. (1994). What is science? Preservice elementary teachers' conceptions of the nature of science. *International Journal of Science Education, 16,* 475–487.

Abell, S., Magnusson, S., Schmidt, J., & Smith, D. (1996, April). *Building a pedagogical content knowledge base for elementary science teacher education.* Symposium for the annual meeting of the National Association for Research in Science Teaching, St. Louis, MO.

Abraham, M. R., & Schlitt, D. M. (1973). Verbal interaction: A means for self evaluation. *School Science & Mathematics, 73,* 678–686.

Akerson, V. L., & Abd-El-Khalick, F. (2002, April). *Teaching elements of nature of science: A year long case study of a fourth grade teacher.* Paper presented at the annual meeting of the American Educational Research Association, New Orleans, LA.

Anderson, R. D., & Mitchener, C. P. (1994). Research on science teacher education. In D. L. Gabel (Ed.), *Handbook of research on science teaching and learning* (pp. 3–44). Washington, DC: National Science Teachers Association.

Annenberg/CPB. (1997). *Minds of our own: Can we believe our eyes* [Math and Science Videotape Collection]. (P.O. Box 2345, South Burlington, VT 05407-2345).

Appleton, K. (1995). Student teachers' confidence to teach science: Is more science knowledge necessary to improve self confidence? *International Journal of Science Education, 19,* 357–369.

Appleton, K. (1997). Analysis and description of students' learning during science classes using a constructivist-based model. *Journal of Research in Science Teaching, 34,* 303–318.

Appleton, K. (2002). Science activities that work: Perceptions of primary school teachers. *Research in Science Education, 32,* 393–410.

Atwood, R. K. (1973). *Elements of science teacher education as abstracted from ERIC-AETS: In search of promising practices in science teacher education.* Paper presented at the annual meeting of the Association for the Education of Teachers in Science.

Australian Science and Engineering Council (ASTEC). (1997). *Foundations for Australia's future: Science and technology in primary schools.* Canberra, Australia: Australian Government Publishing Service.

Bethel, L. (1984). Science teacher preparation and professional development. In D. Holdzkom & P. Lutz (Eds.), *Research within reach: Science education* (pp. 143–158). Washington, DC: National Science Teachers Association.

Biddulph, F., & Osborne, R. (Eds.). (1984). *Making sense of our world: An interactive teaching approach.* Hamilton, New Zealand: University of Waikato, Science Education Research Unit.

Clough, M. P. (2001, November). *Longitudinal understanding of the nature of science following a course emphasizing contextualized and decontextualized nature of science instruction.* Paper presented at the sixth International History, Philosophy, and Science Teaching conference, Denver, CO.

Clough, M. P. (2003, January). *The value of a Research-Based Framework for understanding the complexities in learning and teaching science.* Paper presented at the annual meeting of the Association for the Education of Teachers in Science, St. Louis, MO.

Clough, M. P., & Berg, C. A. (1995). Preparing and hiring exemplary science teachers. *Kappa Delta Pi Record, 31,* 80–89.

Clough, M. P., & Kauffman, K. J. (1999). Improving engineering education: A research-based framework for teaching. *Journal of Engineering Education, 88,* 527–534.

Cox-Petersen, A. M., & Olson, J. K. (2002). Assessing student learning. In R. Bybee (Ed.), *Learning science and the science of learning* (pp. 105–118). Arlington, VA: NSTA Press.

Department of Employment, Education, and Training (DEET). (1989). *Discipline review of teacher education in mathematics and science.* Canberra, Australia: Australian Government Publishing Service.

DeRose, J. V., Lockard, J. D., & Paldy, L. G. (1979). The teacher is the key: A report on three NSF studies. *The Science Teacher, 46,* 31–37.

Donnellan, K. (1982). *NSTA elementary teacher survey on preservice preparation of teachers of science at the elementary, middle, and junior high school levels.* Washington, DC: National Science Teachers Association.

Driver, R. (1997). In Annenberg/CPB (Producer). *Minds of our own: Can we believe our eyes* [Math and Science Videotape Collection]. (P.O. Box 2345, South Burlington, VT 05407-2345)

Feiman-Nemser, S. (1990). *Conceptual orientations in teacher education.* Issue Paper 90-2. East Lansing, MI: National Center for Research on Teacher Education.

Feiman-Nemser, S., & Buchmann, M. (1985). Pitfalls of experience in teacher preparation. *Teachers College Record, 87,* 53–65.

Freyberg, P., & Osborne, R. (1985). Assumptions about teaching and learning. In R. Osborne & P. Freyberg (Eds.), *Learning in science: The implications of children's science* (pp. 81–90). Auckland, NZ: Heinemann.

Fuller, F. F. (1969). Concerns of teachers: A developmental conceptualization. *American Educational Research Journal, 6,* 207–226.

Heikkinen, M. W. (1988). The academic preparation of Idaho science teachers. *Science Education, 72,* 63–71.

Helgeson, S. L., Blosser, P. E., & Howe, R. W. (1977). *The status of precollege science, mathematics, and social science education. 1955–75: Vol. 1. Science education.* Columbus, OH: Ohio State University, Center for Science and Mathematics Education.

Hurd, P. (1983). State of precollege education in mathematics and sciences. *Science Education, 67,* 57–67.

Kennedy, M. M. (1990). Choosing a goal for professional education. In W. R. Houston (Ed.), *Handbook of research on teacher education* (pp. 813–825). New York: Macmillan.

Kennison, C. (1990). *Enhancing teachers' professional learning: Relationships between school culture and elementary school teachers' beliefs, images and ways of knowing.* Unpublished thesis, Florida State University.

Krajcik, J. S., Penick, J. E., & Yager, R. E. (1986, March). *An evaluation of the University of Iowa's science teacher education program.* Paper presented at the annual meeting of the National Association for Research in Science Teaching, San Francisco, CA.

Kuhn, T. S. (1962). *The structure of scientific revolutions.* Chicago: University Press.

Lawson, A. E., Abraham, M. R., & Renner, J. W. (1989). A theory of instruction: Using the learning cycle to teach science concepts and thinking skills. *NARST Monograph,* No. 1.

Magnusson, S., Krajcik, J. S., & Borko, H. (1999). Nature, sources, and development of pedagogical content knowledge for science teaching. In J. Gess-Newsome & N. G. Lederman (Eds.), *Examining pedagogical content knowledge* (pp. 95–132). Dordrecht, The Netherlands: Kluwer Academic.

Mellado, V., Blanco, L. J., & Ruiz, C. (1998). A framework for learning to teach science in initial primary teacher education. *Journal of Science Teacher Education, 9,* 195–219.

Moore, J. J., & Watson, S. B. (1999). Contributors to the decision of elementary education majors to choose science as an academic concentration. *Journal of Elementary Science Education, 11,* 37–46.

National Science Teachers Association. (1984). *Standards for the preparation and certification of teachers of science, K–12.* Washington, DC: Author.

Ohana, C. (1999). *A tangled web: Interactions and structures in university–school collaborations.* Unpublished doctoral dissertation, Iowa State University.

Olson, J. K. (2003, March). *Embedding problem-based learning within a research-based framework for teaching science.* Paper presented at the annual meeting of the National Association for Research in Science Teaching, Philadelphia, PA.

Olson, J. K. (2003). Light students' interest in the nature of science. *Science Scope, 27,* 18–22.

Olson, J. K., Cox-Petersen, A. M., & McComas, W. F. (2001). The inclusion of informal environments in science teacher preparation. *Journal of Science Teacher Education, 12,* 155–173.

Palmer, D. H. (2001). Factors contributing to attitude exchange amongst preservice elementary teachers. *Science Education, 86,* 122–138.

Rowell, J. A., & Cawthron, E. R. (1982). Image of science: An empirical study. *European Journal of Science Education, 4,* 79–94.

Schön, D. A. (1983). *The reflective practitioner: How professionals think in action.* New York: Basic Books.

Schoon, K. J., & Boone, W. J. (1998). Self-efficacy and alternative conceptions of science of preservice elementary teachers. *Science Education, 82,* 553–568.

Shulman, L. S. (1986). Those who understand: Knowledge growth in teaching. *Educational Researcher, 15,* 4–14.

Skamp, K., & Mueller, A. (2001). A longitudinal study of the influences of primary and secondary school, university and practicum on student teachers' images of effective primary science practice. *International Journal of Science Education, 23,* 227–245.

Tamir, P. (1983). Inquiry and the science teacher. *Science Education, 67,* 657–672.

Tobin, K., & Garnett, P. (1988). Exemplary practice in science classrooms. *Science Education, 72,* 197–208.

Tolman, M. N., & Campbell, M. K. (1989). What are we teaching the teachers of tomorrow? *Science and Children, 27,* 56–59.

U.S. Department of Education. (2002). *Meeting the highly qualified teachers challenge.* Washington, DC: Author.

Windshitl, M. (2003). Inquiry projects in science teacher education: What can investigative experiences reveal about teacher thinking and eventual classroom practice? *Science Education, 87,* 112–143.

Young, B. J., & Kellog, T. (1993). Science attitudes and preparation of preservice elementary teachers. *Science Education, 77,* 279–291.

9

Language, Learning, and Science Literacy

Vaughan Prain
La Trobe University

Brian Hand
Iowa State University

Over the last 10 years, there has been strong interest in redefining and expanding definitions of the nature of science literacy as both the central goal and proposed broad outcomes of learning in science (Hodson, 1998; Hurd, 1998; Lemke, 2002; National Research Council, 1996; Norris & Phillips, 2001). These accounts of science literacy have moved beyond a traditional focus on student acquisition of technical conceptions and science terminology to include and emphasize cognitive abilities, reasoning processes, commitment to a science worldview, and communication skills to explain and justify concepts in science to diverse readerships. Although these goals are clearly viewed as outcomes resulting from extended study of science through years of secondary education, there is clearly a range of implications for how the foundations for achieving such goals might be established and promoted in K–6 classrooms.

This chapter suggests that preservice teachers of science in elementary school need a clear conceptualization of the relations between language, learning, and science literacy if they are to understand how these diverse goals might be addressed. The issue of what is meant by "language" in relation to science education is further complicated by growing recognition that science should be understood as a complex mix of different languages, including mathematics, graphic representation, and verbal language (Lemke, 1998, 2002; Russell & McGuigan, 2001). From this perspective, students need to understand the purposes of these different modes in enabling explanation and measurement in science, and also understand how to integrate these modes to represent science ideas, processes, and findings. Although this chapter focuses mainly on verbal language in relation to learning science, it recognizes the importance of the other modes and also comments on their place in science literacy.

The chapter first reviews current understandings of the nature of science literacy as the desired outcomes of learning in science, and then considers the implications of these accounts for broad learning goals in science in the K–6 classroom. It then outlines a framework for conceptualizing the role of language in achieving this learning, drawing on recent research in learning in science related to aspects of this framework. The implications for elementary science teacher education are then explored through discussion of what needs to be included in preservice courses about literacy and education, suggestions on ways to incorporate this language focus into the teaching in these courses, and some proposed methods for supporting elementary teachers' professional development, including recent research on how successful science teachers in elementary schools in Australia address the issue of effective learning in science in relation to the use of language.

CURRENT UNDERSTANDINGS
OF SCIENCE LITERACY

Definitions of science literacy have been broadened in recent years to encompass diverse understandings, activities, values, and effects, as science educators have recognized the complexity of outcomes proposed for this literacy. At the same time, four overarching themes can be identified in this literature. These themes include deeper student knowledge about the nature and methods of science so that students can act and think scientifically in ways that are consistent with the ethos and practices of current professional scientists (Hand, Prain, Lawrence, & Yore, 1999; Hurd, 1998; Lemke, 1998; Norris & Phillips, 2001; Ryder, 2001; Unsworth, 2001; Yore, Hand, & Prain, 2002); communicative ability to explain scientific ideas and their applications to different social, economic, or cultural issues for diverse readerships (Hurd, 1998; Lemke, 1998; Linn, 2000; National Research Council, 1996); a personal lifelong commitment to a science worldview and the values of science inquiry in ways that inform everyday understandings and actions (Cobern, Gibson, & Underwood, 1995; Hurd, 1998; National Research Council, 1996); and the ability to critique science values and their social, economic, and cultural effects (Lemke, 2002).

Many science educators have focused on the need for students to understand the assumptions, methods, reasoning procedures, and representational practices of science inquiry, including subject-specific vocabulary as part of learning science literacy. There is a very large literature on this aspect of the content and methods of science implied in science literacy, as reviewed by many researchers, including Hand, Prain, Lawrence, and Yore (1999), Norris and Phillips (2001), and Ryder (2001). From this viewpoint, as summarized by the National Research Council (1996), students need to "develop a rich knowledge of science" as they become familiar with "modes of scientific inquiry, rules of evidence, ways of formulating questions and ways of proposing explanations" (p. 21). There is also growing recognition of the centrality of new technologies in representing science as a multimodal method of inquiry, entailing verbal, graphic, and mathematical reporting of science findings and investigations (Lemke, 2002; Linn, 2000). At the same time, there is increasing recognition in this literature that contemporary science literacy entails more than familiarity with the procedural and conceptual knowledge of science, and must include a capacity and willingness to contribute to public discussion about the application of scientific principles to social issues. This implies that science literacy should mean broad-based understanding of the procedures and claims of science rather than the understandings of experts in different science fields (Cobern et al., 1995). Contemporary science literacy, therefore, means the abilities and emotional dispositions to construct science understanding and to inform others about these science ideas and to persuade them to take informed action. The National Research Council (1996) supported this broader view of science literacy as the development of educated citizens who can "engage intelligently in public discourse and debate" (p. 13), can "construct explanations of natural phenomena, test these explanations in many different ways and communicate their ideas to others" (p. 20). This viewpoint tends to conceptualize science literacy as an adult attribute, equating science literacy with being scientifically literate in relation to different fields of science rather than having general knowledge about science. However, this definition also implies that experience of science in the early years of schooling should support a process that leads to the development of scientifically literate citizens.

At the same time, within this general redefinition of the nature of science literacy, some science educators have insisted that this literacy must entail, as a core focus, explicit induction into the particular language and representational practices and vocabulary of science as a discipline (Gee, 2002; Norris & Phillips, 2001, 2003; Unsworth, 2001). This perspective asserts that science can only be learned when these aspects of the subject are addressed directly. Norris and Phillips (2001, 2003) contended that the co sential aspect of science literacy relates to particular skills, knowledge, and attitudes associated with responding to and constructing science texts. They claimed that it is possible to identify a "fundamental sense" to science literacy in contrast to a "derived sense" (Norris & Phillips, 2001, p.

3). The fundamental sense refers to knowing why science texts have the history, form, and functions they have, and being able to read and write these texts in the appropriate manner, whereas the derived sense refers to being generally well informed, interested in, and educated about science. They claimed that science education has usually focused on the derived sense, on concept development and science content rather than on the representational practices of science writers. In other words, they asserted the need for a strong disciplinary focus for the subject where the acquisition of a fundamental sense of science literacy must mean that students need to develop skills in effectively "comprehending, interpreting, analyzing and critiquing" science texts (Norris & Phillips, 2001, p. 19), according to the rules of the science community. From this viewpoint, students need to be able to assess the validity, reliability, and creditability of claims made in both print and electronic science texts, as well as construct their own science texts in a manner consistent with the conventions of science representation and reasoning. Clearly, if followed, these prescriptions about the nature of science literacy have significant implications for ways science should be taught in K–6 classrooms, suggesting that science teaching should focus more directly on reading and producing science texts. Science literacy in both its fundamental and derived senses needs to be addressed in the elementary classroom, and the next section proposes a framework within which these different aspects can be addressed. Each sense of science literacy is interdependent, and there are strong psychological and motivational reasons for focusing predominantly on the everyday meaningfulness of science when teaching and learning science in the elementary school. Such a focus is likely to give students a positive meaningful introduction to the aims, methods, and value of this subject.

This very issue, the question of the personal meaningfulness of science values, methods, and concepts for students, is also highlighted in recent accounts of science literacy. Science educators argue that science literacy must entail commitment to the value of a science worldview and its usefulness in explaining everyday experiences. As noted by Hurd (1998), Hodson (1998), Ryder (2001), and many others, this perspective implies that science literacy is a "civic competency required for rational thinking about science in relation to personal, social political, economic problems, and issues that one is likely to meet throughout life" (Hurd, 1998, p. 410). By implication, school science will promote science literacy when students, over time in undertaking science inquiry, make durable, personal, strongly meaningful connections with their own lifeworld, values, and interests. In other words, school experience of science should have the positive effect of encouraging students to identify strongly with the goals, values, and usefulness of this subject area. That such a view needs to be made explicit points to a recognition that such outcomes have not always been the case in terms of science teaching and learning. By implication, there is a strong need to address the issue of effective student motivation in engaging with this subject.

A further recurrent theme in new accounts of science literacy relates to the importance of communication of science to others. From this perspective, students should see the study of science as about sharing new understandings, informing and persuading others about science-based issues, contributing to classroom community knowledge about science, and disseminating science ideas and issues more broadly in the larger community. This emphasis on explaining and justifying science has various implications for the aims, readerships, and diverse kinds of writing and presentation that students should practice in the science classroom in elementary school. Students need to understand traditional ways in which science is reported, but they also need to practice writing about science for varied readerships, using other kinds of writing types such as letters and brochures if they are to achieve the border crossing of readerships implied in these new accounts of science literacy. In addition, such work should foster in students an ability and willingness to contribute to public debate over the application of science to social issues. In this way, new accounts of science literacy emphasize the centrality of communications skills and a commitment to informed and accessible contributions to public debate over the uses of science. Writing and presentational tasks in science classrooms will need to provide opportunities for students to learn how and why they should aim to meet these goals.

Lemke (2002, p. 1) also claimed that science literacy should entail a critical dimension, where students learn how to "challenge the authority of written texts . . . placing them in their larger social contexts, and asking who benefits and how power and privilege are inscribed in them." Lemke (2002, p. 2) indicated that students should learn how to be critical "of the projects and employments of scientific research as an institution." He argued that students never meet scientists, and are not taught about "who does science, where they do it, how resources are found for it, and even why some people choose to do it in the first place" (p. 2). Although he did not specify a particular agenda for critical science literacy in the elementary classroom, he cited other researchers, such as Comber, Cormack, and O'Brien (2001), who clearly favored a critical focus at this level of schooling, where students are expected to identify gender stereotypes and other kinds of bias in texts. Although many of the goals Lemke attributed to critical science literacy are more suited to the secondary classroom, some goals (e.g., increased student understanding of the nature of the everyday work of professional scientists, and why "scientific" findings are used to support or criticize different practices and activities) would seem suitable for coverage in the elementary classroom.

In summary, these new accounts of the nature of science literacy have strong implications for the goals of, and approaches to, science education in the elementary school. They imply that science is both a specialist area of knowledge, with its own particular language (or languages), values, and conventions for representing concepts, processes, and arguments, and that science is also, or should be, part of mainstream experience, and therefore should be made meaningful to children's everyday culture, interests, values, and identities. If elementary teachers of science are to achieve a satisfactory blending of these potentially divergent goals for science education, then they need a clear framework for conceptualizing the role of language in promoting and integrating these different dimensions of science literacy into their classrooms. The next section outlines an overview to achieve these different goals.

A FRAMEWORK FOR USING LANGUAGE FOR LEARNING IN ELEMENTARY SCIENCE

Christie (1981) offered a useful broad framework for conceptualizing the ways in which language might be used to support learning in science consistent with current understandings of science literacy. In seeking to characterize links between language and learning generally, Christie (1981) proposed three main relationships. She claimed that language can be used as a broad resource for learning new content, where use of everyday language enables learners to explore and clarify the meaning of new experiences, technical terms, or new concepts. In this way, the vernacular language of the children—the language of their community—is a major resource for making sense of the "new" literacy of science, its particular vocabulary and language practices, activities, and processes. A second relation entails learning about a language as a system, as in learning the history, typical linguistic structures, grammar and form/function relations of the language, and its usage. In this case, children need to learn how to read and write texts that reflect how science ideas, evidence, and argument are normally represented in the broader science community. Christie (1981, p. 8) purported that students need to understand the "varieties" of expression within a language. Although she was referring to regional and dialect variations in a language, there is also a sense in which this is relevant to science as a language or languages, where there is a diversity of styles, vocabulary choice, and formats for how science is represented in different ways for different contexts. Popular accounts of science ideas and processes on the Internet may use different formats and language to represent this content, in contrast to those used in print and other media accounts; but each representation is still part of the language of science.

The third relation entails learning language, as in learning new vocabulary and how to use a language competently in different contexts for different purposes and audiences. Researchers in science education, over the last 15 years, have advocated strongly the need for each of these possible relations to inform student learning of science, where science is understood, among other things, as a particular language, or more recently, languages, with a history, specific values, discourses, and rationale.

There has been strong advocacy of the role of everyday language as a resource for learning science concepts and processes, with student talk seen as a major way to achieve this goal (Halliday & Martin, 1993; Lemke, 1990; Ogborn, Kress, Martins, & McGillicuddy, 1996; Sutton, 1992, 1996; Wellington & Osborne, 2001). Such talk, arising from inquiry-based learning opportunities, should entail the routine practices of the elementary classroom, including small group and whole class discussion, responses to different kinds of print and nonprint texts, panels, debates, guest speaker talks with question and answer sessions, and individual and group presentations by students of project work, including science findings and recommendations. At a theoretical level, this approach draws on social constructivist perspectives of learning as the process whereby learners use and engage their own linguistic resources and conceptual and everyday categories and frameworks to develop and demonstrate understandings in relation to target concepts and the vocabulary and processes of science (Fensham, Gunstone, & White, 1994; Hand & Prain, 1995).

There has also been strong support for the value of students using writing as a resource for learning science. Researchers have suggested that students should use a diversified range of writing types, both formal and informal, to learn about science, reflect on their learning, and communicate science ideas to others (Boscolo & Mason, 2001; Hanrahan, 1999; Hildebrand, 1999; Hodson, 1998; Prain & Hand, 1996; Rivard & Straw, 2000; Stadler, Benke, & Duit, 2001; Sutton, 1992, 1996). Such writing is viewed as enabling students to achieve border crossing between the language conventions, and values implicit in science and everyday worlds. This perspective does not imply that learners have innate competencies that enable them to learn science vocabulary and concepts without teacher guidance, but rather that writing can be a useful resource, parallel to discussion, to clarify science concepts and practices, to connect new concepts and meanings to past understandings, and to develop critical perspectives. Boscolo and Mason (2001, p. 85) proposed that the act of writing enabled elementary school students to make "systematic connections" in producing a "a meaningful understanding of subject matter," and that tasks requiring students to paraphrase, reword, elaborate, unpack meanings, express uncertainties, analyze comparisons, and reconstruct understandings provided an effective resource for establishing these connections. This approach assumes that the richer the network of correspondences of meaning learners can establish across different contexts and wordings, the more secure their learning will be. Such writing, as noted by Hodson (1998, p. 162), "need not entail very much physical writing at all," but rather should provide opportunities for "stimulating reflective thinking," and build richer networks of associative meaning. Rowell (1997), Hand, Keys, Prain, and Sellars (1998), and Tucknott and Yore (1999) claimed that where students were beginning to learn new concepts, there were pedagogical advantages in opportunities for them to write in diverse ways that led them to connect emerging knowledge and technical vocabulary to their everyday language and past experiences in a two-way process. Michaels and Sohmer (2000) also maintained that teachers needed to provide opportunities for students to build connections between their personal meaning-making processes incorporating their everyday perceptions of the phenomenal world and the official inscriptions and representations of school science knowledge.

This approach to writing in science assumes that students learn when they engage successfully with the demands of communicating to actual readerships, including themselves, for meaningful and varied purposes. These purposes might include brainstorming initial ideas, predicting and speculating about causes and outcomes, re-presenting their ideas in different forms, explaining or justifying their explanations to themselves or others, persuading others to accept their interpretations, constructing textbook explanations for other or younger students, or modifying their views in the light of additional evidence. This diversity of purposes leads to a matching breadth of possible writing types, including informal writing such as notes, maps, observations, and summaries, as well as writing that fits more resolved generic frameworks. From this perspective, students should use their understanding of the rules and methods of language practices beyond science-specific genres, such as debates and discussion.

In identifying strategies to assist teachers to implement writing-to-learn approaches within classrooms, Prain and Hand (1996) proposed a model to help teachers frame a writing task. The model divides the writing task into five components: topic, purpose, audience, genre, and method of text production. The model, which is envisioned as a barrel lock with each segment able to rotate independent of the others, attempts to provide teachers with a greater range of options to integrate writing as a learning tool within science classrooms. In the elementary school, the model would have greatest relevance in the upper grades. The model encourages teachers to explore having students use different writing types, for different purposes, using different audiences, to help them understand science better. Students need to be asked to write for audiences other than the teacher. When interviewed about having to write to someone other than the teacher, students have consistently revealed that they now have to do more explaining than if writing to the teacher. They are familiar with the teacher doing considerable interpretation of their writing rather than achieving greater levels of explanation from their texts themselves. In implementing these writing activities involving different audiences, we have:

- spent time discussing the writing type being used,
- given students opportunities to plan their writings activities in small groups before doing any writing,
- expected students to complete a first draft that is given to the audience for initial evaluation,
- expected students to complete revisions based on feedback from the audience, and
- invited the teacher to evaluate the writing for science content and for language effectiveness.

Possible writing tasks can vary from small informal responses and notes that do not fit any standard writing style to tasks that have strong resolved generic frameworks and expectations. Examples of the first kind of writing task include:

- devising analogies or metaphors to explain a process;
- producing a map or diagram to explain a process or show links between concepts;
- producing a sequence of labeled diagrams to explain a process;
- explaining a diagram or illustration of a machine in terms of the inventor's intention or purpose;
- producing a PowerPoint set of slides as a resource to support an explanatory talk on a topic or process;
- writing a summary/revision notes/mind map of a topic;
- producing a test or a quiz of a topic for other children;
- making predictions, observations, and explanations of experiments and excursions;
- expressing their ongoing reactions to a topic or scientific explanation;
- making personal reflections on how and what students learn in science topics; and
- captioning diagrams where text is not provided.

Examples of the second kind of task include typical genres for science, such as:

Scientific and Verbal Reports

- report and analysis of experiments and observations;
- explanation of a process, structure, or issue, such as health care or causes of disease; and
- history of famous scientists and their achievements including perceived "sidetracks."

Narratives

- stories to show knowledge of a process or a sequence of events, and
- stories to demonstrate multiple viewpoints through multiple narrators and points of view.

Travelogues

- descriptions of travel, such as a tour of the solar system, and
- description of a process or a chain of events.

Guidelines and Instructions

- guides to use of materials,
- survival manual,
- part of a factual text rewritten to explain the topic for younger children,
- designs for safety labels for use at school, and
- step-by-step instructions on a process.

Scripts for Debates or Speech

- debate on a controversial subject such as different uses of the environment,
- positive and negatives issues relating to a topic,
- performance of process,
- scripted dialogue of an interview of a scientist, and
- producing a new script for a commercial video on a science topic targeting a different audience or purpose.

Posters

- demonstration of knowledge at completion of topic,
- ordering of knowledge into appropriate sequence, and
- representation of knowledge not easily produced as written text.

Brochures

- guides for field work, and
- guide for efficient use of resources.

Letters

- persuasive writing on a scientific issue that affects society or the community, and
- explanation of a topic to a friend.

In keeping with a strong communicative focus, students should also write or present science understandings to a diverse, meaningful range of readerships, viewers, and listeners. Recognizing the needs of different audiences encourages students to engage with the science concepts in multiple ways. Traditionally, students tend to participate in a dialogue with their teacher either orally or through text using the teacher's language. Understanding of the science concepts is noted by correct use of the scientific terminology provided by the teacher. Whereas this type of dialogue enables students to engage with science language, there is insufficient opportunity for students to translate science language into forms that they understand. Although students implicitly translate science language into their own understandings, this process is not encouraged in a more explicit manner. Changing the audience for whom students write makes this process of translation more explicit. By asking students to write or present science ideas to peers, younger students, parents, grandparents, or the general public, the students have to become translators of the science language for the audience with whom they are communicating. There is a three-way dynamic set up between the science language of the topic, the student, and the audience (see Fig. 9.1).

For example, a grade 6 class completed three laboratory activities on a topic (Hand & Keys, 1999), and in order to draw their investigations together the teacher asked them to write a report of their activities for their grandparents. The following week was grandparent week, and the students were asked to have them ready for evaluation by their grandparents. Grandparents were provided with an evaluation sheet that asked them to comment on some basic language criteria and the science explanations. These evaluations were returned to the students for use in their final draft of the report. Two important criteria need to be adhered to for successful learning from such writing: the students need to perceive the audience as real; and the audience needs to participate in actual evaluation of the presentation or writing.

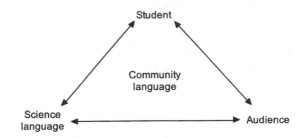

FIG. 9.1. Dynamic relation between science language, student, and audience.

In teaching the subject-specific writing, reading, and viewing practices associated with science, various researchers have proposed a range of strategies. Classroom learning should focus on the discipline-specific structural and functional features of different types of science writing and presentations (Halliday & Martin, 1993; Unsworth, 2001), their subject-specific vocabulary, and the student knowledge required to understand and reproduce these genres (Martin & Veel, 1998; Unsworth, 2001). According to Martin (1999) and others, students will learn effectively the rules and meanings of these particular language practices through the following teaching strategies: detailed analysis of linguistic features of textual examples, joint construction of genres with their teacher, and through an explicit extensive teacher focus on key textual function/form relations and their rationale. Unsworth (2001) argued that students can learn to write scientifically and incorporate multimodal resources into their writing through analyzing the schematic structures and grammatical patterns of sample texts, and then reproducing these functions in their own writing. For Unsworth, the appropriate genres to study in order to promote science literacy in elementary school include procedural recounts, explanations, descriptive reports, taxonomic reports, and expositions and discussions. Unsworth held that students need to understand the "meaning-making systems deployed in images" (p. 111) if they are to understand how to interpret and construct science ideas represented visually in texts. Similarly, students need to understand how different types of writing are used in science as a means to achieve contrasting purposes. In knowing how to integrate visual and verbal meaning systems, students will have started to learn the new multiliteracies of the current and emerging information age. Unsworth further outlined how students in the upper elementary classroom can develop these multiliteracies skills in reading, viewing, and writing through engagement with the topic of the greenhouse effect and climate change. Champagne and Kouba (1999), Scheppegrell (1998), and others also asserted that students need exposure to explicit teaching of these genres so that tacit knowledge can be made more overt.

In summary, the framework proposed by Christie (1981) about the relations between language, learning, and literacy, and confirmed by the research reported in this section on developing science literacy, implies that teachers of science in elementary school need to focus on three broad goals. The first concerns the use of students' everyday community language as a resource for learning science. The second concerns learning about the particular language practices of science, and the third concerns becoming an effective and motivated reader, writer, speaker, listener, and viewer of science in different contexts and for varied purposes. Whereas each of these goals can be addressed successfully by viewing science as a separate area of the curriculum, there is growing recognition that integrating the proposed standards for language learning (IRA/NCTE, 1996) with the relevant standards for science learning (NRC, 1996) could provide additional benefits to science learning in the elementary classroom. The next section considers the rationale for such an approach, and identifies some strategies for achieving this integration.

CONNECTING THE SCIENCE AND LANGUAGE CURRICULA IN THE ELEMENTARY SCHOOL

The standards for language learning (IRA/NCTE, 1996) propose a range of generic interpretive and communicative skills as the goals and outcomes of literacy. The following examples of these standards imply that learners must know how to construct, interpret, and evaluate examples of all the mainstream text types, presumably including both verbal and multimodal informational and science-based texts:

Standard 3: Students apply a wide range of strategies to comprehend, interpret, evaluate, and appreciate texts. They draw on their prior experience, their interactions with other readers and writers, their knowledge of word meaning and of other texts, their word identification strategies, and their understandings of textual features.

Standard 4: Students adjust their use of spoken, written, and visual language (e.g., conventions, style, vocabulary) to communicate with different audiences and for different purposes.

Standard 5: Students employ a wide range of strategies as they write and use different writing process elements to appropriately communicate with different audiences for a variety of purposes.

Standard 6: Students apply knowledge of language structure, language conventions (e.g., spelling and punctuation) media techniques, figurative language and genre to create, critique, and discuss print and non-print texts.

Standard 7: Students conduct research on issues and interests by generating ideas and questions, and by posing problems. They gather, evaluate, and synthesize data from a variety of sources (e.g., print and non-print texts, artefacts and people) to communicate their discoveries in ways that suit their purpose and audience.

Standard 8: Students use a variety of technological and informational resources (e.g., libraries, databases, computer networks, video) to gather and synthesize information and to create and communicate knowledge.

Standard 11: Students participate as knowledgeable, reflective, creative, and critical members of a variety of literacy communities.

Although elementary teachers, in general, are very comfortable with these standards when they teach language, they do not transfer them across to science and mathematics subjects with as much confidence. However, there is clearly overlap between the broad approach of student inquiry in the science standards (NRC, 1996), and some of the specific goals of the language standards. For example, in Standards 7 and 8, student inquiry could be easily integrated with the science standards where students are expected to learn how to use evidence to develop or revise an explanation, communicate science explanations, apply the results of experiments to scientific arguments and explanations, and communicate science ideas to classmates. In engaging in these processes, students will be necessarily addressing the three goals implied in Christie's framework, and therefore participating in activities that promote science literacy.

Integration of science and language implies that learning goals in each curriculum area can be connected more closely, or overlap, including content learning and literacy skills. Various methods of integration have been proposed in recent years (Fogarty, 1991; Peters & Grega, 2000; Roberts & Kellough, 2000), including integration between topics in a subject, between subjects, and within learners' understandings when exploring a range of dimensions to a single topic. Integration can focus on content within broad topics or themes related to different subjects, such as "transport," "the Olympics," "war," or "community"; on concepts common to different topics or subjects, such as the idea of "system," "cooperation," or "cause and effect" within different topics; on genres that are common to different topics or subjects, such as the report or the biography; on skills and learning processes, such as developing cooperative learning skills when working on different topics; or on different learning methods, such as inquiry learning, predicting, and justifying a viewpoint. As noted by many educators (Charbonneau & Reider, 1995; Peters & Grega, 2000; Roberts & Kellough, 2000; Tchudi & Lafer, 1996), there are various benefits in integrating science with other areas such as language, including the following:

- increased student motivation to learn more meaningful topics,
- improved coherence of curriculum planning and implementation,
- fostering of cooperative learning as students justify and explain diverse views and findings,
- increased focus on "big ideas,"
- higher order thinking development through synthesizing understanding across different contexts,
- transfer of knowledge/skills/understandings to new contexts,
- improved student learning through more links across concepts and contexts for learning,
- catering for a diversity of student aptitudes and interests through a variety of learning tasks,
- increased range of resources for teaching key concepts,
- contextualizing science concepts in a meaningful way for learners, and
- covering science topics more thoroughly through links with reading and writing in the language curriculum.

Recent research claims have also been made for the benefits of integrating science into the reading curriculum (Vasquez, 2003), where a group of over 41,000 students raised their standardized test scores in both elementary science and reading, through a range of methods including integration of science topics into the reading curriculum. When learning science is understood in part as learning how to read, write, and view science texts, then there is considerable opportunity for border crossing between language and science lessons in terms of the study and construction of texts. Whereas language education focuses on a broad range of genres, including biographies, scripts, poetry, records, reports, journals, narrative, and factual texts, genres in science usually refer to text types associated with science writing, such as reports, posters, labeled diagrams, concept maps, and observation records. However, as noted earlier in this chapter, students can use a range of genres to explain, explore, and report their ideas on science topics. In this way, reading, writing, and thinking skills associated with language education can be a major focus in a science lesson. For example, skills in science refer to understanding and using science methods, language, and reporting, including the ability to apply appropriate scientific vocabulary to identifying and communicating key concepts, select and record data relevant to tasks, achieve effective verbal and written/pictorial presentation of science topics, and conduct effective research on science topics. These skills are also central to language education where students are expected to develop skills in effective reading, writing, speaking, listening, and viewing, such as skill in summarizing from reading, integrating information from different sources into single reports, and critical skills in identifying the assumptions, purposes, values, omissions, and appeal of particular texts. Some examples of integrating science and language education include:

- Applying language skills (speaking, listening, reading, writing, reporting, researching) to learning a science topic.
- Learning more about key concepts/themes in science through extension language activities such as guided reading.
- Applying science concepts (e.g., orbit, rotation, or system) to other contexts and topics.
- Applying a method in science (e.g., prediction, observation, explanation) to reading or viewing various different kinds of texts.

Student group discussion and written work can assist the development of science concepts, and can also be assessed by the teacher for both content and skills in reading and writing effectively. In this way, an integrated curriculum enables teachers to assess students' work for both its science and language aspects. If grade 3 and 4 students are required to demonstrate their ability to design, organize, and write an informative text in order to meet language standards, then this work can also be assessed for both the student's demonstration of science knowledge and their language education communicative skills, including structural features such as: capital letters, full stops, and commas; textual coherence for reader or listener appeal; and communicative effectiveness.

IMPLICATIONS FOR ELEMENTARY
SCIENCE TEACHER EDUCATION

The foregoing discussion has various implications for the content and pedagogy of courses in elementary science teacher education in relation to literacy, learning, and science, as well as for effective change to teachers' practices and professional development.

Course Content Issues

Elementary science teacher education courses need to focus on ensuring that students are aware of recent broadened definitions of science literacy and their implications for the elementary classroom. These definitions emphasize four contrasting general aspects to science literacy: understanding of the language, methods, values, and history of science; willingness and capacity to communicate science concepts, methods, and findings to others in diverse ways and for diverse readerships; commitment to the value of scientific ways of thinking and acting in relation to everyday issues and problems; and capacity to evaluate the purposes, uses, and findings of science in relation to different topics.

Preservice elementary teachers of science also need to develop a clear conceptualization of the relations between language, learning, and science education. As noted by Christie (1981), language can be viewed as a resource or medium for learning as well as a goal of learning in science education, especially in relation to the acquisition of science literacy. Christie's (1981) perspective implies multiple purposes and outcomes for the use of classroom talk and student writing for learning about science, as well as learning how to write scientifically. Children's talk and writing can serve learning where the children are able to make durable links between their everyday language, concepts, values, and identities, and the target concepts, values, processes, and procedures of science. Preservice teachers of elementary science need to be familiar with, and know how to implement and evaluate, a broad range of talking and writing activities of the kind noted earlier in this chapter, which can serve this "translation" process. In this way, the language resources of the children, including the vernacular language of their community, can be utilized to serve learning the new literacy of science. At the same time, preservice teachers need to know how to teach children to understand and construct the subject-specific types of writing and representation, including electronic texts that are consistent with how science is represented by scientists, and science writing within the broader community. As noted by many authors, this teaching needs to focus on the form–function aspects of the structure of such texts, including multimodal computer-generated texts incorporating visual, verbal, and mathematical modes. Preservice elementary teachers also need to develop an understanding of how the goals, methods, and evaluation strategies of elementary language education can be integrated effectively into approaches to learning science.

Pedagogical Issues

To enable preservice teachers to become familiar with the practical implications of these content issues, there is value in the pedagogy of preservice courses modeling the implementation of these approaches. This modeling can be implemented in various ways, and the following example is offered as one way in which language can be effectively used to address the issues raised earlier. In this instance, preservice teachers undertake the investigation of a topic they will teach to their students, where the teachers monitor their prior and emerging understandings of the topic, and represent these understandings at different stages of the topic in different forms for different readerships. Volkman and Abell (2003) outlined such an approach to the topic of the Moon, where their preservice teachers kept a journal of observations and findings about the Moon, made models of various phases of the Moon, attempted to explain lunar and solar eclipses in relation to their models, investigated the accuracy of Moon representations in children's picture books, and wrote reports on their understandings, incorporating diagrams. They then designed posters to explain these understandings to other faculty and students. Such a program effectively gives the preservice teachers practical knowledge in the effective use of language as both a resource and outcome in learning science literacy, while also building their science expertise and commitment to the value of a science perspective and methods.

Teacher Professional Development Issues

By implication, this chapter also raises issues about how current elementary teachers might change their practices in relation to the role of language in learning science and in developing children's science literacy. Some useful insights into this question are provided by a major research project in Australia that focused on the professional development of elementary teachers in teaching science effectively (Tytler & Waldrip, 2002; Tytler, Waldrip, & Griffiths, 2002). This project required elementary schools to audit their provision of the science curriculum, and then develop action plans for improvement, based on perceptions of individual school needs. The plans were formed through exchange of ideas in regional workshops, external professional development programs, and conference participation. The research project found that changes supporting increased effectiveness in teaching and learning science included the following:

- a higher profile for science in the school, including practices such as "family science events" (p. 24) with community involvement;
- a more coherent coverage of the science curriculum by recognizing and addressing gaps in topics and resources in many schools;
- more teaching of science by linking the subject to language learning through reading, writing, and discussion;
- increased access to resources to encourage more hands-on work;
- improved teacher confidence and knowledge in teaching science;
- increased diversity of teaching approaches, including open-ended investigations; and
- the development of whole-school and mentor-based approaches that supported reluctant teachers to change their approach to, or neglect of, science.

Some of the interviewed science teachers considered that interrelating science with other areas of the curriculum was important, so that science topics might cross over into mathematics and language. They considered that science should be taught in the context of educating the "whole child," and their practices in science "sometimes related to generic critical thinking, and at other times to other learning areas, especially language" (Tytler et al., 2002, p. 14). One teacher in a composite grade K–6 in a small community school organized many excursions for the children and also ran family science nights where the children presented findings to parents. This teacher encouraged the children to put emerging questions on a topic on the board so that they could be revisited over time, and sometimes revised. Another elementary teacher focused strongly on environmental issues, involving community experts, local businesses, and community members. She strongly supported the view that children need to be encouraged to explain ideas "in their own words and supported gradually to develop scientific explanations" (p. 12). Another teacher considered that it was important for her students in grade 4 to discuss contemporary science ideas evident in newspaper reports and other media, as part of understanding the relevance of science to everyday life.

In analyzing the teachers' practices and beliefs, Tytler et al. (2002, p. 14) identified the following as crucial in effective science learning: the teachers' commitment to inquiry-based learning based around conceptually coherent topics; the need for learning to be a collaborative activity; and the need to connect science learning with both "generic critical thinking" as well as language activities. The researchers proposed that for teachers to achieve active student engagement with science in the elementary school their programs needed to focus on both processes and products in learning (as children engage with ideas and evidence and develop deeper understandings about the world) and on learning as both a community classroom activity and as an individual responsibility. The study further confirmed the need for science learning to connect to the children's everyday experiences and be embedded in meaningful communicative contexts. These points reiterate the centrality of the role of children's everyday language as a resource for learning as they engage with learning the practices, thinking skills, and new languages of science.

CONCLUSIONS

This chapter has suggested that preservice elementary teachers of science need a clear conceptualization of the relations between language, learning, and literacy, especially in relation to science, if they are to succeed in addressing the current goals of the science curriculum at this level of schooling. From this perspective, children's everyday language and understandings need to be understood as a major tool for learning the particular languages and practices of science inquiry. It has also been suggested that children will learn to think and act scientifically when they learn how to respond to and construct science texts and texts about science. At the same time, the increasing range of new multimodal texts for engaging in and representing science on the Internet and elsewhere, as noted by Lemke (1998), Unsworth (2001), and others, poses further challenges for students in learning the languages of science for the future, and in developing critical skills in assessing this material. If a major aim of science education is to establish positive attitudes toward science in the elementary school, and also give children some knowledge of how science texts and practices are currently read, written, viewed, reported, and assessed, then there is a need to create strongly motivating and guided opportunities for children to engage with these texts and practices as they begin the process of acquiring science literacy.

REFERENCES

Boscolo, P., & Mason, L. (2001). Writing to learn, writing to transfer. In P. Tynjala, L. Mason, & K. Lonka (Eds.), *Writing as a learning tool* (pp. 83–104). Dordrecht, The Netherlands: Kluwer Academic.

Champagne, A., & Kouba, V. (1999). Written products as performance measures. In J. Mintzes, J. Wandersee, & J. Novak (Eds.), *Assessing science understanding: A human constructivist view* (pp. 224–248). New York: Academic Press.

Charbonneau, M., & Reider, B. (1995). *The integrated elementary classroom: A developmental model of education for the 21st century*. Boston: Allyn & Bacon.

Christie, F. (1981). The Language Development Project. *English in Australia, 58,* 3–10.

Cobern, W. W., Gibson, A. T., & Underwood, S. A. (1995). Valuing scientific literacy. *The Science Teacher, 62*(9), 28–31.

Comber, B., Cormack, P., & O'Brien, J. (2001). Schooling disruptions: The case of critical literacy. In C. Dudley-Marling & C. Edelsky (Eds.), *Where did all the promise go? Case histories of progressive language policies and practices* (pp. 171–193). Urbana, IL: National Council of Teachers of English.

Fensham, P., Gunstone, R., & White, R. (1994). *The content of science. A constructivist approach to its teaching and learning*. London: Falmer.

Fogarty, R. (1991). *The mindful school: How to integrate the curricula*. Palantine, IL: Skylight Publishing.

Gee, J. (2002). *Playing the game: Language and learning science*. Retrieved June 6, 2002, from http://www.educ.uvic.ca/faculty/lyore/sciencelanguage/

Halliday, M., & Martin, J. (1993). *Writing science: Literacy and discursive power*. London: Falmer.

Hand, B., & Keys, C. (1999). Inquiry investigations: A new approach to laboratory reports. *The Science Teacher, 66*(4), 27–29.

Hand, B. M., Keys, C. W., Prain, V. R., & Sellars, S. (1998, April). *Rethinking the laboratory report: Writing to learn from investigations*. Paper presented at the annual meeting of the National Association for Research in Science Teaching, Boston.

Hand, B., & Prain, V. (Eds.). (1995). *Teaching and learning in science: The constructivist classroom*. Sydney, Australia: Harcourt Brace.

Hand, B., Prain, V., Lawrence, C., & Yore, L. (1999). A writing in science framework designed to enhance science literacy. *International Journal of Science Education, 21*(10), 1021–1035.

Hanrahan, M. (1999). Rethinking science literacy: Enhancing communication and participation in school science through affirmational dialogue journal writing. *Journal of Research in Science Teaching, 36*(3), 699–718.

Hildebrand, G. (1999, April). *Breaking the pedagogical contract: Teachers' and students' voices*. Paper presented at the annual meeting of the National Association for Research in Science Teaching, Boston, MA.

Hodson, D. (1998). *Teaching and learning science: Towards a personalized approach*. Buckingham, England: Open University Press.

Hurd, P. (1998). Science literacy: New minds for a changing world. *Science Education, 82*, 407–418.

International Reading Association and National Council of Teachers of English. (1996). *Standards for the English language*. Newark, DE: Author.

Lemke, J. (1990). *Talking science: Language, learning, and values*. Norwood, NJ: Ablex.

Lemke, J. (1998). Multiplying meaning: Visual and verbal semiotics in scientific text. In J. Martin & R. Veel (Eds.), *Reading science. Critical and functional perspectives on discourses of science* (pp. 87–113). London: Routledge.

Lemke, J. (2002, September). *Getting critical about science literacies*. Paper presented at the Language & Science Literacy Conference, University of Victoria, British Columbia.

Linn, M. (2000). Designing the knowledge integration environment. *International Journal of Science Education, 22*, 781–796.

Martin, J. (1999). Mentoring semogenesis: "Genre-based" literacy pedagogy. In F. Christie (Ed.), *Pedagogy and the shaping of consciousness: Linguistic and social processes* (pp. 123–155). London: Cassell (Open Linguistics Series).

Martin, J., & Veel, R. (Eds.). (1998). *Reading science. Critical and functional perspectives on discourses of science*. London: Routledge.

Michaels, S., & Sohmer, R. (2000). Narratives and inscriptions: Cultural tools, power and powerful sense-making. In B. Cope & M. Kalantzis (Eds.), *Multiliteracies: Literacy learning and the design of social futures* (pp. 267–288). London: Routledge.

National Research Council (NRC). (1996). *National science education standards*. Washington, DC: National Academy Press.

Norris, S., & Phillips, L. (2001, March). *How literacy in its fundamental sense is central to scientific literacy*. Paper presented at the annual meeting of the National Association for Research in Science Teaching, St. Louis, MO.

Norris, S. P., & Phillips, L. M. (2003). How literacy in its fundamental sense is central to scientific literacy. *Science Education, 87*, 224–240.

Ogborn, J., Kress, G., Martins, I., & McGillicuddy, K. (1996). *Explaining science in the classroom*. Buckingham, England: Open University Press.

Peters, J., & Grega, P. (2000). *Science in elementary education* (9th ed.). Upper Saddle River, NJ: Merrill.

Prain, V., & Hand, B. (1996). Writing and learning in secondary science: Rethinking practices. *Teaching and Teacher Education, 12*, 609–626.

Rivard, L. P., & Straw, S. B. (2000). The effect of talk and writing on learning science. An exploratory study. *Science Education, 84*, 566–593.

Roberts, P., & Kellough, R. (2000). *A guide for developing interdisciplinary thematic units* (2nd ed.). Upper Saddle River, NJ: Merrill.

Rowell, P. A. (1997). Learning in school science: The promises and practices of writing. *Studies in Science Education, 30*, 19–56.

Russell, T., & McGuigan, L. (2001). Promoting understanding through representational redescription: An illustration referring to young pupils' ideas about gravity. In D. Psillos, P. Kariotoglou, V. Tselfes, G. Bisdikian, G. Fassoulopoulos, E. Hatzikraniotis, & E. Kallery (Eds.), *Science education research in the knowledge-based society. Proceedings of the Third International Conference of ESERA* (pp. 600–602). Thessaloniki, Greece: Aristotle University of Thessaloniki.

Ryder, J. (2001). Identifying science understanding for functional scientific literacy. *Studies in Science Education, 36,* 1–44.

Scheppegrell, M. (1998). Grammar as resource: Writing a description. *Research in the Teaching of English, 25,* 67–96.

Stadler, H., Benke, G., & Duit, R. (2001). How do boys and girls use language in physics classes? In H. Behrendt, H. Dahncke, R. Duit, W. Graber, M. Komorek, A. Kross, & P. Reiska (Eds.), *Research in science education: Past, present, and future* (pp. 283–288). Dordrecht, The Netherlands: Kluwer Academic.

Sutton, C. (1992). *Words, science and learning.* Buckingham, England: Open University Press.

Sutton, C. (1996). Beliefs about science and beliefs about language. *International Journal of Science Education, 18,* 1–18.

Tchudi, S., & Lafer, S. (1996). *The interdisciplinary teacher's handbook: Integrating teaching across the curriculum.* Portsmouth, NH: Heinemann.

Tucknott, J. M., & Yore, L. D. (1999, March). *The effects of writing activities on grade 4 children's understandings of simple machines, inventions and inventors.* Paper presented at the annual meeting of the National Association for Research in Science Teaching, Boston.

Tytler, R., & Waldrip, B. (2002). Improving primary science: Schools' experience of change. *Investigating, 18*(4), 23–26.

Tytler, R., Waldrip, B., & Griffiths, M. (2002). Talking to effective teachers of primary science. *Investigating, 18*(4), 11–15.

Unsworth, L. (2001). *Teaching multiliteracies across the curriculum: Changing contexts of text and image in classroom practice.* Buckingham, England: Open University Press.

Vasquez, J. (2003, March). *The guaranteed way to reduce standardized test scores—throw out the science.* Paper presented at the National Science Teachers Association National Convention, Philadelphia, PA.

Volkman, M., & Abell, S. (2003, June). Seamless assessment. *Science and Children,* pp. 41–45.

Wellington, J., & Osborne, J. (2001). *Language and literacy in science education.* Buckingham, England: Open University Press.

Yore, L., Hand, B., & Prain, V. (2002). Scientists as writers. *Science Education, 86*(5), 672–692.

10

A Standards-Infused Approach to Curriculum: A Promising Interdisciplinary Inquiry

Katherine C. Wieseman
Western State College

Hedy Moscovici
California State University, Dominguez Hills

A bifurcated emphasis on language and mathematics literacies in U.S. elementary schools, the use of students' standardized test results, and the implications arising from these practices has led district level and school-based administrations to send messages to teachers about the relative value of learning in different curriculum areas. The content of many of these messages is that a curriculum stressing language and mathematics performance is the prime focus of elementary schooling (National Science Teachers Association [NSTA], May/June 2002). Within such an educational context, science learning is likely to occur only within an interdisciplinary orientation to learning. Further complicating the teaching of elementary science is the widely acknowledged finding that elementary teachers who lack confidence in their abilities to teach science tend to avoid spending time on science (e.g., Abell & Roth, 1992; Czerniak & Lumpe, 1996; Smith & Neale, 1989). Consequently, in elementary science teacher education programs, there tends to be an emphasis on teaching science using interdisciplinary approaches.

This chapter presents and examines tensions related to the use of interdisciplinary orientations in elementary science teacher education. The tensions are presented in composite stories (Barone, 1988; Connelly & Clandinin, 1988; Manen, 1990) to encourage the reader to become immersed in the rich complexity of the situations at hand and to become a partner in a disciplined inquiry process rather than to act as a passive bystander. The stories are based on three key assumptions: Teachers' beliefs systems and decisions in classroom practice are grounded in their life histories and experiences, both personal and educational (Pajares, 1992; Connelly & Clandinin, 1988; DeCourse, 1996); learning is a social process of making sense of experiences, and is based on individuals' constructed extant knowledge (Von Glasersfeld, 1989; Tobin, Tippins, & Gallard, 1994); and teachers need to be encouraged to question content-related assertions, as well as established social norms (Giroux & Simon, 1989).

The next section outlines the educational context of this report, and the second section briefly describes themes salient to elementary science teacher preparation drawn from the educational and research literature about interdisciplinary orientations. The third and fourth sections each present a story depicting tensions experienced by preservice teachers involved in elementary science education courses promoting an interdisciplinary inquiry orientation for learning. The stories are set in educational settings in which teachers are held accountable for aligning what they teach to state or district standards. The chapter closes with a discussion of the implications of possible tensions that preservice elementary teachers can encounter as they consider which approach to curriculum to employ. Distinguishing between teachers' approaches to curriculum holds promise for developing different views of the future of science teacher preparation at the elementary level and in elementary school science. The distinction may prove particularly useful in institutions where science education is being preempted by views of learning revolving around narrow interpretations of standards that lead to visions of schooling and/or learning focused strictly on language literacy and mathematics literacy instruction.

THE EDUCATIONAL CONTEXT

Standards are broad statements about the essential content knowledge in school subjects that all students should know by the completion of grade 12 schooling in the United States. This essential content knowledge appropriate for grade level bands (e.g., K–4 and 5–8) or specific grade levels is communicated in more detailed learning outcome statements. Students' achievement and accomplishments with respect to these standards are being assessed through, what has become in many states, high stakes[1] standardized tests aligned to the standards. Teachers are using a kaleidoscope

---

[1]This term refers to situations where test results are used to reward or sanction schools, teachers, and/or students according to test performance (e.g., school funding linked to school test performance, teacher promotion dependent on class test performance, and student grade advancement dependent on personal test results).

of approaches to curriculum to respond to accountability issues emerging from the emphasis being placed on the standardized test results.

In the first story, "No Easy Journey!" recounted in the third part of this chapter, the preservice teachers' field experience takes place in a district in which the school administration requires teachers to indicate which standards are targeted in every lesson. In the second story, "Will the *Open Court* Police Chastise Me?" an emergency permit teacher completing a teacher education credential program must address standards and implement a time-consuming, district-mandated, prescribed curriculum for reading and writing. The stories serve to illuminate how preservice teachers approach planning and curriculum decisions as a response to perceived accountability issues.

A "Standards-Infused" Approach

In these stories, standards were not the starting point or triggers of thinking about planning or classroom practice. Instead, what is labeled a "standards-infused" approach was used. The basis of this approach is inquiry in an interdisciplinary context—a holistic view of curricular objectives and standards. Higher order thinking is necessary to infuse standards of different content areas within a curriculum that recognizes teacher and students' interests and knowledge. The teacher identifies a problem-solving situation, issue, or theme that encompasses multiple standards from several content areas and promotes investigation and experimentation. In a standards-infused approach, the triggers for what is taught and learned are problems, issues, themes, and big ideas that transect content areas. Decisions about targeted standards leave open pathways for individual teacher design, students' interests, and creativity. The standards are in the backdrop of a curriculum planning process, and thinking is not restricted to consideration of the standards of one content area only. The teacher must be knowledgeable and motivated to diagnose, analyze, synthesize, evaluate, and create a curriculum that can immerse both the teacher and students in learning (Tchudi & Lafer, 1996).

A Traditional Approach to Standards

Another more traditional way of thinking about standards accountability issues is based on the idea of a standards-based approach to curriculum, one in which a teacher identifies particular standards in one content area as triggers for what will be taught and learned. The teacher might sequence students' learning experiences based on addressing aspects of the standards in the order presented in curriculum documents. The teacher selects resources that support this decision, and sequences learning opportunities to promote conceptual understanding with respect to the targeted standards. In this approach to curriculum planning, the standards constitute the curriculum.

Standards-infused and standards-based approaches to curriculum are two of a kaleidoscope of responses used by teachers to approach curriculum and resolve accountability issues.

WHAT DO WE ALREADY THINK AND KNOW?

No consensus on a definition of curriculum integration exists in the educational community (Czerniak, Weber, Sandmann, & Ahern, 1999; Grossman, Wineburg, & Beers, 2000; Hurley, 2001). The construct, interdisciplinary curriculum, refers to a number of positions on a continuum of curriculum possibilities, ranging from a discipline-based focus to a more holistic focus on an idea, an issue, a learner's question, or an essential question chosen by the teacher (DeCourse, 1996; Drake, 1991; Fogarty, 1991; Goodlad & Su, 1992; Jacobs, 1989; Martinello & Cook, 2000; Maurer, 1994; Post, Ellis, Humphreys, & Buggey, 1997; Tchudi & Lafer, 1996). Lederman and Niess (1997) used metaphors to distinguish between curricula that connect yet maintain the integrity of the disciplines involved (chicken noodle soup), and curricula in which disciplines are blurred into a seamless whole (creamy tomato soup). Hurley (2001) suggested that single continuum models do not fully represent the diverse aspects of the nature of curriculum integration. Interdisciplinary refers to curricula holistic in nature that have as their focus an idea, theme, issue, or essential question that can and will be investigated from multiple perspectives.

Regardless of the range of interpretations of integrated curricula, and despite a paucity of empirical research (Czerniak et al., 1999; Grossman et al., 2000), interdisciplinary orientations to learning are thought to have value and merit across all levels of schooling, including preservice elementary teacher education (Akins & Akerson, 2000; Beane, 1991; Carter & Mason, 1997; Collins et al., 1999; Grossman et al., 2000; Hurley, 2001; Minstrell, 2000; Roberts & Kellough, 2000; Saam, 2000; Sterling, 2000; Wood, 1997; Zoller, 2000). According to Czerniak et al. (1999) and Grossman et al. (2000), the ambiguity associated with the construct is a reason for the limited research base. They called for research to illuminate what actually happens with interdisciplinary curricula used in the classroom, and to provide insight into the effectiveness of interdisciplinary curriculum compared to traditional, discipline-based curriculum.

The research literature (e.g., Akerson & Flanagan, 2000; Collins et al., 1999), reviews of research (e.g., Carter & Mason, 1997; Czerniak et al., 1999; Hurley, 2001; Wineburg & Grossman, 2000), and other educational literature (e.g., Roberts & Kellough, 2000; Wood, 1997) have suggested numerous merits to interdisciplinary orientations, as well as cautions (Roth, 2000). The merits have included:

- increased student and elementary preservice teacher achievement (Collins et al., 1999; Czerniak et al., 1999; Hurley, 2001; Renyi, 2000);
- development of elementary students' higher level thinking skills (Akins & Akerson, 2000; Wood, 1997);
- relating of higher order conceptual skills to problem-solving situations (Collins et al., 1999; Zoller, 2000);
- rich understandings of the connections between public policy or social meaning, different disciplines, and personal lives (Akins & Akerson, 2000; Beane, 1991; Carter & Mason, 1997; Grossman, Wineburg, & Beers, 2000; Saam, 2000; Sterling, 2000; Wood, 1997);
- student ownership of learning process (Roberts & Kellough, 2000);
- "intellectual pleasure" and a sense of personal accomplishment (Minstrell, 2000, p. 472);
- suitability for students with special needs due to greater flexibility in scheduling and use of alternative methods and reporting techniques (Wood, 1997); and
- reframing of teachers' ideas of science to be more consistent with the ideas held in the scientific community (Sterling, 2000).

Obstacles to successful use of an interdisciplinary orientation have also been reported. The obstacle most cited by teachers across all grade levels is time—the extensive time requirement for basic planning (Carter & Mason, 1997; Roberts & Kellough, 2000) and for working through a personal learning process of conceptual change regarding new methodologies of pedagogy (Sterling, 2000; Wood, 1997). Preservice elementary teachers' previous experiences in negotiating a learning culture based on respect for diversity, and teacher willingness and openness for collaboration, also have been identified as powerful influences on the use of interdisciplinary approaches to science learning (Akerson & Flanigan, 2000; Wieseman, Moscovici, Moore, van Tiel, & McCarthy, 2001). Team approaches to planning and teaching can reduce some of the problems, but not having students in common can compromise these efforts (Sterling, 2000). Management issues associated with increased movement and noise challenge some teachers (Roberts & Kellough, 2000).

The next two sections present two composite stories, "No Easy Journey" and "Will the *Open Court* Police Chastise Me?" Each story presents tensions experienced by elementary preservice teachers using an interdisciplinary inquiry orientation for science learning.

"NO EASY JOURNEY!"

This story is based on the individual voices of Bella, Kelly, Ellen, Amanda, and Regina, five preservice teachers, in a course that encourages inquiry-oriented learning within an interdisciplinary context. They are enrolled in two one-credit courses, *Methods of Teaching Elementary Science* and *Methods of Teaching Elementary Social Studies*, which are taught at a local elementary school.

The Context of the Story

Desiring to make efficient use of the total of 40 contact hours for the courses, and to increase the opportunities for preservice teachers to make connections between disciplinary perspectives, the professor combined the courses into one interdisciplinary methods course. Thirty-seven percent of the course time was designated as *Adventure Buddies*, a field experience in which the preservice teachers used pedagogy that they experienced as learners in class sessions modeled by the professor and/or elementary teachers. During *Adventure Buddies*, the preservice teachers interviewed the elementary children they were to teach, and observed the professor and elementary teachers team-teach lessons modeling inductive discovery, deductive discovery, problem-based learning, concept attainment, and the generative learning model (Osborne & Freyberg, 1985). They then taught the lessons they had planned, and reflected with the professor and/or elementary teachers after each teaching episode.

Three themes transect the preservice course curriculum: (a) What is inquiry? (b) What is the nature of each of the disciplines, science and social studies? When and how is it appropriate to integrate the two? (c) How is the teaching philosophy meshing with what is happening in the course, and why? During the course preservice teachers examined topics, including:

- Influences on learning: Inquiry in science and inquiry from social studies perspectives;
- "Place" of inquiry in the nature of science and social studies;
- Safety in an inquiry-oriented classroom;
- Discovery learning (deductive vs. inductive, demonstration, discussion); and
- Conceptual change theory and the generative learning model.

Inquiry-oriented learning within an interdisciplinary context was defined in this course as structured and guided inquiry (Windschitl, 2001) centered round a big idea, problem, theme, or topic. It was embodied into the learning opportunities in the course, and was described by the professor when the preservice teachers were examining what was happening in course sessions and the reasons why they were happening. These forms of inquiry were prioritized because the professor viewed 40 contact hours as a limitation to effective learning. She was also concerned that the preservice teachers did not engage in "busy work"—merely philosophizing about an educational idea or construct, or completing a course assignment they perceived not explicitly applicable to the actual teaching of science and social studies. Finally, she wanted to be sensitive to their request that class time be allocated to planning the lessons for their forthcoming field experience.

The professor, at the beginning of the course, introduced the preservice teachers to several approaches for planning curriculum aligned to state content standards (see Fig. 10.1).

The Participants in the Story

Bella, Kelly, Ellen, Amanda, and Regina are from 20 to 30 years old. All but Bella had teaching experience in other settings and Kelly was the only one who had experiences with inquiry prior to the methods course. Their memories were largely of teacher-centered, factual learning based on discipline-specific learning experiences. These experiences, combined with concerns about classroom management, contributed to a range of opinions about the appropriateness of an inquiry-oriented interdisciplinary approach to planning, within which standards/benchmarks are infused.

Regina, who was disinterested in and uncomfortable with science, needed tight structure and teacher control, and so was skeptical. Ellen, whose previous learning experiences were text-based and emphasized rote learning, and whose emotional comfort zone required her to have a solid knowledge base in order to teach a concept, was ambivalent. Bella and Amanda, who both enjoyed and valued science, were advocates. Kelly was a leader of the advocates.

From interviewing teachers about their approaches to curriculum planning and thinking about issues related to standards, Bella, Kelly, Ellen, Amanda, and Regina concluded that teachers often use the third, fourth, and fifth approaches to integration portrayed in Fig. 10.1. In class sessions dedicated to enhancing participants' pedagogical content knowledge, the preservice teachers experienced the first and second approaches. As revealed in conferences with the professor while planning for the field experience episodes, Bella, Amanda, and Regina favored the third and fourth approaches, whereas Kelly and Ellen tended toward the first approach.

Without consistent coaching, Bella, Amanda, and Regina tended toward what Moscovici and Nelson (1998) called "activitymania," or string-

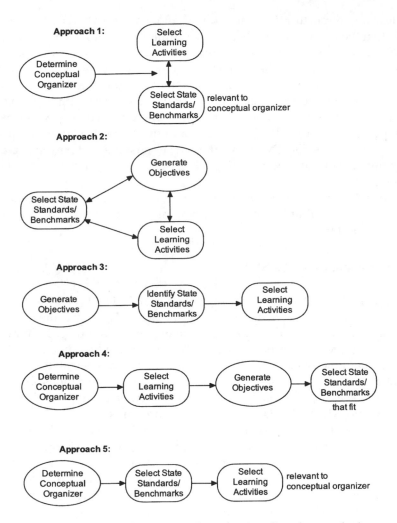

FIG. 10.1.   Approaches to curriculum planning aligned to standards.

ing together activities related to a topic that they thought would be interesting and fun for elementary children to do without regard to conceptual scaffolding, higher order thinking, or connections to standards or benchmarks. High in their thinking was planning a fun multimodal experience incorporating visual, auditory, tactile, and—if possible—kinaesthetic aspects of learning.

On the other hand, Kelly and Ellen were conscientious to scaffold learning experiences to address an essential question derived from the children's interests. As teaching partners for a "Let's Meet in the Playground!" teaching episode of the field experience, they framed their lessons around the question "How is a playground equipment's general shape or design related to its functional characteristics?" For example, the triangular shape of the swing set provides stability so a person swinging will not roll or tip it over. They developed this question by analyzing questions about playgrounds, in which their children had indicated an interest in exploring. Kelly and Ellen kept a close eye on benchmarks in science, geography, physical education, math, and reading/writing relevant to the essential question as they devised learning activities and sequenced them. In addition to having children make observations while swinging, they built miniature models of swing sets using different shapes, and kept a journal of their observations. They compared and contrasted appearances of playgrounds around the world and how children used them. When they played games that children in other countries might play on their playgrounds, the idea was reinforced that shapes are always present and play a role.

The remainder of the story focuses on the transactions during the last course session. Italics provide contextual information.

The Story

The last course session has arrived! Bella, Kelly, Ellen, Regina and Amanda are seated around a table in the elementary classroom where the course has been taught. Like teachers congregated in a teachers' lounge reveling and/or venting about what's happening in their classrooms or in the school, the five are evaluating what they have learned in the course.

Bella: I liked the interdisciplinary inquiry approach. The way we taught the lessons sure mimicked how people approach problem solving, how they try to come up with solutions to real-life problems. It's not that the questions we were asking necessarily represent every-day problems. But once you are faced with a question, you try to find an answer and you draw on everything that you know to come up with the best possible solution. It was a lot more work on me as the teacher, but . . .

Kelly [interrupts]:   Just don't make it activitymania. Interdisciplinary teaching using inquiry wasn't easy! I worked so hard at keeping the lesson focused, together, and intertwined. Figuring out which benchmarks were relevant to our big question took time and thought too. But it's the way to teach! Students really remembered what we had done and talked about!

Ellen [blurts]:   You have to do so much prep before you get into it, but it makes such a difference if you have your resources, have read through them, and know as much as you possibly can before you do it.

The more prepared you are, the more able you are to handle children shooting questions at you.

Concurrent dyad conversations become audible as the others voice feelings of unease and uncertainty when elementary children had posed questions they had not anticipated.

Amanda [above the din of the conversations]:   This new orientation to learning has forced me to think so differently. I don't even want to lecture because the kids are so charged up to get their opinions out and get involved. No way you could do this with a lecture! When I was in Catholic school, I just sat there and looked at the teacher as they talked at me.

Regina: Education is changing, but it's sad that it's still so lecture-based. What we've been using in this course is incredible. Rather than just sit regurgitating information, our first graders were in a very excited emotional state. I really got to know what kids thought and to build on their thinking. Like building a modular home, you lay the foundation down and build from bottom up.

Amanda: And it is so important to relate what children are learning to their outside world. Otherwise they think, "Who cares? This has nothing to do with me." When I was a kid in school I never got to take it outside! I never saw connections between what we read and listened to in one subject to another.

Ellen: Yes, it makes the children have a mission. Learning becomes meaningful if they're guiding the learning, and they truly are! One of the problems that I had in facilitating discussions was that they would take off somewhere where I didn't want them to go, but sometimes you have to let that happen. The children were learning how to communicate among themselves, rather than have everything directed back to the teacher. This builds ownership. [with exasperation] When do they get this opportunity in school to talk about their ideas? It always seems so structured. It wasn't easy letting go, but at the same time it was cool. Everyone keeps saying, and I agree, that good teachers are flexible, they let children "do" to learn and step aside to act as a guide.

Amanda: Another plus—if a teacher is really thoughtful and careful during planning, an interdisciplinary approach is a way to hit lots of standards, even in one lesson. What we teach and children learn needs to be aligned to the standards. Remember what the teachers told us about how they have to turn in their plans with the standards identified every Monday? . . . How our plans too had to have the standards clearly indicated for the principal? Then from the interviews we found out they certainly paid attention to standards but didn't let them drive how lessons and units were organized. Remember how they said there was never enough time, so an interdisciplinary approach allowed them to integrate science and social studies with other subjects?

Regina: Was I surprised to hear all this! [hesitantly] Even though I think an interdisciplinary approach is incredible I wonder whether it's always the best choice. I think some topics are specific to a subject. Don't you get confused when planning more extensive interdisciplinary units, like we did the curriculum and assessment course? I wasn't sure if my unit had a clear focus. I felt like I was going in so many directions just trying to learn the content that I lost sight of the big picture. I wonder if children will learn less in the subject area concepts if you try to integrate everything. I know I did.

Amanda [voice above a general mumbling]:   Sure, I have felt confused and unsure, but I still think children learn more because they are interested in what's happening. They can make more connections than if the focus is just on one subject. Interdisciplinary and inquiry are how we live our everyday lives.

Ellen: Confusion is a natural part of a learning process. [with a tone of uncertainty] What I wonder is, "How much of a solid basis of knowledge in each discipline does one need to be successful using an interdisciplinary approach?" All my preping was so I would understand the different concepts. Then I could understand the big picture and keep a focus. That's why for me an interdisciplinary approach is not easy. Maybe that will change with more experience; I hope so.

"WILL THE *OPEN COURT* POLICE CHASTISE ME?"

This story focuses on Diane, an emergency permit teacher (EPT) in grade 1 at a low socioeconomic elementary school in Los Angeles. Emergency permits are granted to preservice teachers who are completing the requirements in an accredited teacher education credential program while teaching during the day. Being an EPT has advantages and disadvantages. The EPTs come to class sessions, such as the elementary science methods course, with real-life questions because they are teaching while attending courses in pedagogy. However, they are often tired during the evening classes. They tend to be very reflective, always looking to improve their practices, and constantly questioning whether they are doing the best job possible. In this sense, reflection becomes a two-edged sword, especially because they can be fired without reason and they fear losing their positions.

During her science methods course, Diane learned about the interdisciplinary environmental education curricula: *Project Wild, Aquatic Wild,* and *Project Learning Tree.* She thought that her first graders would be very excited with the tasks and challenges in these ecological enrichment activities. She also believed that her first graders could achieve at a much higher level than they were using a newly mandated reading and writing program, *Open Court.* On the other hand, she had to follow this program to the minute. Otherwise, who knows what the *Open Court* "Police" would do, especially because she was working under emergency permit.

The *Open Court* program is a prescribed curriculum for reading and writing being implemented in low achieving elementary (and some middle) schools in the district. Its goal is to improve students' scores on standardized tests. All teachers across the district implementing the program are expected to be in perfect synchrony, that is, on the same page at any given date. Teachers not in perfect synchrony are given a coach to help them get on track. Otherwise, they are fired or moved to another grade level or school. School administrations and the *Open Court* Police enforce perfect synchrony in content coverage through visits to classrooms.

A large quantity of materials, such as color-coded books, are provided by the editor to ensure that teachers are using the "right" materials and are not getting confused. For example, a "recommended/required model" lesson for grade 1 (as written in the curriculum) requires 180 minutes daily to complete the activities. The lesson is divided, with from 55 to 65 minutes of the green color-coded "Preparing to Read" activities (33% of the total time), 20 to 30 minutes of the red color-coded "Reading and Responding, Comprehension Strategies/Skills" activities (14%), 50 minutes of the blue color-coded "Integrating the Curriculum" activities (28%), and 45 minutes of the special mats "Independent Work Time" activities (25%).

Not only was Diane uncomfortable with the nature of the *Open Court* program, she was especially disgruntled with the rigidity imposed by the editor that was enforced by her school's administration and the *Open Court* Police. With afternoons dedicated to teaching math, no time existed in the school day for science or social studies, even though the district has science standards. The principal did not want teachers to teach science. Instead, the principal's first goal was to raise standardized test (SAT 9) scores in reading, writing, and mathematics. As in the previous story, italics provide contextual information.

The Story

Diane [with frustration]:   What should I do? As if dealing with the district standards and this constraining *Open Court* curriculum is not enough! Now I need to decide if I should use the enrichment materials that I received from my methods course. Should I just pretend that I have not received these materials, and use them later when I have my job secured? Should I just pretend that I do not see the bored and miserable faces of my students? Sometimes I don't know who is more bored—me or my first graders! [in a sly tone]: Maybe I can be sneaky and integrate all three (*Open Court*, standards, and the environmental education materials) in my lessons, like my science methods professor suggested. This way I could integrate standards from science and social studies with language arts and math. Maybe this would give me the best of everything. It is worth a try.

Diane decides to risk integrating these pieces together into a multi-day lesson she calls "What's hiding around us?" After outlining her ideas, she approaches her methods professor.

Diane [full of nervous excitement]:   After all children love to discover things. They love mysteries. The idea is to teach my first graders about diversity—of soil particles, of soil kinds, of objects in different soils, of trees and forests. Wow! I am so wired up! They will learn about diversity in general, diversity in the habitats of living things, and diversity in the environment. It should be exciting! And all this is in the California content standards. I can use *The Forest* by S. T. Shrew, as well as use materials from *Project Wild* and *Project Learning Tree*. I will use parts of the units on *Animals* and *Machines in our Garden* from *Open Court*. I will ask my first graders to bring relevant materials from home. We will use the KWL[2] tool to have them write and decide what to learn. Just think— having the children collect soil samples from around the school, I can see all their faces lighting up, can't you?

Diane puts her plan into action. With sparkles in their eyes, her students collect soil samples and analyze them. They compare and contrast trees in the neighborhood. They present this information to the whole class using drawings and whole sentences. The students start bringing things related to their studies to show to classmates. They explore the effects of changing the environment such as cutting the trees. They read stories and poems. Parents start expressing delight with their children's pleasure and interest in the subject. Diane is having a wonderful time with her students, and then . . . a surprise visitor appears who makes her panic.

Diane [to her elementary methods course classmates]:   I had made space for students' artefacts from home, but I didn't expect as many as

---

[2]An acronym for "What do you already KNOW, What do you WANT to learn, and What have you LEARNT?"

came in. Books, pieces of natural wood, wood miniatures, pictures of forests (the Redwood forest, the temperate rainforest on the Olympic Peninsula in the State of Washington, tropical forests all over the world). Of course when I said to bring soil samples from home, I did not expect the high amount nor the variety. Some students brought more than four samples. So much stuff! Here they were learning about diversity, all by themselves! Parents even volunteered for the short fieldtrips in the neighborhood, and they enjoyed learning together with the first graders.

On the third day of the multi-day lesson, a representative of the *Open Court* Police, Ms. Davis, arrived at my door. She walked in just as I was reading *The Forest* to my first graders who were gluing the drawings of the insects that they found in their soil samples on their ecology posters. My heart crashed to my feet! I swallowed hard, took a deep breath, and continued reading pretending that she was not there. Ms. Davis visited for a week and took a lot of notes. In the afternoons and after school she asked me questions, "Where did you go on the field trip you mentioned today in class, and what did the children do? Could you show me the books and other artefacts that students brought from home?" By the end of the week, she told me that she had not seen such a well-done integration. My students were more ready to learn than those in any of her other classroom visits. She mentioned that my students' interest in science motivated them with reading and writing, and complimented me for the elegant way in which I addressed standards from all the subjects in this unit. On the last day of her visit she actually took pictures of the students during the read aloud exercise and of their project presentations.

Story Epilogue: Today Diane is a resource teacher at her elementary school where she assists teachers who stumble. She is an appreciated colleague and a valuable role model. From the experience related in the story Diane learned that for students to learn they needed to be challenged and become active participants in the learning process. She also learned that a good thematic curriculum intertwines standards from various content areas "like the notes played by different instruments in a good piece of music."

In 2003, the state of California implemented a statewide science test for fifth graders. This mandatory test assesses student achievement on science standards for grades 4 and 5. Elementary principals and teachers are in a panic. Diane wonders whether and how the low socioeconomic schools in the Los Angeles area will address science learning standards for the grades 4 and 5 and continue to use the *Open Court* program. She "at least was lucky to be introduced to interdisciplinary thematic units that integrate so elegantly and naturally various content areas in an inquiry adventure for teachers and students alike." She knows that she "is prepared to face the new challenge."

IMPLICATIONS FOR THE FUTURE
OF ELEMENTARY SCIENCE EDUCATION

The tensions depicted in the two stories (see Table 10.1) challenge the teacher education community to think about the nature and purpose of elementary teacher preparation in general and elementary science teacher preparation in particular. An examination of the tensions experienced by these preservice teachers in relation to a broader educational discourse (Akerson & Flanagan, 2000; Carter & Mason, 1997; Roberts & Kellough, 2000; Sterling, 2000; Wineburg & Grossman, 2000; Wood, 1997) suggests that the tensions are not unique.

In general, during their teacher preparation, elementary teachers are not likely to be required to balance and resolve perceived real-world issues and challenges associated with an interdisciplinary orientation to learning (Hillman, Bottomley, Raisner, & Malin, 2000). The stories in this

TABLE 10.1
Preserve Teachers' Tensions

Network of Issues Needing Resolution

- Whose interests—participants in the classroom or outsiders—are most important when deciding learning goals?
- How do I (teacher) get student commitment to learning? I want my students to be "charged up" and have "glittering eyes."
- Which intelligences and thinking skills are most important to encourage and nurture?
- Who is the learner here—students and teacher as inquirer and/or students as direction-followers and recipients of dispensed knowledge?
- Will making connections between subject areas really lead to more meaningful/relevant learning and a richer understanding? It seems a lack of connection between subject matter and a potpourri of activities can lend more to fragmented learning and understanding.
- What knowledge is of most worth—factual, integrated? Is the way of knowing in one discipline really more important than another?
- How closely should learning mimic real-life problem-solving situations?
- What solutions are there for competing multiple directives and accountability pressures—standards, district mandates, expectations of achievement test results, students' interests and needs, and teacher's desires? There have to be some.
- If an interdisciplinary orientation is to be successful, how solid does my (preservice teacher and teacher) knowledge base in each discipline first need to be?
- I do not feel confident in my knowledge base in the separate disciplines. Maybe an interdisciplinary orientation is the solution to this discomfort, right?

chapter, as well as research by others (e.g., Akerson & Flanigan, 2000; Grossman et al., 2000; Hillman et al., 2000), suggest that formal experiences with interdisciplinary orientations may play a critical role in an elementary teacher's understandings of and attitudes toward science and interdisciplinary teaching and learning.

The research reported here, and that of others such as Akins and Akerson (2000), Wineburg and Gossman (2000), Zoller (2000), and Wieseman et al. (2001), suggest that an interdisciplinary inquiry orientation can promote meaningful learning when an appropriate curriculum organizer is chosen and the learning experiences encourage the learner to use higher order thinking skills relevant to solving problems in real-life situations. We, among others (e.g., Renyi, 2000; Wieseman & Moscovici, 2001; Wood, 1997), purported that interdisciplinary orientations that encourage inquiry and student ownership of their learning are inviting to a wide variety of students. The students are "charged up" to engage in a learning experience. Some students may be attracted by the tactile sensations of the materials with which they are working, or kinesthetic motion like the body swinging to and fro. Other students may be hooked as a result of the inclusion of an artistic and aesthetic view of the topic of the inquiry, such as their immediate surroundings. Another set of students might be drawn to question and search for relations between factors—for example, search for a relation between the diversity and distribution of trees and soil type. The essential element contributing toward student engagement and ownership appears to be deep and documented, not superficial and informal, inquiry.

Where does the standards reform movement of the 1990s fit in? Standards tend toward a disciplinary orientation, evidenced by their organization in content areas with the labels of science, math, and so on; by the nature of knowledge communicated through the language of the standards; and by the identities and interests of the organizations that coordinated their construction. A teacher aligning a learning experience with a standard/substandard benchmark in one content area thus perpetuates a disciplinary orientation. Courses, even those in these stories, have titles that connote a disciplinary orientation and content specificity. The preservice teachers in the *No Easy Journey!* story were enrolled in science methods and social studies methods courses. The EPT, Diane, was enrolled in a science methods course. Interdisciplinarity became part of the framework of these methods courses through the values and decisions of the professors concerned.

The introduction to the chapter proposed that distinguishing between approaches to curriculum used by teachers, such as standards-infused and standards-based approaches (see Table 10.2), could be useful for developing different views of elementary science teacher preparation and elementary school science, particularly in institutions in which science education is preempted. The promise of a standards-infused approach to curriculum, as characterized in Table 10.2, emerges as a result of positive experiences with a curriculum design process that feels natural, even though tension may be experienced. When the results in the classroom validate risks taken and allay the associated tensions, subsequent curriculum planning is more likely to follow an integrated approach.

What was the curriculum design process used? The preservice teachers focused on developing problem- or theme-based units that provided multiple entry points for constructing conceptual understanding. The standards were in the backdrop of the curriculum planning process, and

TABLE 10.2
Characteristics of Standards-Infused and Standards-Based Approaches

Standards-Infused	*Standards-Based*
• Focus on problem-solving situation, issue, or big idea that transects content areas.	• Focus on the nature of knowledge of one content area (big ideas, concepts, methodologies, values, beliefs).
• Holistic approach to planning involving: —Problem-solving situation, issue, or big idea that transects content areas, promotes inquiry, and recognizes teacher and students' interests and knowledge identified. —Resources selection, sequencing of learning opportunities, and identification of multiple standards from multiple content areas (backdrop of planning).	• Linear approach to planning involving —Grade level standards or benchmarks in one content area are identified as entry points, and triggers what will be taught and learned. —Selection of resources and sequencing of learning opportunities for understanding with respect to entry point.
• Perception that use of thinking skills and inquiry is the priority, solution-building and conceptual connections are the goals.	• Perception that factual learning is the priority and goals are standards/benchmarks.
• Curriculum developed by classroom teacher using curriculum organizers broad in scope to provide multiple entry points for learning and promoting inquiry.	• Greater tendency toward using a prescribed curriculum, which may/may not promote inquiry and may/may not meet needs of diverse learners.
• Activitymania (Moscovici & Nelson, 1998) avoided.	• Activitymania (Moscovici & Nelson, 1998) can be basis of selecting hands-on activities
• Purpose of hands-on activities is to raise interest and stimulate thinking, so they need to be well sequenced and coherent to capture children's attention and excite them (e.g., use of questions such as *What's hiding?*) and to meet needs of diverse learners.	• Purpose of hands-on activities is to fit with and reinforce a benchmark or targeted concept.

teacher thinking considered standards of multiple content areas. Hands-on activities were sequenced to provide opportunities for constructing conceptual connections, and required the use of higher order thinking skills and solution building. In the first story, Kelly and Ellen were consistently successful in infusing consideration of children's needs and interests in relation to their conceptual organizer, an essential question focused on relations between form and function. They identified benchmarks from several content areas pertinent to both the organizer and childrens' interests. In the second story, Diane and her first graders investigated the essential question "What's hiding around us?" This curriculum experience infused (a) science and social studies standards advancing the big idea of diversity, math standards in numeracy and data analysis, and reading and writing standards; (b) a passion for nature; (c) students' curiosity and interest in their surroundings; (d) available local resources; and (e) the mandated *Open Court* program.

Even though the vision of the standards reform movement of the 1990s advocates open-ended and authentic inquiry, and a classroom dynamic that is fluid, what can happen in classrooms is selection of and adherence to a linear, text-based, prescribed curricula. This tendency has been familiar in elementary school science (e.g., Abell & Roth, 1994; Tobin, Briscoe, & Holman, 1990). Problems reported in the literature associated with this continue to be played out in classrooms. Curiosity and inquisitiveness for scientific questions are often not priorities. Although leery of the kind of science that they were exposed to as learners in the K–16 educational system, many elementary teachers (preservice and inservice) continue to operate a text-based, prescribed curriculum. It seems to function as a life jacket for teachers feeling insecure and experience emotional discomfort because their science discipline knowledge is limited and/or they are uncertain how and why to use inquiry in their classrooms.

Grossman, Wineburg, and Beers (2000) claimed that the "existing literature is . . . comprised of idealized descriptions of programs and how to put them in place, and is almost entirely devoid of descriptions of what actually happens when theory meets school practice" (p. 9). In order to better understand the "disorderly state of the art" there is a need for more discourse in how teachers, both preservice and inservice, approach curriculum with respect to standards, what actually happens in the classroom in relation to curriculum planning, and why.

CONCLUSIONS

This chapter has outlined two case studies of preservice teachers attempting to implement standards-infused, integrated curriculum that incorporates science. However, a number of questions remain: What are the short-

and long-term outcomes of such a curriculum for student learning? How worthwhile are courses for preservice elementary teachers that model interdisciplinary inquiry orientations? What learning opportunities within interdisciplinary teacher preparation courses are viewed by preservice teachers as worthwhile, and why? What tensions do elementary teachers associate with disregarding or reinterpreting discipline-focused requirements made by administration and other controlling bodies, and how do they resolve these tensions?

The intent of this chapter has been to encourage the reader to become a partner in a disciplined inquiry process regarding situations experienced by preservice teachers who are thinking about curriculum, teaching, and student learning. Distinguishing between approaches to curriculum, such as standards infused and standards based, can hold promise for developing different views of the future of elementary science teacher preparation. A standards-infused approach may be particularly useful in institutions where science education is being preempted. In this approach, accountability with respect to standards is present, although thinking about standards is part of the planning process. The promise of standards-infused approach to curriculum may be tempered, however, as teachers must be knowledgeable and motivated to diagnose, analyze, synthesize, evaluate, and create curriculum that can immerse both the teacher and students in learning (Tchudi & Lafer, 1996). As a reader, what are your insights, wonderings, and tensions?

REFERENCES

Abell, S. K., & Roth, M. (1992). Constraints to teaching elementary science: A case study of a science enthusiast student teacher. *Science Education, 76*, 581–595.

Abell, S. K., & Roth, M. (1994). Construction science teaching in elementary school: The socialization of a science enthusiast student teacher. *Journal of Research in Science Teaching, 31*(1), 77–90.

Akerson, V. L., & Flanigan, J. (2000). Preparing preservice teachers to use an interdisciplinary approach to science and language arts instruction. *Journal of Science Teacher Education, 11*(4), 345–362.

Akins, A. T., & Akerson, V. L. (2000, January). *Connecting science, social studies, and language arts: An interdisciplinary approach.* Paper presented at the annual meeting of the Association for the Education of Teachers in Science, Akron, OH.

Barone, T. E. (1988). Curriculum platforms and literature. In L. E. Beyer & M. W. Apple (Eds.), *The curriculum: Problems, politics, and possibilities* (pp. 140–165). New York: State University of New York Press.

Beane, J. (1991). The middle school: The natural home of integrated curriculum. *Educational Leadership, 49*(2), 9–13.

Carter, C. C., & Mason, D. A. (1997, March). *A review of the literature on the cognitive effects of integrated curriculum.* Paper presented at the annual conference of the American Educational Research Association, Chicago, IL.

Collins, A., Bercaw, L., Palmeri, A., Altman, J., Singer-Gabella, M., & Gary, T. (1999). Good intentions are not enough: A story of collaboration in science, education, and technology. *Journal of Science Teacher Education, 10*(1), 3–20.

Connelly, F. M., & Clandinin, J. D. (1988). *Teachers as curriculum planners: Narratives of experience.* New York: Teachers College Press.

Czerniak, C. M., & Lumpe, A. T. (1996). Relationship between teacher beliefs and science education reform. *Journal of Science Teacher Education, 7,* 247–266.

Czerniak, C. M., Weber, W. B., Sandmann, A., & Ahern, J. (1999). A literature review of science and mathematics integration. *School Science and Mathematics, 99*(8), 421–430.

DeCourse, C. B. (1996). Teachers and the integrated curriculum: An intergenerational view. *Action in Teacher Education, 18*(1), 85–92.

Drake, S. M. (1991). How our teams dissolved the boundaries. *Educational Leadership, 49*(2), 20–22.

Fogarty, R. (1991). Ten ways to integrate curriculum. *Educational Leadership, 49*(2), 61–65.

Giroux, H. A., & Simon, R. (1989). Popular culture and critical pedagogy: Everyday life as a basis for curriculum knowledge. In H. A. Giroux & P. L. McLaren (Eds.), *Critical pedagogy, the state and cultural struggle* (pp. 236–252). New York: State University of New York Press.

Goodlad, J. I., & Su, Z. (1992). Organization of the curriculum. In P. W. Jackson (Ed.), *Handbook of research on curriculum* (pp. 327–344). New York: Macmillan.

Grossman, P., Wineburg, S., & Beers, S. (2000). When theory meets practice in the world of schooling. In S. Wineburg & P. Grossman (Eds.), *Interdisciplinary curriculum. Challenges to implementation* (pp. 1–16). New York: Teachers College Press.

Hillman, S. L., Bottomley, D. M., Raisner, J. C., & Malin, B. (2000). Learning to practice what we teach: Integrating elementary education methods courses. *Action in Teacher Education, 22*(2a), 91–100.

Hurley, M. M. (2001). Reviewing integrated science and mathematics: The search for evidence and definitions from new perspectives. *School Science and Mathematics, 101*(5), 259–268.

Jacobs, H. H. (Ed.). (1989). *Interdisciplinary curriculum: Design and implementation.* Alexandria, VA: ASCD.

Lederman, N. G., & Niess, M. L. (1997). Integrated, interdisciplinary, or thematic instruction? Is this a question or is it questionable semantics? *School Science and Mathematics, 97*(2), 57–58.

Manen, M. (1990). *Researching lived experience: Human action for an action sensitive pedagogy.* New York: SUNY.

Martinello, M. L., & Cook, G. E. (2000). *Interdisciplinary inquiry in teaching and learning* (2nd ed.). Upper Saddle River, NJ: Prentice-Hall.

Maurer, R. E. (1994). *Designing interdisciplinary curriculum in middle, junior high and high schools.* Boston: Allyn & Bacon.

Minstrell, J. (2000). Implications for teaching and learning inquiry: A summary. In J. Minstrell & E. H. van Zee (Eds.), *Inquiring into inquiry learning and teaching in science* (pp. 471–496). Washington, DC: American Association for the Advancement of Science.

Moscovici, H., & Nelson, T. H. (1998). Shifting from activitymania to inquiry. *Science and Children, 35*(4), 14–17, 40.

National Science Teachers Association (NSTA). (2002). News & bulletins [column 3]. *NSTA Reports!,* May/June, p. 24.

Osborne, R., & Freyberg, P. (1985). *Learning in science: The implications of children's science.* Auckland, New Zealand: Heinemann.

Pajares, M. F. (1992). Teachers' beliefs and educational research: Cleaning up a messy construct. *Review of Educational Research, 62*(3), 307–332.

Post, T. R., Ellis, A. K., Humphreys, A. H., & Buggey, L. J. (1997). *Interdisciplinary approaches to curriculum: Themes for teaching.* Upper Saddle River, NJ: Merrill.

Renyi, J. (2000). Hunting the quark: Interdisciplinary curricula in public schools. In S. Wineburg & P. Grossman (Eds.), *Interdisciplinary curriculum. Challenges to implementation* (pp. 39–56). New York: Teachers College Press.

Roberts, P. L., & Kellough, R. D. (2000). *A guide for developing interdisciplinary thematic units* (2nd ed.). Upper Saddle River, NJ: Prentice-Hall.

Roth, K. J. (2000). The photosynthesis of Columbus: Exploring interdisciplinary curriculum from the students' perspectives. In S. Wineburg & P. Grossman (Eds.), *Interdisciplinary curriculum: Challenges to implementation* (pp. 112–133). New York: Teachers College Press.

Saam, J. (2000, January). *Teachers' wisdom: Bringing unique perspective to the integrating of middle level mathematics and science.* Paper presented at the annual meeting of the Association for the Education of Teachers in Science, Akron, OH.

Smith, D. C., & Neale, D. C. (1989). The construction of subject matter knowledge in primary science teaching. *Teaching and Teacher Education, 5*(5), 1–20.

Sterling, D. R. (2000, April). *Strategies enabling interdisciplinary teacher-teams to develop and implement standards-based teaching plans.* Paper presented at the annual meeting of the National Association for Research in Science Teaching, New Orleans, LA.

Tchudi, S., & Lafer, S. (1996). *The interdisciplinary teacher's handbook: Integrated teaching across the curriculum.* Portmouth, NH: Boynton/Cook.

Tobin, K., Tippins, D., & Gallard, A. J. (1994). Research on instructional strategies for teaching science. In D. L. Gable (Ed.), *Handbook of research on science teaching and learning* (pp. 45–93). New York: Macmillan.

Tobin, K., Briscoe, C., & Holman, J. R. (1990). Overcoming constraints to effective elementary science teaching. *Science Education, 74*(4), 409–420.

Von Glasersfeld, E. (1989). Cognition, construction of knowledge, and teaching. *Synthese, 80*(1), 121–140.

Wieseman, K. C., & Moscovici, H. (2001, March). *Tales from the field: Dilemmas in science learning through interdisciplinary approaches.* Paper presented at the annual meeting of the National Association for Research in Science Teaching, St. Louis, MO.

Wieseman, K. C., Moscovici, H., Moore, T., van Tiel, J., & McCarthy, E. (2001, January). *Stories from the field: Challenges of science teacher education based on interdisciplinary approaches.* Paper presented at the international conference of the Association for the Education of Teachers of Science, Costa Mesa, CA.

Windschitl, M. (2001, March). *Independent inquiry projects for pre-service science teachers: Their capacity to reflect on the experience and to integrate inquiry into their own classrooms.* Paper presented at the annual meeting of the National Association of Research in Science Teaching, St. Louis, MO.

Wineburg, S., & Grossman, P. (Eds.). (2000). *Interdisciplinary curriculum. Challenges to implementation* (pp. 1–16). New York: Teachers College Press.

Wood, K. E. (1997). *Interdisciplinary instruction: A practical guide for elementary and middle school teachers* (2nd ed.). Upper Saddle River, NJ: Prentice-Hall.

Zoller, U. (2000). Teaching tomorrow's college science courses—Are we getting it right? *Journal of College Science Teaching, 29*(6), 409–414.

11

The Role and Place of Technological Literacy in Elementary Science Teacher Education

Alister Jones
University of Waikato

One of the difficulties in writing about technology in science education is the perceptions that people have of technology are frequently associated with computers or educational technology (Cajas, 2001; Jones & Carr, 1992). In fact, many undergraduate and graduate courses in science education have sections on technology. However, these are often about using computers and the like to teach science concepts or processes, which represents a limited view of technology. Technology has played a central role in human societies. For instance, Bybee (2000) explained how the Newseum, a journalism museum in Virginia, conducted a survey of American historians and journalists to determine the top 100 news stories of the 20th century. He noted that in the top 100 headlines in the 20th century, an estimated 45% were directly related to technology. This ranking of news stories seems to justify increased emphasis on technological literacy, because it clearly represents what the public reads, hears, and values relative to technology. Yet, as Bybee noted, for a society deeply dependent on technology, and particularly in this so-called knowledge age, people are largely ignorant about technological concepts and processes, and the factors that underpin technological development and innovation. In the past, this has been ignored in the educational system. This has led to a society that, generally, knows little about technology. To add to this lack of understanding, there are those who often equate the theme of technology education with the use of computers in schools. The use of computers, as one of many educational technologies, provides important tools for the enhancement of learning across all curriculum areas, but it should not be equated to technology education or limit technology in science education to just the use of computers in the teaching and learning of science. Also, the lack of general notions of technological literacy is compounded by the other misconception that technology is simply applied science. As Bybee (2000) highlighted, a new understanding of technology must be established. In attempting to develop a better understanding of technology in science education, this chapter describes technology, looks at its relation with science, examines teachers' and students' perceptions of technology, explores approaches to introducing technology into the science classroom and some difficulties that may be experienced, and considers ways in which the introduction of technology can enhance student learning in both science and technology. It concludes with considerations of the implications for science teacher education.

Technology in science education and the interdependence of scientific and technological literacy are becoming much more prominent in recent science education literature. For example, there are special issues in technology education in the journals of *Research in Science Education* (2001) and *Journal of Research in Science Teaching* (2001). Many issues raised in the special issue of *Research in Science Education* (Jones, 2001) have direct relevance to thinking about teaching and learning in science. For example, the relation between science and technology, teacher development, enhancing pedagogical content knowledge, learning in early childhood settings, the situated-ness of learning, and the enhancement of formative interactions have direct implications for science education.

Technology can be utilized in a variety of ways in science education. However in using it, it is important to have a clear concept both of the nature of science and the nature of technology. Too often in the past a limited view of technology in science has limited both the learning of science and the learning about technology. An analysis of both the nature of science and the nature of technology shows that there is a complex relation between the two. It is therefore important that some of this complexity is apparent in science curricula. Unfortunately, in the past, a simplistic relation of technology as applied science has held sway. It is time for a reevaluation (Cajas & Gallagher, 2001). Recent reforms in the area of scientific literacy have also included calls for students to have a better understanding of technology (Cajas, 2001).

In the past, technology has been marginalized in general education for all and was often not seen as an area worthy of study by the more able students. However, the new production of technology curricula in Australia, the United Kingdom, the United States, Canada, Hong Kong, and New Zealand emphasizes the importance of students developing a broad level of technological literacy (Jones, 2001).

REASONS FOR TECHNOLOGY EDUCATION

A prime motivation for wanting technology in the curriculum for all students is its place as a major activity in today's culture, which is of course similar to the reason for teaching science and other subjects. There are three basic purposes articulated for technology education: to improve the economy of the country, for its intrinsic value as part of individual development, and as an introduction to culture. McCormick (1992) explicated these in more detail, but a number of other authors have given different names to cover the same ideas (Jones & Carr, 1993; Layton, 1994). The most compelling reason for studying technology is that it is a major, and some would argue, a determining feature of today's world. As part of culture, young people need to be introduced to it so they can understand its nature and participate in it at some level. Technology therefore stands alongside other ways to represent culture (e.g., science, mathematics, music, etc.). This makes it more difficult to represent and for people to agree on its nature, but it is no less an aspect of culture than the other major areas of knowledge and understanding that are represented as subjects. If technology is indeed a determining feature of the world, then it follows that young people, as future citizens, need to understand how it shapes the world. Technology is a value-laden activity and citizens need to understand and control many of the decisions that are made. Therefore education must prepare them to do this by dealing with the technical, social, ethical, political, and economic issues that underlie technological processes. This will allow them to take part as active citizens of their society.

DEFINING THE NATURE OF TECHNOLOGY

Consideration of the nature of technology indicates that technological knowledge and practices are socially constructed and context dependent, and where human mental processes are situated within their historical, cultural, and institutional setting (Wertsch, 1991). Therefore, technology is an activity that involves not just the social context, but also the physical context, with thinking being associated with and structured by the objects and tools of action. Technology is thus based within a philosophical, historical, and theoretical context (Mitcham, 1994). It is its characteristic as an *activity*, as well as a body of knowledge, that is salient. Technological activity makes the idea of *practice* most central, and hence the importance of technological practice. Technological practice is primarily about *doing* technology, as well as studying it and creating technological knowledge. This does not deny that those who do technology create knowledge either through technological activity or in a theoretical fashion, or that there is unique technological knowledge. The uniqueness of technological knowledge, processes, and skills have not always been recognized in general education, although literature in the area is increasing (e.g., Bucciarelli, 1996; Layton, 1993; Kline, 1987; McGinn, 1978; Pacey, 1983; Pinch, Hughes, & Bijker, 1987; Staudenmaier, 1985; Wajcman, 1991). Cultural authenticity of technology leads to the need to reflect the work of communities of practice of technology, which brings up questions such as:

- What are the communities of practice (assuming there is not just one)?
- What are the technologies that these communities practice in?
- Is there anything coherent about or common to these communities?
- Is this too restricted to only professional views of culture?

The *multiple communities of practice* (as is the case in many areas of activity, including science) are defined by the nature of the products, or outcomes of the activity, or some general body of expertise, and even by their role in that activity. Thus, there are *technologies*, not just *technology*. These technologies are not definitively defined but include the likes of food, textiles, electronics, automobile industry, and mechanical engineering, and are not exclusive (e.g., many mechanical engineers work in the food industry; megatronics is a combination of technologies).

This multiple communities view implies that there is likely to be much that is not common across the technologies. Common elements would be some general processes like "design" and perhaps "fault finding" (although the extent to which they are the same processes is debatable), and general concepts like "systems," "product development," and "innovation." The difficulty of seeing commonality is, however, of the same order as it is for science with its multiple disciplines and quite different conceptual structures and methods of investigation. Therefore, it is just as reasonable for technology educators to look for commonalties in the same way as science educators do.

COMPONENTS OF TECHNOLOGY EDUCATION

Technology education is concerned with developing student technological literacy through the exploration and solving of complex and interrelated technological problems that involve multiple conceptual, procedural, societal, and technical variables (Jones, 1997). The key aspects can then be seen as technological knowledge and understanding, technological processes or capability, and the relation between technology and society. The technical aspects relate to particular techniques and skills.

Technological Knowledge and Understanding

It is crucial that technology education does not become a purely process-orientated one, where knowledge is not deemed to be important (e.g., "design cycle" or "design and make"). Technology in the school setting has often been limited to the design aspects, including constraints and compromises. However, technological knowledge and principles have been underplayed. It is impossible to undertake a technological activity without some technological knowledge and the utilization of other knowledge bases. It is crucial that the use of other knowledge bases is valued as an essential part of a technological activity. Each technological area will have different knowledge bases that are important. It is therefore vital that students have an understanding of a range of technologies and how they operate and function. Technological activity does not occur in isolation from people and their needs and values. Different ideas contribute to technology and it is important for students to have the opportunity to investigate how these ideas influence technology, including the direction of technological developments. Students need to develop an understanding of the principles underlying technological developments such as aesthetics, efficiency, ergonomics, feedback, reliability, and optimization. The understanding of systems is essential in developing knowledge in technology. Students should investigate technological systems and analyze the principles, structure, organization, and control of systems. These knowledges and principles will of course be dependent on the technological area and context in which the students are working.

Technological Processes or Capability

Technological activity arises out of the identification of some human need or opportunity. Within the identification of needs and opportunities, students use a variety of techniques to determine such things as consumer preferences, using market research, or similar strategies. Most technological developments occur through modifications and adaptations of existing technologies. In the history of technology, just as in the history of science, attention is paid to the large leaps forward rather than the incremental changes that are involved in technological development.

Students should investigate existing technology in order to propose modifications, innovations, and adaptations to meet needs and opportunities. It is crucial in a technological activity that students develop implementation and production strategies to realize technological solutions. Part of this development involves students in developing and generating possible ideas that may lead to a number of solutions and strategies in order to realize these ideas. To achieve a desired solution, students learn how to manage their time, resources, and people. Outcomes suggested and or produced should be related back to the previously identified needs or opportunities, but not necessarily be constrained by them. Students should communicate their designs, plans, and strategies and present their technological outcomes in appropriate forms. Part of this process is the devising of strategies for the communication and promotion of ideas and outcomes, such as advertising, marketing, pricing, and packaging. Throughout the technological activity, students should continually reflect on and evaluate the decisions they are making. Research indicates (Jones, 1997) that this reflection and evaluation is essential if students are to realize their technological outcomes. Students should not only appraise their own outcomes, but also those from the past, present, and other cultures.

Technology and Society

The understanding of the complex relation between technology and society is essential to technology and technology education. Too often in the past in general education or even in science education, attention has been focused on the impacts of technology on society rather than on different views people have about technology and the ways these are influenced by their beliefs, values, and ethics. Past technological developments have been shaped by people making decisions based on their own and others' opinions and interests. It is essential that students be given the skills and opportunities to investigate the basis of these past influences, and explore the impacts they have had, and continue to have, on technological innovation and development. Such a level of critical awareness of the way in which the lived world is constructed is crucial to developing in students a sense of empowerment. People cannot initiate change if they do not understand the frameworks on which any change is dependent. In many ways, understanding the relation between technology and society is about not only learning the "rules of the game," but also being in a position to critique these rules and feel empowered to change them if they are found to be inappropriate.

Areas of Technology

The practice of technology in the world outside the classroom covers a diverse range of activities from agriculture through to the production of synthetic materials and electronics. Technology education must reflect this diverse practice and not limit itself to designing and making with a limited range of materials. The development of technology education must reflect the relevant areas of technology. Therefore, it is essential that a range of technological areas in the teaching and learning of technology be introduced. Each technological area has its own technological knowledge and ways of undertaking technological activity. So it is important that students experience a range of technological areas and contexts to develop an understanding of technology and technological practice. Theories of learning also point to the fact that the more students can work in a number of contexts and areas, then the more likely they are to develop effective knowledge about technology and transfer this knowledge to other contexts and areas (Jones & Carr, 1993; Perkins & Salomon, 1989). These areas may include materials technology, information and communication technologies, electronics and control technologies, biotechnology and biorelated technologies, structures and mechanisms, process and production technologies, and food technology.

Many activities that have been called technology, such as technology challenges (building bridges to take particular weights), have not acknowledged the importance of technological concepts in the designing and construction process. But, it is also important to remember that a bridges unit that acknowledges the different knowledge bases from science and technology (e.g., structural support systems, material properties, failure analysis, and social impact) can provide a powerful learning environment.

Technology for students can range from ways they can enhance their own environment, such as dealing with concrete understanding of particular materials, through to developing understanding of more abstract technological principles and processes, and the interaction of technology with society.

These components are reflected in a number of technology curricula, particularly in New Zealand, Hong Kong, and the United States. The *Standards for Technological Literacy: Content for the Study of Technology* (TfAAP, 2000) reflects these components of technology by emphasizing the nature of technology, technology and society and design, and producing and assessing products and systems for all students. The standards highlight the importance of technological knowledge and processes, as well as technology and society. The areas of study are medical technologies, agricultural and related biotechnologies, energy and power technologies, information and communication, transportation, manufacturing, and construction technologies.

THE RELATION BETWEEN SCIENCE
AND TECHNOLOGY

The relation between science and technology is a complex one. Discussions about this relation have often been fruitless because a too simplified image of that relation was used (de Vries, 2001). The "technology as applied science" paradigm is well known. Defenders of this paradigm had no difficulties in showing examples in which this idea applied well. There is scarcely any doubt that the transistor would not have been invented in the Bell Labs without the use of solid-state physics. However, at the same time, others could come up with equally valid examples for rejecting the technology as applied science view. They could, for instance, suggest the example of the hot air engine that was invented at a time when engineers' knowledge of thermodynamics was limited. So valid cases could be used both for defending and for rejecting the technology as applied science paradigm. As de Vries (2001) noted it is important to distinguish between different types of technology because the technology as applied science paradigm only applies to some technologies.

As Jenkins (1994) and Layton (1991) noted, science and technology can in some cases be inextricably linked. For example, the laws of physics can limit technological innovation, and scientific activity can be constrained by factors such as commercial advantage. However, even in these instances, the purposes of science and technology are different. For the scientist, the purpose is developing a greater understanding of the natural or perhaps the made world. The purpose of a technologist is to intervene in that world and to change it in some way. This means that technological solutions will often be specifically situated, whereas scientific solutions are usually thought to be more generalizable.

Layton (1991) and Gardner (1995) maintained that scientific information needs to be reworked and translated into different forms for use in technology and technological practice. Technology has a knowledge base in its own right and is not subservient to other forms of knowledge. Technological practice also relies on accessing multiple knowledge bases and transforming these for a particular technological purpose. Jenkins (1994) and Gardner (1995) provided numerous examples to indicate that technology cannot be merely understood as applied science. For example, engineering concepts often need to be developed or utilized in the realization of a technological solution. These concepts are often a long way removed from an understanding of the physics theory. Engineers use concepts such as leakage inductance, effective sensible heat ratio, solar heat gain, ergonomics, functionality, and disability glare, in order to design features and enable the use and viability of different materials. Jenkins indicated that the role of scientific knowledge in technological development is frequently overstated or misunderstood. Even in electronics, the role of science has been overstated (Jenkins, 1994, p. 12); for example, the diode valve required no new scientific knowledge for its discovery and although science contributed initially to radio, it made little contribution to the technology from then on.

Much technology was in place before there was a scientific understanding. For example, iron extraction was taking place before an understanding of redox reactions. Gardner (1995) pointed out that in some cases innovations have worked even when the inventor has a faulty scientific theory to account for it (e.g., hot air balloons based on a concept of rising smoke, or heat treating food to get the air out). Consequently, scientific knowledge is not always needed for technological innovation. However, Gardner made the point that the case of the steam engine shows that technology can develop from science, make progress without science, and also contribute to science. From a historical perspective, therefore, Gardner (1995) clearly showed that technology cannot be understood merely as applied science.

De Vries (2001) suggested that histories of industrial research laboratories can offer opportunities for studying the complex relations between science and technology. A good insight of these relations is relevant for shaping sound concepts of science and technology in both science education and technology education. De Vries derived three different interaction patterns from the history of industrial research laboratories (in particular the Philips Natuurkundig Laboratorium), namely, science as an enabler for technology, science as a forerunner of technology, and science as a knowledge resource for technology.

TEACHERS' PERCEPTIONS OF TECHNOLOGY

Teachers' concepts and practices have shown strong links with the initiation and the socialization of teachers into subject subcultural settings (Goodson, 1985). Therefore, teachers have a subjective view of the practice of teaching within their concept of a subject area (Goodson, 1985). This view is often referred to as a subject subculture, and leads to a consensual view about the nature of the subject, the way it should be taught, the role of the teacher, and what might be expected of the student (Paechter, 1991). Because technology is a relatively new curriculum area, teachers' awareness of their own conceptualizations of technology as a learning area is limited (Jones & Carr, 1992). There is a consequent lack of a technology subject subculture, so other subjects' subcultural impact on technological classroom practice becomes very complex. There are a multitude of subcultures impacting on technology education in a variety of ways, depending on the teachers' subject backgrounds, concepts of technology, and their concepts of learning and teaching both within technology and generally. Paechter (1991) also pointed out that teachers' beliefs about what was important for students to learn in their existing subject were transferred to technology education.

Jones and Carr (1992), in a study that looked at teachers' perceptions of technology and technology education, found that all the science teachers who were interviewed saw technology education in terms of applications of science. In terms of teaching, technology was perceived to be a vehicle for teaching science and often something extra to the conceptual development in science. There was concern expressed about nonscience teachers incorporating the scientific aspects of technology into their lessons. All science teachers made comments similar to the following:

> It [technology] would fit best in science . . . it depends what aspects of technology you are going to explore and what for . . . I try it [technology] as an application [of science]. If it is from a sociological point of view it is better to explore it in social studies . . . but from the scientific aspect it would be better through science rather than having social studies teachers trying to teach the science of technology.

At the time of the study in both the elementary and middle school settings, teachers were trying to integrate computers into their classrooms. In the elementary school, there was one computer per class and at the middle school the computers were located in a resource area. Many teachers at the elementary and middle school viewed technology in terms of computers. For these teachers, technology meant using computers or other information technology to solve problems. Although they might have been aware of the scope of technology as a subject, they tended to focus on computing, as for example, stated by one teacher, not using pen or paper but using computers to solve problems. Many teachers saw technology education as students using technology as part of their learning experiences and this was, in the main, using computers: "We use technology across the curriculum. For example, some of the science that we do can be done on the computer or on calculators. So technology is coming into the science program."

Teachers also mentioned problem solving in relation to finding out about the way things work, "Using examples of technology as a way of finding out how things work."

When talking about technology, teachers have mentioned problem solving both in the context of using computers and finding out how things work. Technology is seen as a mechanism for solving a problem or as a vehicle for approaching a particular type of problem solving: finding out how things work.

A group of teachers who did not emphasize computers in their perceptions of technology education emphasized the links between science and technology. Technology was perceived to be closely linked to science, and the teachers were trying to make the science courses more relevant by including technology. Some of these teachers had attended courses that emphasized this approach:

> Technology would be a more practical way of teaching science, more things to aid them, more equipment . . . maybe children would work in a more individual way. . . . Technology is about science. They go together. Technology is the practical awareness or help with science—make things to understand science things. Technology goes along with science. Children are interested in how things work . . . it would involve using various apparatus or going into more depth in their own way. Children in this school have very enquiring minds and want to find out how things work.

Although teachers in elementary schools teach a range of subjects, they appear to take a special interest in particular areas. This is apparent when one group of teachers places an emphasis on computers and technology education, whereas others talk about the close relation between science and technology.

Moreland (1998) reported that although the teachers she worked with stated that they needed to learn more about the teaching of technology, they felt they had enough skills and understanding to be teaching technology and could do it in the classroom. One teacher with a science strength set the students applied science tasks (design a hot balloon after studying flight). Technological principles were not involved. The task criteria were in terms of why things happened and included a narrow focus of outcomes. Northover (1997) noted that all the science teachers she worked with viewed technology as being applied science and technology as skills and skill development. The teachers went for minimal change and added technology into existing programs rather than developing new ones or new learning outcomes. She found that these teachers generally expressed an interest in technology education and commented on the motivational aspects of technological activities. Teachers often saw changes in their perceptions of technology and technology education as a means of better understanding the curriculum document. However, they did not see the importance of developing a coherent technological knowledge base to their own learning and teaching practice. The dominant science subculture in schools proved to be a powerful conservative influence. Teachers who evidenced a changed view of technology and biotechnology at earlier stages throughout the teachers' development oftentimes by the end had reverted back to their initial perspective. In fact, in cases where teachers did make changes to their perceptions initially, the cognitive dissonance set up by the disparity between their views and their practice was often resolved in the end by reverting to a previously held view.

The strategies developed by the teachers in their classrooms when implementing technological activities were often positioned within that particular teacher's teaching and subject subculture. These subcultures are consistent and often strongly held. The subcultures had a direct influence on the way the teachers structured the lessons and developed classroom strategies. Teachers developed strategies to allow for learning outcomes that were often more closely related to their particular subject subculture (e.g., science or language) than to technological outcomes. Teachers entering areas of uncertainty in their planned activities often reverted to their traditional teaching and subject subculture. Their views of assessment, their expectations of the students, and their views of learning influenced possible learning outcomes identified by teachers in technological activities. Teachers often drew on learning areas with which they were comfortable in order to identify possible learning outcomes rather than draw on technological outcomes.

These research studies highlight that, for technological learning outcomes to be seen as desirable for students, it is crucial for teachers to have a clear understanding of both technology and technology education.

TECHNOLOGY IN A NATIONAL SCIENCE CURRICULUM: THE EXAMPLE OF NEW ZEALAND

Technology is mentioned in a number of science curricula around the world. This section looks briefly at the New Zealand curriculum to see how technology is being incorporated into science programs for all students. The term *technology* is employed throughout the New Zealand science curriculum in a number of ways. In the introduction to the science curriculum, science and technology are expressed as if they are interchangeable terms (Bell, Jones, & Carr, 1995). In much of the curriculum, technology is mentioned as a context for teaching or introducing scientific concepts. For example, in the "physical world" strand, students are expected to investigate how everyday technology works in order to explore how physical phenomena are incorporated into these technologies. However, it is not made clear what aspects of technology should be focused on, nor is it stated how technology is in fact being conceptualized.

In the early years of the science curriculum, the primary reason for introducing technology is for the purpose of clarifying and demonstrating a scientific principle. At higher grades, the curriculum focus shifts to a more general investigation of the relation between science and technology. Emphasis is placed on acknowledging and understanding how technological advances have aided or enabled the development, or major rethinking of, scientific ideas (e.g., the way recent laser technologies have impacted on theories of atomic collisions). The way in which scientific principles have provided crucial knowledge for technological development and advance is also highlighted (e.g., the development and uses of genetic finger printing). When there is a focus on learning how technological artifacts function, this is in terms of scientific principles only, with technological and other knowledge bases crucial to the successful functioning of technological artifacts, systems, and environments ignored. Technological knowledges, including technological concepts and principles, are not acknowledged. Rather, the principles behind technological innovation are perceived to be only those belonging to science.

There is some opportunity within this aim to see how technological developments impact on scientific knowledge, and vice versa. This opportunity is constrained to those technologies fitting the applied science notions of technological developments. There is also opportunity for the exploration of the effect of technological development of society. However, it is specifically stated that the means of such an evaluation should be through the application of scientific knowledge.

This curriculum, although attempting to introduce technology into science, has generally focused on technology as applied science and portrayed a very limited view of technology and therefore impeded the potential for learning both science and technology. However, there was a definite attempt to include the theme of science, technology, and society (STS). Fensham (1987) identified 11 dimensions or aspects of STS learning. These are the relation between science and technology; technocratic/democratic decision making; scientists and socioscientific decisions; science/technology and social problems; influence of society on science/technology; social responsibility of scientists; motivation of scientists; scientists and their personal traits; women in science and technology; social nature of scientific knowledge; and characteristics of scientific knowledge (scientific methods, models, classification schemes, tentativeness). The STS movement began due to a combination of factors, including a growing concern during the 1960s that science education had become divorced both from its social origins and from the social implications of scientific endeavor. This was often expressed as the "social relevance of science" (Fensham, 1987, p. 1). There was also a push for science education to become more technology related. For example, Hurd (1991, p. 258) expressed this movement as a result of the "call for the reconceptualisation of science teaching to bring it into harmony with the ethos of modern science and technology." Solomon (1988, p. 379), in fact, stated, "STS has emerged as a discipline with a discernible history and development." Previously in New Zealand, and in other countries, an STS focus has often been an "add on" in the teaching of science. It is important to note at this point that whereas technology as conceptualized within STS is in practice very much aligned to "applied science," this is in direct conflict with early warnings whereby the following was stated:

> Technology, in the context of STS courses, will mean neither the intricacies of microelectronics and mechanisms, nor those "big machines" that are the general public's equation to technology. Our students need to see technology as the application of knowledge, scientific and other, for social purposes. They will have to recognise some of the pressures for innovation (once called the "technological imperative") and its cultural dependence. Technology may be either small scale or large, but it is always the process of producing social changes as well as the result of social needs. (Solomon, 1988, p. 381)

Therefore, it is important when introducing technology that the students are made aware of the interaction of technology and society, rather than just the application of scientific knowledge for a purpose.

USE OF TECHNOLOGICAL CONTEXTS TO TEACH SCIENCE CONCEPTS

Research in science education that explored the use of technological applications for the teaching of science suggests such contexts do have a positive effect on students' learning of scientific principles and concepts (e.g., Jones & Kirk, 1990; Rodrigues, 1993). This research is in keeping with international research findings on the importance of context in student learning (Brown, Collins, & Duguid, 1989; Hennessey, 1993; Lave, 1991; Perkins & Salomon, 1989).

However, care must be taken that the technological context used is appropriate to the students and the scientific content, and it is presented as an integral part of the learning experience rather than an add-on for the sake of sparking interest. For example, Jones and Kirk (1990) qualified their statement regarding how a technological focus enhances the learning of science concepts for most students, by stating the need for the context to be linked to suitable teaching sequences and the context integrated into the lesson sequence rather than being used for illustrative purposes.

Research was carried out by Jones (1988) into the effect of introducing technological applications on students' concepts of physics. Using such applications as earthquake monitoring systems and baby breathing monitors, it was found that the students indicated that these technological applications helped them to remember scientific concepts involved. However, no change was recorded if the applications were used as an add-on either at the beginning or end of a lesson. The students also commented that the use of such technological contexts also provided frameworks for the construction of further scientific concepts to those specifically targeted. Another important outcome from this research was the significant increase in the students' level of confidence, interest, and enjoyment in science generally. This was a factor noted by both the students and their teachers.

Rodrigues (1993) explored the role and effect of context on female students' learning of oxidation and reduction. Using such technological applications as breathalyzers, and hair perming and coloring systems as contexts, Rodrigues found that not only did students become more interested in the scientific concepts of oxidations and reduction, but they also showed a large increase in the number and quality of classroom interactions both with each other and the teacher. These interactions took many forms, including direct questioning and discussion centered on both the functions and use of the application and the scientific concepts involved (Rodrigues, 1993). The researcher's observations were that the students appeared to take "control of their learning" (Rodrigues & Bell, 1995, p. 807) and teacher, student, and researcher statements all appeared to suggest that the students experienced an increased conceptual understanding of redox reactions.

There is, therefore, a significant body of research that supports the use of technological applications in science education. It would appear that student learning in science can be greatly enhanced by using technological applications in order to increase their understanding of scientific concepts and principles, as well as increasing their enjoyment of science generally. An STS approach allows for an understanding to be developed of the impact of applied science technologies on society and an understanding of the relation between society and science.

However, science curricula generally portray a narrow view of technology. Such a narrow view relies on a concept of technology as very much focused on applied science. As has been stated elsewhere (Bell et al., 1995), the treatment of technology as embedded in science is cause for concern because it means that other forms of knowledge, including technological knowledge, which are all essential for technology, are not apparent. It also excludes many technological innovations and developments that have no direct links to science as a discipline.

SOLVING TECHNOLOGICAL PROBLEMS
IN SCIENCE CLASSROOMS

There have been many attempts to introduce technological problem solving in science classrooms. However, extensive classroom observations undertaken in science classrooms when technology problems have been introduced have shown that the science classroom culture and student expectations appeared to strongly influence the way in which students carried out their technological activities (Jones, 1994). The students in the science classrooms involved in this research enjoyed carrying-out technological problem solving and their teachers reported considerable enthusiasm for these activities. Subject subcultures were a major influence on students' expectations of classroom practice, with regard to both themselves and their teacher (Jones, 1997). For example, throughout the technological activity that was situated in a science classroom and timetabled slot, the students played by the "rules" of the science classroom. Their perceptions of the activities they were to be involved in were significantly affected by prior concepts of "project" work in science. The focus throughout the unit was therefore primarily in terms of collecting information to present to the class. They often did not continue with explorations of wider social issues, because they did not see this as relevant to their notions of science. Jones and Carr (1993) and Jones, Mather, and Carr (1995) indicated that the majority of students in the science classroom limited themselves to using science resources even though they had been encouraged by the teacher to use outside resources. The solutions that the students sought were often in terms of traditional solutions utilized in their prior experiences of the science classroom. When questioned, these students clearly stated that they could have done more toward solving their problems, but they consciously limited themselves to what they considered to be appropriate within the science classroom. For example, they made a simplistic model of a circuit rather than considering the factors involved in developing an actual circuit to solve a particular problem. Consumer benefits, costing, and use of other materials were often ignored when they were actually vital to developing an appropriate solution. Students often stated that they learned scientific concepts when undertaking the technological activity, and appeared to view this as the legitimate learning outcome for the activity.

There are numerous examples from the research by Jones et al. (1995) of students having difficulty translating knowledge taught in alternative subject areas to technological problems, including the translation of science concepts into technological solutions within a science classroom. Formal science knowledge needs to be reconstructed, integrated, and contextualized for practical action (Layton, 1991). In applying abstract knowledge, therefore, there needs to be an intermediate translationary step. The trans-

fer of knowledge in a usable form, from one context to another, is a difficult process and one that will need to be taught to students. In fact, it was
found that there was very little evidence of transferring science concepts
to full technological solutions. McCormick (1992) also noted that the student's inability to transfer was an obstacle in technological activities.
Transfer assumes that students have been taught for the understanding of
when and how to use that knowledge.

When technological problem solving is introduced into science classrooms, students are interested, enjoy the experience, and in many cases,
learn some scientific concepts. There is very little evidence of transfer of
scientific knowledge to technological solutions and little understanding of
the processes involved. The technological process adopted by the students
is somewhat fragmented and appropriate solutions are not forthcoming.
The culture of learning in science classrooms does not appear to lend itself
to helping students develop technological capability or technological literacy. The introduction of technological problem solving into science classrooms needs careful consideration if technological literacy is a desired
learning outcome in science. The main conclusion is that present science
classroom cultures need to be understood as greatly affecting performance in technological problem solving.

Research shows that if technology is perceived as simply applied science, then classroom practice tends to ignore economic, social, personal,
and environmental influences, as well as needs and constraints (Gardner,
1995; Jones & Carr, 1993; Jones et al., 1995). This perception will therefore
limit students' learning of technological concepts and practices. Because
of the limiting effect of alternative subject subcultures, teachers must be
very clear of the technological outcomes they are working toward and not
allow these to be overridden by outcomes from other areas. There must
also be an understanding of the subject subcultural effects on both teachers and students if technological outcomes are to be achieved. It is in no
way suggested that technological outcomes and science outcomes cannot
be met alongside each other in practice. However, the difference in the
power of the science subculture over the still emerging technology subculture, especially in a designated science classroom, indicates the need for
extreme caution when attempting to do this.

IMPLICATIONS FOR ELEMENTARY SCIENCE
TEACHER EDUCATION

Technology has a major impact on society, and the technology used in society is a reflection of the values that people hold. Both scientific and technological literacy are important as a part of a general education for all, yet

there is a general lack of understanding of technology in society. Also, within both education and in society, there are a number of misconceptions about technology. This can make it difficult to introduce technological ideas in a preservice science teacher education course, and for elementary teachers to introduce technology education to their classrooms.

Technology can be utilized in a variety of ways in science education, but in doing so it is important to have a clear concept of both the nature of science and the nature of technology. Both need to be clearly and equally addressed in a preservice science and technology course. Too often in the past a restricted view of technology in science has limited prospective teachers' learning of science, and their learning about technology. However, as recent reforms in the area of scientific literacy have included calls for students to have a better understanding of technology, there are pressures to include technology in preservice science courses.

The same is also true in inservice courses for teachers. If practicing teachers are being required to include technological perspectives in their science teaching, then in inservice programs a crucial starting point is exploring the characteristics of science and the characteristics of technology to highlight both the connections and the differences. This has been found to be crucial if a balanced view of technology is to be introduced to students. The next step in these inservice programs is for the focus to move to consider what knowledge and process is unique to technology. This is done to illustrate that technology cannot be considered a subset of science.

Technology cannot be viewed as simply applied science, but technology has a technological knowledge base, technological procedures, tools and techniques, and a strong interrelation with society. If technology is included in an elementary science teacher education course, then it is important that prospective teachers are aware of the potential technology learning outcomes for students, and not limit students' learning about technology to just applied science notions.

When STS approaches are included in preservice science courses, technology is often portrayed as the application of science ideas and their impact on society, rather than showing that technological developments impact on scientific knowledge and vice versa.

The introduction of technological applications as a context to teach science concepts has been shown to be a powerful means for students to enhance their learning. Students also enjoy the science and show more interest in it. This, however, does not happen by accident. Prospective teachers must be shown how to achieve this meaningfully. It is important that technological applications are not included as an add-on, but are used to provide a context for the science.

It cannot be assumed that students will transfer the ideas they learn in science to technological applications. If one of the outcomes for science education is for students to apply their developing knowledge to technological problems and innovation, then prospective teachers have to be taught how to help students do this. This requires a specific focus on linking science knowledge to technological applications in science methods courses. However, a preferred option is for preservice elementary teacher education programs to include methods courses in both science and technology. Whether or not this occurs in part depends on curriculum priorities in both education systems and universities.

Helping elementary teachers introduce technology as a new subject into their curriculum requires teachers to make considerable adjustments in both their knowledge base and teaching practice. For a start, their knowledge of technology is usually very limited, and the addition of a new subject into the curriculum requires major adjustments to curriculum priorities and practices.

CONCLUSIONS

This chapter has shown that students need to develop an effective understanding of technology if they are to gain an understanding of the world in which they live. Technology has often been interpreted as applied science or related to computers in education. The introduction of STS and technological applications can enhance the learning of science concepts and increase students' interest and motivation. However, if technology is taught as a subset or as subservient to science, then this will be detrimental to students' learning a clear understanding of technology. It is important that teachers and students develop an understanding of technology and science as two areas that can interact but are also distinct in nature. Technology is a discipline in its own right (Mitcham, 1994) and is not a subset of other learning areas. For example, technological knowledge is not reducible to science, mathematics, or social studies learning areas. Science must not be seen as a gatekeeper for students undertaking further work in technology, because this will limit students' learning in both fields. If teachers want to develop technological outcomes in their classrooms, then it is important that both teachers and students develop an understanding of technology, technology education, and technological practice. Equally, science teacher educators need to appreciate these issues and plan carefully how they deal with science and technology in methods courses.

REFERENCES

Bell, B., Jones, A., & Carr, M. (1995). The development of the recent national New Zealand science curriculum. *Studies in Science Education, 26,* 73–105.

Brown, J. S., Collins, A., & Duguid, P. (1989). Situated cognition and the culture of learning. *Educational Researcher, 18*(1), 32–42.

Bucciarelli, L. (1996). *Designing engineers.* Cambridge, MA: MIT Press.

Bybee, R. (2000, April). *Achieving technology literacy: A national imperative.* A presentation for a Government Industry Dialogue, the Technological Literacy and Workforce Imperative. Washington, DC.

Cajas, F. (2001). The science/technology interaction: Implications for science literacy. *Journal of Research in Science Teaching, 38,* 715–729.

Cajas, F., & Gallagher, J. (2001). The interdependence of scientific and technological literacy. *Journal of Research in Science Teaching, 38,* 713–714.

De Vries, M. (2001). The history of industrial research laboratories as a resource for teaching about science–technology relationships. *Research in Science Education, 31,* 15–28.

Gardner, P. (1995). The relationship between technology and science: Some historical and philosophical reflections. Part 2. *International Journal Technology and Design Education, 5*(1), 1–33.

Goodson, I. F. (Ed.). (1985). *Social histories of the secondary curriculum.* Lewes: Falmer.

Fensham, P. (1987, December). *Relating science education to technology.* Paper prepared for the UNESCO Regional Workshop, Hamilton, New Zealand.

Hennessey, S. (1993). Situated cognition and cognitive apprenticeship: Implications for classroom learning. *Studies in Science Education, 22,* 1–41.

Hurd, P. (1991). Closing the educational gaps between science, technology and society. *Theory into Practice, 30,* 252–259.

Jenkins, E. (1994). *The relationship between science and technology in the New Zealand Curriculum.* Wellington: Educational Forum.

Jones, A. (1997). Recent research in student learning of technological concepts and processes. *International Journal of Technology and Design Education, 7*(1–2), 83–96.

Jones, A. (1988). *Student perceptions of physics and the introduction of technological applications.* Unpublished doctoral dissertation, University of Waikato, Hamilton, New Zealand.

Jones, A. (1994). Technological problem solving in two science classrooms. *Research in Science Education, 24,* 182–190.

Jones, A. (2001). Developing research in technology education. *Research in Science Education, 31,* 3–14.

Jones, A., & Carr, M. (1992). Teachers' perceptions of technology education—implications for curriculum innovation. *Research in Science Education, 22,* 230–239.

Jones, A., & Carr, M. (1993). *Analysis of student technological capability* (Vol. 2). Working Papers of the Learning in Technology Education Project. Hamilton, New Zealand: Centre for Science and Mathematics Education Research, University of Waikato.

Jones, A., & Kirk, C. (1990). Introducing technological applications into the physics classroom. Help or hindrance to learning? *International Journal of Science Education, 12*(5), 481–490.

Jones, A., Mather, V., & Carr, M. (1995). *Issues in the practice of technology education.* Hamilton, New Zealand: Centre for Science and Mathematics Education Research, University of Waikato.

Kline, R. (1987). Science and engineering theory in the invention and development of the induction motor, 1880–1900. *Technology and Culture, 28*(2), 283–313.

Lave, J. (1991). Situated learning in communities of practice. In L. B. Resnick, J. M. Levine, & S. D. Teasley (Eds.), *Shared cognition: Thinking as social practice, perspectives on socially shared cognition* (pp. 63–82). Washington, DC: American Psychological Association.

Layton, D. (1991). Science education and praxis: The relationship of school science to practical action. *Studies in Science Education, 19,* 43–49.

Layton, D. (1993). *Technology's challenge to science education.* Buckingham, England: Open University Press.

Layton, D. (1994). A school subject in the making? : The search for fundamentals. In D. Layton (Ed.), *Innovation in science and technology education* (Vol. 4, pp. 11–28). Paris: UNESCO.

McCormick, R. (1992). The evolution of current practice in technology education. In R. McCormick, P. Murphy, & M. Harrison (Eds.), *Teaching and learning technology* (pp. 3–14). Wokingham, England: Addison Wesley.

McGinn, R. (1978). What is technology. In P. Durbin (Ed.), *Research in philosophy and technology* (Vol. 1, pp. 179–197). Greenwich, CT: JAI Press.

Mitcham, C. (1994). *Thinking through technology. The path between engineering and philosophy.* Chicago: University of Chicago Press.

Moreland, J. (1998). *Technology education teacher development: The importance of experiences in technological practice.* Unpublished master's thesis, University of Waikato, Hamilton, New Zealand.

Northover, B. (1997). *Teacher development in biotechnology: Teachers' perceptions and practice.* Unpublished master's thesis, University of Waikato, Hamilton, New Zealand.

Pacey, A. (1983). *The culture of technology.* Oxford, England: Basil Blackwell.

Paechter, C. (1991, September). *Subject sub-cultures and the negotiation of open work: Conflict and co-operation in cross-curricular.* Paper presented to St. Hilda's conference, Warwick University, UK.

Perkins, D., & Salomon, G. (1989). Are cognitive skills context bound? *Educational Researcher, 18,* 16–25.

Pinch, T., Hughes, T., & Bijker, W. (1987). *The social construction of technological systems.* London: MIT Press.

Rodrigues, S. (1993). *The role and effect of context on learning form six oxidation and reduction concepts by female students.* Unpublished doctoral dissertation, University of Waikato, Hamilton, New Zealand.

Rodrigues, S., & Bell, B. (1995). Chemically speaking: A description of student–teacher talk during chemistry lessons using and building on students' experiences. *International Journal of Science Education, 17*(6), 797–809.

Solomon, J. (1988). Science technology and society courses: Tools for thinking about social issues. *International Journal of Science Education, 10*(4), 379–397.

Staudenmaier, J. (1985). *Technology's storytellers: Reweaving the human fabric.* Cambridge, MA: MIT Press and the Society for the History of Technology.

Technology for All Americans Project. (2000). *Standards for technological literacy: Content for the study of technology.* Reston, VA: International Technology Education Association.

Wajcman, J. (1991). *Feminism confronts technology.* Sydney, Australia: Allen & Unwin.

Wertsch, J. (1991). *Voices of the mind: A sociocultural approach to mediated action.* Cambridge MA: Harvard University Press.

12

Assessment and Elementary Science Teacher Education

Christine A. Harrison
Kings College, London

Assessment has become a major focus in some educational systems, and is proving to be a challenge for teachers and teacher educators. This chapter examines the issues of assessment in science and in science teacher education. Prospective elementary teachers need to gain insights into important aspects of assessment in school science—especially in the context of high-stakes systemwide testing. Teacher educators also need to be aware of how findings about assessment impact on their preservice and inservice programs. The early discussion in the chapter examines assessment in terms of facilitating learning using science instances, but many aspects also apply to preservice science and methods courses, as well as inservice courses.

ASSESSMENT IN FORMAL LEARNING CONTEXTS

Schools and other educational institutions are places where students go to learn and where teachers create experiences in which learning can take place. Assessment helps teachers and students focus on learning because it provides both a system and a process where both parties can consider and plan for the next steps to take. In this way, assessment plays a vital role in learning because it helps sort out the direction, pace, and intensity of the learning and so guides teachers with their planning and preparation.

Although the main purpose of assessment in schools and similar organizations must be to facilitate and drive learning, assessment plays a much broader and pervasive role. This is because assessment data are also used for accountability issues and for certification. These functions are necessary for evaluation and comparison purposes, and for providing individuals with documentation to take them to further stages of education or employment, but it is essential that these purposes do not hijack the primary purpose of assessment, which is to promote and direct learning. This is difficult to focus on at some stages of the educational process because the need to gain certification or the necessity to show class or school performance in a good light can sidetrack teachers into abandoning assessment for learning in favor of "teaching to the test." In order to understand why this action is not only unnecessary but also counterproductive, first think about how people learn and how teachers can support learning.

How Do Students Learn?

Contemporary understanding on the ways that students learn centers on constructivist ideas (see also chap. 6) where the main premise is that students respond to new information or problems by having to build models in their heads of how the parts fit together and how things work. In essence, they make connections between the new knowledge and the knowledge that they already hold in their heads and so make sense of the new situation. In this way, the new knowledge is retained because it links with an existing knowledge system and the model of understanding grows as more information is added. Memory becomes rather like a filing cabinet in that the filing system enables the learner to know where to look for specific information. As the knowledge grows, filing systems can become outdated and so the learner needs to rebuild or replace systems on occasion and this can be quite a threatening and unsettling time for the learner because they have to reframe their thinking and resort their previous knowledge within the new system that they form. It is evident that transformation of new ideas can only be achieved in light of what a learner already knows and understands, so the reception of new information and ideas depends on the existing knowledge and understanding. So improved understanding comes through analyzing and transforming any new information into the framework formed, which in turn allows learners to revisit and check the model inside their heads.

Research into learning science has shown that learners make sense of the world around them before they are introduced to the science ideas of the classroom and that they resist change to their naïve models of how the world works (Driver, Squires, Rushworth, & Wood-Robinson, 1994). For example, many students believe that a plant feeds by taking in chemicals from the soil rather than through photosynthesis. In some cases, despite this incorrect view of plant nutrition, they can correctly answer questions on a test about the process of photosynthesis, such as naming chlorophyll, writing out the word equation, or describing the starch test on leaves. So teaching needs to do more than deal with surface knowledge; it needs to probe, explore, and develop students' understanding. Moreover, assessment needs to provide the necessary facilitating tools and opportunities. Whereas the starting point in this process lies with the teacher, learners have a complementary role: Unless learners make their thinking explicit to others, they cannot become aware of the need for conceptual modification. Consequently, a teacher's role in the classroom is to set up activities that encourage learners to talk about their current understanding. Once out in the open, these ideas need to be challenged, reshaped, and repackaged and the key to fostering this is flexibility. Good formative assessment responds to the evidence that is revealed through the activities engaged in by learners, by recognizing the current state of understanding and selecting the next steps needed to take learning forward. Formative assessment fashions the next learning experience.

For teachers, formative assessment can be a tricky business because it is reliant on reacting to whatever evidence emerges from the learners. As such, it requires teachers to organize learning opportunities rather than focus on the presentation and management of activities. The ongoing collection and use of evidence to shape the learning experience requires a flexible approach that has the capacity to evolve in different ways. It is much easier to plan and present a series of activities on a specific topic where the aim is coverage of the scientific ideas, rather than planning activities that cater to learners' needs as and when they arise.

FORMATIVE ASSESSMENT TOOLS

If assessment for learning is to function in the classroom, then teachers need to develop and hone their skills. Three important teaching skills in this area are questioning, feedback on specific pieces of work, and supporting peer and self-assessment (Black, Harrison, Lee, Marshall, & Wiliam, 2002; Black & Wiliam, 1998).

Questioning

If teachers want to get below the surface knowledge that learners hold and persuade their students to reveal understanding, then the type of questioning that they use matters. Questions such as "What units do I measure force in?" or "How many chambers are there in the heart?" or "What is the test for hydrogen gas?" seek surface knowledge that the learner can simply recall and often warrant answers that are either right or wrong. Questions that seek understanding, and particularly misunderstanding, force the learner into using higher cognitive skills and, as a consequence, utilize longer answers in trying to demonstrate their state of understanding. Answers are rarely completely right or completely wrong, but have parts that need unraveling and reflecting on to reveal the complexity of their understanding.

An example of such a question would be, "What happens to the electric current as it passes through a bulb?" Many learners would incorrectly explain that the current is "used up by the bulb" and this might prompt the teacher to connect an ammeter into the circuit before and after the bulb to show that the current remains the same after passing through the bulb. This challenge to students' understanding is one step forward, but other learning experiences need to be put in place so that this new information can be put in place and made sense of, in order for it to be sufficiently resilient to move the learning forward. Such activity requires discussion, opportunity to think and reflect, and especially a chance for the learner to ask questions and discuss the emerging ideas with others. Minstrel and van Zee (2003) called this type of work "scientific argumentation," where teachers encourage learners to generate new knowledge from observations, and to trace the reasoning that turns this new information into conclusions and understanding. Such practice begins with elicitation questions, and ideas are then developed through several rounds of class discussion to negotiate a shared understanding of meaning. In this discussion, questions are used to force learners, in making sense of new data, to generate new understanding, and to encourage learners to justify their reasoning. Further questioning might encourage learners to try their new knowledge in different contexts and to develop their metacognitive processing by helping them monitor and become aware of their own learning.

To encourage scientific argumentation, the teacher needs a set of good questions that evoke discussion, a series of activities that challenge misunderstanding or support the development of new understanding, and classroom practices that enable these to function successfully. Considerable work has been done over the last 30 years in documenting many of the typical misconceptions that students hold about science (e.g., Driver, Guesne, & Tiberghien, 1985; Osborne & Freyberg, 1985; Wandersee, Mintzes, & Novak, 1994; Wood, 1998) and this can be used as a starting point to develop searching questions that challenge students' understanding. For example, students' understanding about gravity can be very mixed. Whereas most would respond to the question, "Why does a pencil fall when you drop it?", with the answer, "Because of gravity," their uncertainty about the concept of gravity starts to show when you go on to ask them questions like: "Why does a cloud not fall?" "Why do the Sun and Moon not fall?" "Is there gravity on the Moon?" "What is gravity?"

When Sneider (Sneider & Ohadi, 1998) asked such questions of 11 year olds, she received varied responses, most of which suggested problems with understanding of gravity. Before teaching, she asked them, "Does gravity act in space where there is no air?", and only 13 out of the 48 children correctly answered "yes." To help move the children's ideas forward, the teaching program included activities in which the children observed the curved path of balls rolling off tables, looked at pictures of the trajectories of cannon balls fired from guns, and studied why satellites orbit the Earth. After these experiences, more children answered the gravity–space question correctly, justifying their answer with statements like, "This is why the planets go round the Sun," or "Since the Earth keeps the Moon in orbit, there must be gravity in space." However, only 23 out of the 48 children were able to answer the question correctly after the teaching, which demonstrates how difficult it is to move children's ideas forward and how tenaciously they hold onto their deep-seated ideas about the world around them.

Giving an Answer That "Fits."   Finding questions that seek understanding is not an easy task. The process is not helped by learners who can become adept at satisfying questions with answers that fool teachers into believing they hold similar ideas to the teacher—when in fact they do not. Wiliam (2001) referred to these situations as "fit" rather than "match." For example, a group of my preservice teachers came to my house for dinner. At the time, my daughter was about 8 years old. Being inquisitive prospective teachers, one asked my daughter what she had been studying in school that week. My daughter told them that she had been studying the water cycle and proceeded to draw out for them a perfect replica of the typical water cycle found in the textbooks, correctly spelling the difficult words such as condensation and evaporation. I remember the quiet gasp from my group of preservice teachers, because they were teaching similar ideas to 14-year-olds in secondary schools, and they clearly thought my daughter was a genius. Although I do believe that my daughter is bright, I was pretty sure that the detailed water cycle diagram had been pulled from memory and was not the culmination of understanding of the various processes involved. So to check this, I asked my daughter to explain how the water in her diagram moved from the lake to the clouds. She answered, "The water evaporated," I remember that one of my preservice teachers said, "Wow!" So I asked my daughter, "But how does the water get there? What does water do when it evaporates?" She thought hard and eventually looked me straight in the face, searching for clues, and said quizzingly, "Giant, invisible straws?" Because my daughter's first answer "fitted" with my understanding, I could readily have believed that she had the same understanding of the water cycle as I did. However, her answer to the follow-up question showed that her conceptual understanding did not "match" mine. Indeed, I believe that she had no model in her head to explain how water moved from the lake to the clouds, but desperately tried to find a model to satisfy the situation (and a demanding mother!).

Working with individuals to try and move understanding forward is difficult, but attempting this with a whole class can be problematic unless the teacher utilizes the diversity and richness of ideas in the class to challenge and develop everyone's understanding. In both Sneider's (2003) and Minstrell and van Zee's (2003) work, the management and facilitation of the classroom discussion played an important part in its success. Hearing what peers say can sometimes challenge or consolidate a student's ideas more than what the teacher says. This is partly due to the language that peers use, which allows others easier access to the idea. It is also because students feel that it is possible to challenge a peer's ideas rather than those espoused by a teacher, whom they see as an expert. Students need to be encouraged to articulate their ideas without fear of ridicule from their peers. They also need to be active participants in terms of their learning, and try to think through answers rather than waiting and allowing others in the class to do this for them.

Wait Time. In class question and answer sessions, where the teacher is working with the whole class, "wait time" is important. Wait time is the time between asking a question and taking an answer. Rowe's (1974) study found that the average wait time in elementary science classes in the United States was 0.9 seconds. This is an incredibly small amount of time—barely enough time to recognize what the question is asking, never mind thinking through the answer.

In a more recent project, the wait time of secondary school mathematics and science teachers was typically from 1 to 2 seconds, even after they had been made aware of Rowe's study and had discussed the need to lengthen this time (Black & Harrison, 2001). When the teachers did manage to increase the wait time, it had a profound effect on the classroom discourse, which in turn enabled the teachers to develop their formative assessment practices. The first student response to increased wait time was that more students began to answer the questions, and the answers were generally longer than they had been previously. From an assessment point of view, this means that there is a greater data input, which can help the teacher make clearer judgments about the next learning steps. Also, answers given by students were more often challenged by classmates, and more alternative explanations were offered. This resulted in a more inclusive dialogue in which ideas were debated and negotiated by the class with their teacher, rather than answers simply being validated as correct or incorrect by the teacher's response to answers from individuals. The aim was not for right answers, but for students to reveal their ideas, and then to accept that these ideas might be challenged, refined, or built on by answers from other students.

The teachers who were involved in fostering these changes in their classroom quickly became aware of the need to find questions that support and promote the more open discussions that may result from increased wait time. So starter questions like "What do we mean by friction?", which perhaps only a few students might attempt, evolved into "Some people think that friction is the opposite of slipperiness. What do you think?" This more open question, which asks for opinion on a statement about friction, is much less threatening than the former friction question that seems to ask for a definition. So, children feel capable of attempting an answer, because it is acceptable to answer "yes" or "no" and then give a reason to justify the decision. To encourage students to challenge or negotiate answers, teachers would ask, "How does Ian's answer fit in with your ideas?" or "What could we add to Claire's answer?" or "What do we think of Mary's answer?" The skilled teacher uses these decisions to encourage the individuals in the class to think through and agree on their answer to the question, and at the same time to receive data on the way that several in the class are thinking about the concept of friction and force.

Feedback on Specific Pieces of Work

An essential part of formative assessment is feedback to the learner both to assess their current achievement and to indicate what the next steps in their learning trajectory should be. Butler's (1988) study looked at the type of feedback that students received on their written work. In a controlled experimental study, she set up three different ways of providing feedback to learners—marks, comments, and a combination of marks and comments. The study showed that learning gains were greatest for the comments only group, with the other two treatments showing no gains.

It is easy to understand and to justify comment-only marking practice if the events are observed when students' work is returned to them in classrooms. Their first reaction on getting work back is to compare marks with peers—they rarely read comments. Reading and responding to comments is not encouraged because teachers do not usually give students time to read comments that are written on work and probably few, if any students, return to consider these at home. Further, comments on work are often brief and are not specific (e.g., "Details?"). This makes it difficult for students to understand what is needed to improve a piece of work. Also, the same written comments frequently recur in a student's book, suggesting that students do not respond to the comments and teachers fail to check that students are making the improvements that the comments suggest. In other words, the feedback system does not work. Teachers need to give time to formulating detailed and specific comments that learners can understand and respond to. Teachers also need to make time in lessons for students to read and respond to such comments.

For feedback to be effective, students need to know what they should do to improve their work and should also be encouraged and supported to make the necessary changes. It is not simply a matter of not giving a mark or grade. It centers more on finding the best way to communicate to the learners about what they have achieved and what they need to work on next. It is also about engendering behaviors in the learners that lead them to take action on the feedback, and about providing support systems that foster this approach.

The key to effective feedback is providing sufficient acknowledgment of what has been achieved alongside advice on what next steps to take. Some teachers use a technique called "two stars and a wish" to achieve this in a concise but informative way. So a piece of work is assessed by noting two features that demonstrate the strength of the work or perhaps relate to an area that a specific student is currently targeting, and then adding one aspect that might improve the piece. Learners can then begin to recognize for themselves what they have done well and understand where they need to put their efforts next. It also helps if the comment can be personally directed and written to encourage thinking. Examples might look like:

> James, you have provided clear diagrams and recognised which chemicals are elements and which are compounds. Can you give a general explanation of the difference between elements and compounds?

> Gita, you have identified the anomalous results and commented on the accuracy of your experiment but what can you say about the sample size?

Teachers also need to demonstrate to their learners that improvement is the main purpose of assessment, and opportunities for students to follow up feedback comments should be planned as part of the overall learning process. A particularly valuable way of doing this is to devote some lesson time to redrafting one or two pieces of work, so that emphasis can be put on feedback for improvement within a supportive environment. This can change students' expectations about the purposes of class work and homework, as well as provide polished pieces of work that they can be proud of. Implementation of such practices can change the attitudes of both teachers and students to written work: The assessment of students' work will be seen less as a competitive and summative judgment, and more as a distinctive step in the process of learning.

Although feedback benefits all students, it is particularly important that low achievers receive appropriate feedback, because they usually have greater difficulty than other learners in understanding the concept of quality that the teacher requires. The problem for the teacher is two-fold. First, they need to select the specific improvements that are needed, and second, they must find the language to explain how the leaner might approach the remediation work. However, it is only the learner who can close the gap between the current level of performance and the new level that the teacher sets; so cooperation and a willingness to try is required. For some low achievers, feedback has been a painful experience where their shortcomings have been revealed and they have retired hurt from the learning process. Such learners often exhibit a type of behavior that Dweck (1986) termed "learned helplessness," where their self-esteem has been damaged by the assessment process and they are unwilling to engage in meaningful learning work. Teachers need to handle such learners carefully and find strategies to engage them once again with the learning process. In such a situation, marks seem to aggravate the problem and so this provides another reason for teachers using qualitative, rather than quantitative, data in their feedback to students.

Clearly, there are time implications relating to comment-only marking because good comments take time to write. Teachers need to make decisions about their marking strategies and consider which pieces of work offer opportunity for students to reveal understanding, because it is these that are worth the time spent on careful consideration and marking by comment. Other pieces that perhaps are checking terminology, consist of a few recall questions, or are preparation for class work can be dealt with in other ways (e.g., self-checking or peer checking). In this way, teachers can create the time they need to give useful feedback to their students about their work. There will also be occasions when a good proportion of the class is having similar problems with a piece of work and, in this instance, oral feedback and discussion with the class is possibly a more appropriate strategy. It is also worth considering giving oral feedback to individuals about their work. Whereas such an approach means that further discussion can arise that will aid the teacher in making judgments and support the learner in taking the next steps, it can be time consuming if a teacher tries to do this regularly to each member of a class. Effective teachers take decisions about when, to whom, and how to provide feedback, and balance the needs of individuals with those of the whole class.

Peer Assessment and Self-Assessment

Learning has to be done by students—it cannot be done for them. The teacher provides guidance, but if students are to learn with understanding, then they have to be able to guide their own efforts. This means they need to understand the goal (i.e., they should know what they are meant to achieve and have some idea of what counts as good achievement). They can then assess where they stand in relation to the goal and so orient their efforts to move toward achieving it. They can be helped to do this by teachers who choose and explain feasible goals, who convey the criteria for achievement, and who can then guide their progress. A prior condition here is that teachers, in addition to having elicited evidence about where students are in relation to the learning goals, can be clear about those goals, can adapt them to the capacities of the students, and can convey them clearly.

There is good evidence that, as students are helped to assess themselves, they learn more effectively. Students can also help one another in such ways of working. Their communication with one another can use shared language forms, the achievements of some can strengthen the credibility of the exercise to others still struggling, and the feedback of a group to a teacher can command more attention than that of an individual. Thus, both self-assessment and peer assessment are important components of assessment for learning. Peer assessment allows students to begin to judge their own work by recognizing similar or different qualities in the work of their peers. For example, students' learning can be enriched by marking their own or another's work because this helps students develop the meaning and sense of the quality criteria in a given piece of work. This will then enable them to seek such quality within their own work. It is not only low attaining students seeing what more able students can do that benefit here. In a creative writing class that I was observing, where students were peer assessing one another's work, the most able student in the class said that she liked to do peer work because "By reading my partner's work, I now realise what I do in my own writing that makes it work. Before, I knew that I was good at English, but now I know why I am good and what I should try and do more of." So, formative assessment is not simply remediation, it is helping learners recognize their strengths and weaknesses and then working on both of these aspects. It is also uniquely valuable because students may accept, from one another, criticisms of their work that they would not take seriously if made by their teacher (Black et al., 2002); and they are in a language and form that students themselves would naturally use and can readily comprehend. In practice, peer assessment turns out to be an important complement to self-assessment and a means by which the necessary skills can be developed.

Self-assessment will only happen if teachers help their students, particularly the low achievers, to develop the skill. This can take time and practice. One simple and effective idea is for students to use "traffic light" icons, labeling their work green, amber, or red according to their confidence levels with specific pieces of work. These labels serve as a simple means of communication of students' self-assessments. The teacher can then reorganize the working groups so that the students that are greens and ambers can pair up to deal with problems between them, whereas the red students can be helped by the teacher to deal with their deeper problems. For such peer group work to succeed, many students will need guidance about how to behave in groups (e.g., in listening to one another and taking turns), and how to judge quality (e.g., familiarizing themselves with criteria). Techniques, such as the "two stars and a wish" approach described in the feedback section, can also be used in developing these behaviors, particularly as they can be used to explore how the criteria for quality can be achieved within different student's work. Again, peer assessment through collaborative group work provides the vehicle through which self-assessment practices can develop.

An essential aspect is to encourage an emphasis on, and a progressive move on the part of the student, toward improvement, so that the self-assessment skills they learn within one activity can be modified and utilized in a later one. This allows students to build their short-term goals into medium and eventually long-term goals. To ensure this, students should be encouraged to focus on the aims of their work, assess their own progress, and improve their work to meet these aims as they proceed. They will then be able to guide their own work, and so become independent self-regulated learners. What is clear is that peer assessment and self-assessment not only make unique contributions to the development of students' learning, but they secure aims that cannot be achieved in any other way.

TESTS

Although the focus on assessment so far has centered on fashioning classroom practice to develop formative assessment practices, tests and quizzes have a role too. Tests provide a stopping point where the teacher and learners can check on knowledge gains. They allow a measure to be applied to the learning and this is sometimes needed for accountability issues or for contribution toward certification. However, the aftermath of tests can also be an occasion for formative work. Peer marking of test papers can be helpful, as the range of acceptable answers can be studied. It is also particularly useful if pupils are required to first formulate a marking scheme, an exercise that focuses attention on criteria of quality relevant to their productions. After peer marking, teachers can reserve their time for discussion of the questions that give particular difficulty; those problems encountered by only a minority might be tackled by peer tutoring.

A further idea has been introduced by research studies (Foos, Mora, & Tkacz, 1994; King, 1992), which have shown that pupils trained to prepare for examinations by generating and then answering their own questions outperformed comparable groups who prepared in conventional ways. Preparation of test questions calls for, and so develops, an overview of the topic so that students come to understand the essential components for conceptual understanding.

These developments challenge common expectations. Some have argued that formative and summative assessments are so different in their purpose that they have to be kept apart, and such arguments are strengthened by experience of the harmful influence that narrow "high stakes" summative tests can have on teaching. However, it is unrealistic to expect teachers and students to practice such separation, so the challenge is to achieve a more positive relation between the two.

Teachers need to use tests sparingly to maintain the emphasis on assessment for learning rather than on assessment of learning. This is especially so in countries like England, where testing by external agencies now occurs at ages 7, 11, 14, 16, 17, and 18. With high stakes testing occurring so frequently in the school lives of the learners, it is quite difficult to escape the tyranny of the testing regime and gain a good balance between formative assessment, which promotes learning, and summative assessments, which encourage "teaching to the test." There are many well-intentioned but misguided people, particularly politicians, who believe that testing is supposed to raise standards of achievement. However, there is also much anecdotal evidence that high stakes testing damages many students' motivation for learning; pupils are finding school stressful and feel the pressure of constant testing. In order to find evidence about these very different views of the effects of testing and to seek firm evidence as to the impact of testing on motivation, a systematic review of research on these matters was carried out by Harlen and Deakin Crick (2002). The following are the most pertinent findings that evolved from this review:

- High stakes tests can become the rationale for all that is done in classrooms, permeating teacher-initiated assessment interactions.
- When passing tests is high stakes, teachers adopt a teaching style that emphasizes transmission teaching of knowledge, thereby favoring those students who prefer to learn in this way and disadvantaging and lowering the self-esteem of those who prefer more active and creative learning experiences.

- Lower attaining students are doubly disadvantaged by tests. Being labeled as failures has an impact on how they feel about their ability to learn, lowers further their already low self-esteem, and reduces the chance of future effort and success.
- Repeated practice tests reinforce the low self-image of the lower attaining students.
- Tests can influence teachers' classroom assessment and tend to squeeze out formative assessment. Pupils interpret classroom assessment as purely summative, regardless of the teacher's intentions.

SUMMARY

The main function of assessment is to foster learning. Formative assessment is the ongoing feedback that teachers give to learners during the learning process. Its intention is not to measure or grade or level, but to inform, support, and develop the learning. Such strategies engage teachers and students in evolving ways of evaluating understanding as it develops, and so provides the impetus for taking the next steps in the learning process. It requires teachers to actively seek student understanding and to respond to learner's needs. It also involves helping the learner develop self-assessment strategies so that they become less reliant on their teacher and become self-regulated learners. Such behaviors require teachers to plan carefully the learning environment and experiences of their classes so that collaborative learning thrives and the common aim is for improvement.

Perhaps one way to think of how to develop effective formative practice is to think of it as a coaching role. If a swimming coach wants to improve the quality of a student's performance, then the way forward is not to provide more competitions for the swimmer to take part in. Rather, the coach diagnoses the areas that require work and improvement. So the swimmer's breathing techniques or leg strength or arm action might need to be improved. Only when the coach feels confident that improvements have been made in one area while maintaining and perhaps enhancing others, will the coach allow the swimmer to race again. It is the judicial balance between formative assessment for learning and summative testing for measurement that will improve the standard of performance.

IMPLICATIONS FOR ELEMENTARY SCIENCE TEACHER EDUCATION

The foregoing discussion has particular implications for tertiary teacher education, especially regarding the dilemma of formative assessment for learning, and accreditation to grade performance in a course. There are also specific implications for three main areas of elementary science teacher education: science courses as part of the teacher preparation program, prospective elementary teachers developing appropriate assessment practices as part of their teaching repertoire, and teacher professional development—helping those who have been teaching awhile review and change their practice as necessary.

Formative Assessment Versus Accreditation

Many tertiary institutions follow a traditional pattern of semester courses taught via a combination of lectures, tutorials, and workshops. At the end of the semester, each student must be given a result, usually a letter grade—sometimes a percentage—which determines whether the student has "passed" the course to be accredited toward an award. There is often the requirement that more than one summative assessment task be used to determine the student's grade, but to avoid overassessment, the number of assessment tasks may have an upper limit, such as three in a 14-week semester. Appeal procedures usually allow a student to challenge a result that they feel does not reflect their learning, so professors need to retain records of marking systems and criteria. Frequently, there is also a requirement that formative assessment feedback be provided to students so they can gauge their progress (usually interpreted as an interim grade); some institutions even demand that all formal assessment tasks be graded using the official grading system. Comments are seen as an explanation as to why a better result was not given.

Within such a context, a culture may develop among students and professors that works against the principles of assessment for learning outlined earlier: All assessment is seen as summative, with perhaps some formative feedback provided in pieces of work submitted early in semester. Students consider tasks that do not contribute to a grade or result as optional, and many do not do them. The emphasis is on "What do I have to do to pass?", rather than "What do I need to do to learn?" This has become exacerbated where standards have been introduced for teacher accreditation, assessed using testing regimes.

Fortunately, many professors have found ways of working within such restrictive systems, usually in preservice methods courses. The descriptions of practice in, for example, chapters 8, 13, and 14 are illustrations. Useful strategies include peer and self-assessment, multiple submission opportunities, project and problem-based learning, and portfolios. An overriding consideration for methods professors is that they should model, where possible, assessment practices that prospective teachers should use in their own classrooms—not always easy when there is an institutional demand that each assessed item must include a result.

Science Courses

Many teacher education programs include required science courses taught by science faculty. Unfortunately, professors in science faculties are not always aware of issues such as assessment for learning, and use traditional forms of assessment in science, where testing tends to dominate. Prospective teachers consequently have a traditional view of assessment in science reinforced, making the task of their acquiring different assessment priorities and strategies more difficult.

It is essential that education faculty liase with their science colleagues so they are aware of such dilemmas, and then motivate them to develop alternative approaches to assessment. Science courses that are team taught by both science and education professors, or by an education professor are preferable, but such arrangements are institution dependent. The latter arrangements also allow for pedagogy to be dealt with alongside science content, providing a more meaningful learning experience for the prospective teachers (see chap. 3).

Developing Appropriate Assessment Practices

If prospective teachers of elementary science are to develop assessment strategies in line with those discussed here, then these strategies need to be consistently explicit in their courses. That is, such assessment practices need to be included in the content of general pedagogy and methods courses, as well as in the practice within those courses and within the practicum—not just in an isolated science methods course. To achieve this, a high level of coordination within a program is necessary. The desire for prospective teachers to see effective assessment practices during their practicum is of particular concern, because many supervising/mentor teachers remain unclear about assessment to enhance learning, and may even be constrained by high stakes testing regimes to teach to the test. This highlights the need for ongoing teacher professional development.

Teacher Professional Development

For a teacher to change assessment practices, some major shifts in views of learning and teaching, and assessment, may be needed. This is not an easy exercise for most, and it requires more than an intellectual awareness of the need to change—especially if there are external pressures from testing regimes. Teachers need to see that proposed changes have benefits in terms of student learning (and perhaps in test performance), and are both possible and manageable in their classroom. Further, many require support when trying to make changes. A mentor, or critical friend working in the classroom with them, can make a considerable difference.

CONCLUSIONS

The perspectives about assessment discussed in this chapter present a departure from traditional practice in many schools and universities, and illustrate how current trends toward external testing regimes can be counterproductive to student learning—despite the avowed intention that they are supposed to improve learning. It is essential that elementary science teachers and university professors are aware of these assessment issues, can use them in their own teaching, and are passionate about communicating their ideas to others. More importantly, when students, parents, and politicians see that student learning can be enhanced by effective teaching and assessment practices without extensive testing regimes, there is hope that a trend away from widespread testing toward formative assessment for learning will develop.

REFERENCES

Black, P., & Harrison, C. (2001). Feedback in questioning and marking: The science teacher's role in formative assessment. *School Science Review, 83*, 43–49.

Black, P., Harrison, C., Lee, C., Marshall, B., & Wiliam, D. (2002). *Working inside the black box: Assessment for learning in the classroom*. London:: King's College.

Black, P., & Wiliam, D. (1998). Assessment and classroom learning. *Assessment in Education, 5*(1), 7–71.

Butler, R. (1988). Enhancing and undermining intrinsic motivation: The effects of task-involving and ego-involving evaluation on interest and performance. *British Journal of Educational Psychology, 58*, 1–14.

Driver, R., Guesne, E., & Tiberghien, A. (Eds.). (1985). *Children's ideas in science*. Milton Keyne, England: Open University Press.

Driver, R., Squires, A., Rushworth, P., & Wood-Robinson, V. (1994). *Making sense of secondary science: Research into children's ideas*. London: Routledge.

Dweck, C. S. (1986). Motivational processes affecting learning [Special issue: Psychological science and education]. *American Psychologist, 41*(10), 1040–1048.

Foos, P. W., Mora, J. J., & Tkacz, S. (1994). Student study techniques and the generation effect. *Journal of Educational Psychology, 86*(4), 567–576.

Harlen, W., & Deakin Crick, R. (2002). *A systematic review of the impact of summative assessment and tests on students' motivation for learning* [EPPI-Centre Review]. Research Evidence in Education Library. Issue 1. London: EPPI-Centre, Social Science Research Unit, Institute of Education.

King, A. (1992). Comparison of self-questioning, summarizing, and note taking-review as strategies for learning from lectures. *American Educational Research Journal, 29*, 303–323.

Minstrell, J., & van Zee, E. H. (2003). Using questioning to assess and foster student thinking. In J. M. Atkin & J. E. Coffey (Eds.), *Everyday assessment in the science classroom* (pp. 61–74). Arlington, VA: National Science Teachers Association Press.

Osborne, R., & Freyberg, P. (Eds.). (1985). *Learning in science: The implications of children's science*. Auckland, New Zealand: Heinemann.

Rowe, M. B. (1974). Relation of wait-time and rewards to the development of language, logic, and fate control: Part II—Rewards. *Journal of Research in Science Teaching, 11*, 291–308.

Sneider, C., & Ohadi, M. M. (1998). Unravelling students' misconceptions about the earth's shape and gravity. *Science Education, 67*, 205–221.

Wandersee, J. H., Mintzes, J. J., & Novak, J. D. (1994). Research on alternative conceptions in science. In D. L. Gabel (Ed.), *Handbook of research on science teaching and learning* (pp. 177–210). New York: Macmillan.

Wiliam, D. (2001). *Level best? Levels of attainment in national curriculum achievement*. London: Association of Teachers and Lecturers.

Wood, D. (1998). *How children think and learn: The social contexts of cognitive development* (2nd ed.). Oxford, England: Oxford University Press.

III

CONTEXTUALIZED PRACTICE AND PROFESSIONAL DEVELOPMENT

The chapters in this part examine aspects of elementary science teacher professional development, with particular foci on preparing prospective teachers, and issues related to educational contexts for professional development. In chapter 13, van Zee discusses her approach to teaching prospective elementary science teachers about and through inquiry, using a multi-pronged view of inquiry investigated through a "teacher as researcher" approach. She explains how her students inquire into natural phenomena, their own learning, children's learning related to their own teaching, and the methods instructor's teaching practices. Akerson and Roth McDuffie also argue in chapter 14 that using a "teacher as researcher" approach is a means of teaching prospective elementary science teachers about inquiry. However, their focus is more on the inquiry process itself, in the form of educational research—action research into their own trials of a selected aspect of pedagogy. These two chapters provide similar yet contrasting views of ways of incorporating the "teacher as researcher" approach into elementary science methods courses.

In chapter 15, McGinnis explores the emergence of research into the culture of schooling, and looks at how the intersecting cultures of science, the classroom, and the school influence a neophyte teacher and shape her perspectives of herself as a teacher of elementary science. The influence of various aspects of culture on prospective elementary science teachers is explored in chapter 16 through the use of cases in a preservice course. Bryan and Tippins describe two examples of cases that provide opportunities for prospective teachers to explore different perspectives and influences on science teaching. In chapter 17, Jones and Edmunds examine the role of specialist science teachers in elementary schools, raising issues about generalist teachers teaching science, teacher professional development for generalists wishing to become specialists, and teacher preparation for prospective teachers on a science specialist career track. The section concludes with a perspective on future directions in elementary science teacher education.

13

Teaching "Science Teaching" Through Inquiry

Emily H. van Zee
University of Maryland

What does it mean to teach science through inquiry? If teachers should teach science through inquiry, and if they should first experience the ways they are expected to teach, then what might it mean to teach "science teaching" through inquiry? These questions have emerged from my experiences during the past decade in designing activities and assignments in courses for prospective teachers (van Zee, 1998a, 1998b, 2000; van Zee & Roberts, 2001; van Zee, Lay, & Roberts, 2003). I write from the perspective of an instructor inquiring into my own teaching practices (Cochran-Smith & Lytle, 1993; Hubbard & Power, 1993, 1999; Shulman, 1998). This chapter reflects on the meaning of teaching through inquiry and ways in which such teaching can form the basis for courses on methods of teaching science in elementary school.

WHAT DOES IT MEAN TO TEACH SCIENCE THROUGH INQUIRY?

There are many views of teaching science through inquiry. The following sections summarize studies presented in the literature, my country's recommendations, and my own perspective.

Studies of Inquiry Approaches to Learning and Teaching

Researchers have examined inquiry-based science instruction from many theoretical perspectives (Keys & Bryan, 2001; Minstrell & van Zee, 2000). Dewey (1900) and Schwab (1962) were early advocates for inquiry approaches to science learning and teaching. Exemplary inquiry practices have been described in journals (Flick, 2000; Hammer, 1995; Keys & Kennedy, 1999; Roth & Bowen, 1995) and books (Layman, Ochoa, & Heikkinen, 1996; National Research Council, 2000; Tippins, Koballa, & Payne, 2002). Settings include college science courses (McDermott, 1990; Smith & Anderson, 1999), teaching methods courses (Gess-Newsome, 2002), professional development (Erickson, 1991; Luft, 2001), museums (Allen, 1997), Web sites (Cennamo & Eriksson, 2001), families (Barton et al., 2001), as well as K–12 classrooms.

Research indicates that inquiry teaching can produce positive results (Anderson, 2002; Chang & Mao, 1999), including with special education students (Mastropieri, Scruggs, & Boon, 2001). Major issues include understanding factors that help or hinder implementation of inquiry-based practices (Appleton & Kindt, 2002; Brickhouse & Bodner, 1992; Crawford, 1999; Lederman & Niess, 1998; Tobin & McRobbie, 1996; Windschitl, 2003), making science accessible to all students (Fradd & Lee, 1999; B. White & Frederiksen, 1998; Yerrick, 2000), utilizing technology (Edelson, 2001; Songer, Lee, & Kam, 2002), modifying curricula (Huber & Moore, 2001), and assessing progress (Briscoe & Wells, 2002). Also critical is understanding how to better engage students in developing scientific arguments and explanations (Driver, Newton, & Osborne, 2000; Herrenkohl, Palinscar, DeWater, & Kawasaki, 1999; Hogan, 1999; Kuhn, 1993; Polman & Pea, 2001).

Experienced teachers have contributed insightful analyses of their own inquiry-based science teaching practices in journal articles (Iwasyk, 1997; Roberts, 1999; Simpson, 1997; Scott, 1994), book chapters (Saul et al., 1993; Saul & Reardon, 1996), and books. Doris (1991), for example, presented transcripts of conversations, copies of students' writings and drawings, and her own reflections on her students' investigations of their world. Gallas (1995) reported ways in which she engaged children in "science talks" and built curriculum around their questions and theories. Pearce (1999) described the many ways he nurtured inquiry in his classroom. P. Whitin and D. J. Whitin (1996) discussed the wonders that children generated while watching birds.

Interest in creating communities of inquiry (Wells, 1993, 2001) extends well beyond science education. McGilly (1994), for example, described "communities of learners" where "students engage in a continual, largely self-directed process of inquiry and knowledge building and share their knowledge with other members of the 'learning community' to which they belong" (p. 18).

My Country's Recommendations

In teaching methods of teaching science in elementary school, I emphasize recommendations published by my country's National Research Council (NRC; 1996) in the *National Science Education Standards*. This document articulates several aspects of inquiry: what students should be able to *do* as scientific inquirers (inquiry as a capability to be developed), what students should know *about* scientific inquiry (inquiry as subject matter content to be understood), how teachers can *teach* (inquiry as an instructional approach), and how teachers can *learn* (inquiry as a way to generate knowledge about science learning and teaching).

The Science Content Standards include both the abilities necessary to do scientific inquiry and understandings about such inquiry. Students in grades K–4, for example, should develop the ability to "ask a question about objects, organisms, and events in the environment, plan and conduct a simple investigation, employ simple equipment and tools to gather data and extend the senses, use data to construct a reasonable explanation, and communicate investigations and explanations" (NRC, 1996, p. 122). They also should develop understandings about scientific inquiry such as "scientists develop explanations using observations (evidence) and what they already know about the world (scientific knowledge)" (NRC, 1996, p. 123).

Teaching Standard A requires that "teachers of science plan an inquiry-based program for their students" (NRC, 1996, p. 30). An elaboration of this standard states that "inquiry into authentic questions generated from student experiences is the central strategy for teaching science" (p. 31). Many examples of inquiry-oriented classrooms are provided in a second document, *Inquiry and the National Science Education Standards* (NRC, 2000).

In addition to standards for science teaching, the U.S. National Research Council (1996) articulated standards for professional development. These standards refer both to the learning of science content and to the generation of knowledge about science and science teaching. Professional Development Standard A states: "The professional development of teachers of science requires learning science content through the perspective and methods of inquiry" (p. 59). Professional Development Standard D states that "professional development for teachers of science requires building understanding and ability for lifelong learning. Professional development activities must . . . provide opportunities to learn and use the skills of research to generate new knowledge about science and the teaching and learning of science" (p. 68).

These recommendations suggest to me that I should engage the prospective teachers in my courses in learning both science content and science pedagogy through inquiry.

My Perspective

Inquiry approaches to learning and teaching seem to lie along a continuum of learning environments that vary in learner self-direction (see Fig. 13.1). At one end of the continuum is direct instruction, in which the

| Direct Instruction | Inquiry Approaches to Teaching and Learning | Learning during Everyday Life |

Teacher decides:	Teacher and students decide:	Learner decides:
Topics	Topics	Topics
Tasks	5Es: Engage, Explore	Tasks
Texts	Explain, Extend	Resources
Tests	Evaluate	Self-assessments

FIG. 13.1.   Continuum of learning environments.

teacher decides what topics to teach, tasks to require, texts to use, and tests to administer to measure learning. At the other end is learning during everyday life, in which the learner decides what topics to explore, tasks to undertake, resources to access, and self-assessments to make to monitor learning. Between these two extremes lie various approaches to inquiry learning and teaching, with the degree of directiveness depending on the students, the teacher's intent, and the situation. I think of myself as shifting back and forth along this continuum in many ways during each class and envision the prospective teachers doing the same as they teach. Within this framework, I try to provide them with many opportunities to generate and explore their own questions. My goal is to help the prospective teachers to build their capacities for engaging themselves, students, colleagues, friends, and families in sustained inquiries throughout life.

As indicated in Fig. 13.1, I design inquiry-based instruction around the well-known strategy called the five E's: engage, explore, explain, extend, and evaluate. These seem to be useful guidelines for organizing instruction, sometimes reiteratively within a session, sometimes spread over several class meetings, with evaluation ongoing rather than relegated to the end. Derived from earlier reform efforts (Bybee, 1997), the 5E's have become the instructional model recommended by the county in which we place the prospective teachers for field experiences. The county also has developed a kit curriculum based on many of the curricula developed with National Science Foundation support by the Educational Development Center, Lawrence Hall of Science, and National Science Resources Center. Thus, I am fortunate that the prospective teachers in my courses are likely to find reform-based curricula in place in their mentor teachers' classrooms, and therefore available for them to use when I ask them to prepare and teach a lesson drawn from a science curriculum (see Fig. 13.2).

In both research and teaching, I have emphasized engaging students in talking together about what they think. My reading assignments include some of the case studies of student and teacher questioning that several teachers and I collaborated in developing (Iwaysk, 1997; Kurose, 2000; van Zee, Iwasyk, Kurose, Simpson, & Wild, 2001). In addition, I am beginning

Journal #6

Unfortunately I have not had very many opportunities to observe science in my mentor teacher's classroom because it is not a subject that is given much attention. However, my mentor teacher was excited that I offered to teach the next Science Unit; Rocks, Sand, and Soil, to her first grade class. She had me begin my lesson that same day, which was Monday, April 8, 2002.

I began the first lesson by informing the students that we had a new unit to begin studying for Science. I told the students that it was Rocks, Sand, and Soil. I then asked the students what they thought we would be learning about in this unit and as a class we made a list. The list is as follows: worms, rocks can break, sand (soft and hard), plants, building with rocks, building with sand. After (it) seemed the students had brainstormed all the possibilities, I began taking out materials from the Science Kit that we would be using and asking for the students to add on to the list. The remainder of the list was: magnifying glass to look closer, jewelry, funnel (used for sand), digging, boxes for worms, collect samples, reading about rocks, sand and soil, clay, seeds, measure and weigh, trays to collect stuff, sandpaper, and strainer.

The students were so excited and were trying to peek into the Science Kit to get a glimpse of what else I would pull out. I played on the student's enthusiasm and built up the mystery by keeping the box closed and slowly pulling out the materials. Afterwards, we went outside as a class and the students were given the task of finding three rocks that were completely different from each other. We took about ten minutes to do so. We sat outside and shared our rocks, describing the texture, color, size, and shape of the rocks. We then tried to find rocks that would fit into specific boxes (we only had two different sizes because I did not have time to prepare.) The students enjoyed the activity and seemed motivated. I was impressed at how well the students were able to describe their rocks to me. I asked the students if they wanted to continue studying rocks and describing them and all of the children were enthusiastic.

The factors that fostered science learning were that it was: hands on, learning by doing, enthusiastic teacher, working in groups, accomplishing a goal, getting out of the classroom, observation, tangible end product, and brainstorming. I feel that my students learned a great deal during this assignment and had fun learning science. I am excited to continue this unit with the students and hopefully my next lesson during the full week in the schools will go just as well.

FIG. 13.2.   Prospective teacher's journal about using the science curriculum in her placement classroom.

to use case studies of student inquiry in physical science that my colleague, David Hammer, and I are developing in collaboration with elementary and middle school teachers, with support from the National Science Foundation.

WHAT MIGHT IT MEAN TO TEACH "SCIENCE TEACHING" THROUGH INQUIRY?

As a new instructor for a course on methods of teaching science for prospective elementary school teachers, I used "teacher as researcher" as the guiding metaphor in the design of activities and assignments (van Zee,

1998a, 1998b). My vision of teachers as researchers was based on my experiences as a participant in a research site created by a high school physics teacher, Jim Minstrell (Minstrell, 1989; van Zee & Minstrell, 1997a, 1997b). With funding from various sources, Minstrell was able to continue teaching physics while exploring physics learning with the help of his students, his colleagues, and interested university researchers. I also participated in an experimental teacher education program at the University of California at Berkeley where prospective science and mathematics teachers were welcomed into ongoing educational research programs at the inception of their preparation to teach (Lowery, Schoenfeld, & B. White, 1990) and where I learned various ways of probing understanding (R. White & Gunstone, 1992). In addition, I learned to listen closely to students while assisting in a coordinated program of physics education research, curriculum development, and instruction (McDermott, 1990, 1996). I also drew on my earlier experiences teaching science in middle school.

My course on methods of teaching science in elementary school meets for 2 hours once a week. The prospective teachers spend 2 days each week in their placement classrooms and 2 days on campus in courses on methods of teaching language arts, mathematics, reading, science, and social studies. They also spend 2 full weeks in the schools, one near the beginning and the other near the end of the semester. In this chapter, I have drawn examples from the spring 2002 class ($n = 23$; mostly middle-class, White females of college age). Four levels of inquiry occur in this course. The first level involves the prospective teachers in inquiring into natural phenomena. The second involves them in inquiring into their own learning. The third involves them in inquiring into children's science learning and their own teaching. The fourth involves my inquiring with them into my own teaching practices.

Level 1: Prospective Teachers Inquire into Natural Phenomena

Teaching science through inquiry is challenging, particularly if one has not experienced this approach to instruction oneself. Therefore, I devote a portion of my course to engaging the prospective teachers in a sustained inquiry into natural phenomena. Like many other instructors, I choose to engage prospective teachers in inquiries about the sun and the moon (Abell, George, & Martini, 2002; Duckworth, 1987; McDermott, 1996). This is a readily accessible context for prospective teachers to experience what it means to develop explanatory models. Some have studied the phases of the moon in college astronomy courses taught by lecture, at the direct instruction end of the continuum shown in Fig. 13.1. Presumably all have learned about the moon at the other end of the continuum by noticing the sun and the moon in the sky during their everyday lives. I call this assignment a "Conversation about the Sun and the Moon" to emphasize the importance of ongoing discussions in learning science. I also provide case studies written by teachers who have engaged their students in learning about the sun and the moon through inquiry (Iwasyk, 1997; Kurose, 2000; Roberts, 1999).

We begin in an open-ended way, close to learning during everyday life on the continuum shown in Fig. 13.1. I invite the prospective teachers to "Look at the sky daily. Enjoy what you see. If you happen to see the moon, record your observations. If you do not see the moon, record that too." After several weeks of such informal observing, I ask the prospective teachers to design explorations based on questions they generate about what they have been seeing. For example, one group of students wondered, "Will we always see the same orientation of the spots on the moon?" Another group asked, "What pattern would we be able to see by looking at the moon at the same time every night?"

After many weeks of watching the sun and the moon, we summarize the patterns observed and develop a model to explain these. To do this, we go outside on a day when the moon is visible (or, if necessary, work with a light representing the sun in a darkened room). I invite the prospective teachers to hold up ping-pong balls so that they match the shape of the sunlit portions of the balls in their hands to the shape of the sunlit portion of the moon in the sky. Then I challenge them to move their balls to reproduce with the balls the changing shapes of the moon that they have been seeing in the sky. This is close to the direct instruction end of the continuum in Fig. 13.1 as I lead the prospective teachers through the details of the logic of this explanatory model. Yet, there is opportunity for discussion and making sense, in terms of this model, of the questions they generated and have been exploring in the small groups.

We extend the model by considering where the earth is heading if one is looking at a third quarter moon (toward the moon's position in space, which lies approximately in the orbit of the earth around the sun) and by estimating how soon the earth will get there (in about 3½ hours). There are many ways to work this problem if one assumes a circular orbit for the earth around the sun and knows rough estimates of the distance between the sun and the earth and between the earth and the moon. This provides a challenging example of how to integrate science and mathematics for older students.

Throughout this sustained inquiry, the prospective teachers evaluate their progress by completing weekly handouts and by writing a paper summarizing their understanding of the moon's changing phases. I provide explicit directions for the paper but leave many opportunities for them to make their own choices. For example, I direct them to make connections to issues raised in the readings but invite them to discuss which-ever issues they find most interesting.

Level 2: Prospective Teachers Inquire into Their Own Learning

For many prospective teachers in my courses, the sustained inquiry into the causes of the moon's changing phases seems to be the first time they have been asked to formulate questions, to design ways to collect relevant data, and to make sense of their findings. Some become deeply engaged in this process by seeking information from books and the Internet and by involving family and friends, as well as classmates, in their data gathering and discussions. Thus, watching the moon provides a context within which they can watch themselves learning science.

On the first day of class, I ask the prospective teachers to write re-sponses to a questionnaire that documents their initial knowledge about the sun and the moon, questions they have about the sun and the moon, explanations for day and night and the changing phases of the moon, de-scriptions of the nature of scientific explanations, and definitions of in-quiry approaches to teaching and learning. During subsequent weeks, they respond to handouts that track their growing understanding. Some also reflect on their learning about the moon in their journals.

On the tenth week of class, the prospective teachers document their current knowledge by writing responses to the same questionnaire they had answered on the first day of class. I assure them that they are not ex-pected to have understood everything there is to know about the moon, that their task is to document progress. However, I feel responsible for communicating "right answers" at a minimum level and talk individually with any whose responses indicate they still seem to believe that the phases of the moon are caused by the shadow of the earth.

The prospective teachers use the evidence from their pre- and postinstruction questionnaires and class work to write a paper that re-flects on changes in their understandings about the phases of the moon, the nature of scientific explanations, and inquiry approaches to learning and teaching. Thus, by doing this assignment, they are learning to *do* sci-entific inquiry, learning *about* the nature of scientific inquiry, and learning about inquiry as a method of instruction. One of the prospective teachers wrote:

> I think that the outside experiment where we ourselves modeled the phases of the moon with the ping-pong balls attached to pencils also caused my change in thinking. This activity, while led by the teacher, really focused on students modeling the world around them. When we went outside, you did not say, "Stand here with your arms like this," but instead allowed our own

inquisitiveness to lead us to understanding by asking us to "Stand so that your model (the ball) matches the phase of the moon." This altered my thinking about how a teacher should structure inquiry learning and how the way you structured the questions led to our own individual understanding. (prospective teacher, conversation about the sun and the moon, spring, 2002)

My belief is that prospective teachers can deepen their understanding of inquiry teaching by reflecting on such instances of their own learning.

Level 3: Prospective Teachers Inquire into Children's Learning and Their Own Teaching

The primary way in which I teach science teaching through inquiry is to guide the prospective teachers in generating and exploring questions about pedagogical issues that interest them. Like Akerson and Roth McDuffie (see chap. 14), I engage prospective teachers in learning how to do research as an integral part of learning how to teach. As discussed later, the prospective teachers undertake pedagogical inquiries during group investigations, individual research projects, and development of a personal framework for science teaching and learning.

Group Investigations.   During the first third of the course, the prospective teachers formulate and explore pedagogical inquiries in small groups. They formulate these inquiries in collaboration with experienced teachers participating in a teacher researcher group, the Science Inquiry Group (van Zee, 1998a; van Zee et al., 2003). By bringing the two groups together in the early part of the course, I provide evidence for the prospective teachers that graduates of the course and other experienced teachers value both my focus on research and my unexpected ways of teaching.

The group investigation includes participating in a Research Festival, a meeting held after school jointly with the Science Inquiry Group (SING). After the SING teachers describe their studies, each meets with a small group of the prospective teachers. In this way, the SING teachers mentor the prospective teachers in planning an inquiry-based science lesson they will later teach in the SING teacher's classroom. With mentoring from their SING teacher, each small group of prospective teachers also formulates a question about learning and teaching science that they later explore in the SING teacher's classroom.

After visiting their SING teacher's classroom and then practicing in my class, each group goes back out to their SING teacher's classroom to teach and research with children. During the next class session, each small group develops interpretations of the data they have collected and constructs a poster and report to present their findings. The poster and report summarize their research question, relevant literature, methods of data collection and analysis, claims with evidence supporting their claims, and implications for instruction. Finally, at a second Research Festival, each small group presents their findings to their classmates and the SING teachers.

During spring 2002, the small groups formulated pedagogical inquiries that focused on gender issues, the use of prior knowledge in making predictions, student questions about a topic, and initial student knowledge. For example, one group asked, "How do first graders perceive wind? What do they already know about wind?" For data, they audiotaped instruction, took digital pictures, and collected student responses on a worksheet. They formulated the claim that "First graders know that wind moves things. For example, these children knew the effects of wind and how it moves things like leaves on trees by blowing them." For evidence, they provided a transcript:

Prospective teacher:	Wait, I have a question . . . could you *see* wind?
Class:	No
Prospective teacher:	You can't see wind?
First grade student 1:	Yes, almost, because you could see what it's doing.
Prospective teacher:	What does it do? [Name], what does it do?
First grade student 1:	Knock down trees.
Later excerpt:	
Prospective teacher:	[Name] remember when you told me wind moves leaves?
First grade student 2:	Yeah
Prospective teacher:	How does it move leaves?
First grade student 3:	Air . . . it blows. (Report of group investigation, prospective teachers, Spring 2002).

One of the group's conclusions about implications for instruction was: "In order to keep students engaged in learning science, teachers should build upon the background knowledge of students and challenge students to look deeper into the subject by further analyzing their observations."

These experiences in teaching and researching provide opportunities for each group of prospective teachers to formulate pedagogical inquiries about issues that interest them, to collect data within a context in which they are doing the teaching, to develop claims based on these data, and to consider the implications of their findings. For many, this is their first experience formally "teaching" in a classroom; they are forming images of themselves as teachers of science with support from peers and science enthusiasts. By teaching and researching together, each group shares a context within which to reflect on many issues raised in the readings.

Individual Research Projects. After completing the group investigations, the prospective teachers begin planning individual research projects that they undertake as interns in their placement classrooms. These projects follow a process similar to the group investigations. In the individual projects, the prospective teachers formulate questions about science learning and teaching that interest them and plan ways to explore these issues while teaching lessons selected from their mentor teachers' science curriculum. They first practice with peers in class and then teach the lessons in their placement classrooms while collecting data such as tapes of instruction, copies of student work, and so on. Finally, they develop interpretations of these data in class and present their findings in posters at a Research Festival. They also write formal reports that serve as artifacts of their teaching and research in their teaching portfolios.

During spring 2002, the prospective teachers individually explored a wide variety of issues. These included gender effects, motivation, the use of models, conceptual understanding, discourse, and aspects of inquiry learning and teaching. For example, one prospective teacher wondered how the context that a teacher provides students affects student understanding, involvement, and enthusiasm for science learning. She and a colleague had the opportunity to teach the same lesson to two classes. For one class, they gave detailed directions and for the other, more open-ended instruction. Their data sources included student journals, video recordings of instruction, and notes taken by their mentor teachers. Her expectation was that "students with more guidance and direction . . . would be more likely to demonstrate understanding, be involved through active participation, and show more enthusiasms for science learning than students who were not given a clear introductory context to a science activity." Her expectation was not confirmed, however. One of her claims was that "by intentionally not guiding the students with specific questions to answer, students are more able to creatively come up with their own system of collecting and interpreting their own data." For evidence she presented the following:

> For example, during the morning when class "B" was just given the vague statement "record any observations you want to make about your shadow outline and anything you think might affect it today," one group asked "Joe" to lie down on top of his shadow outline and see if it was his height (it

turned out to be slightly larger). None of the groups from class "A" demonstrated such initial creativity. Additionally, the other class [B] thought to
draw pictures of what they were observing during the beginning of the experiment while the other class [A] only answered the questions provided on
their paper. (prospective teacher's research report, spring 2002)

My belief is that prospective teachers will learn how to teach science by
designing and conducting such pedagogical inquiries as they teach science in their placement classrooms.

Development of Personal Frameworks for Science Teaching.   I want
prospective teachers to leave my course with a firm commitment to teaching science in ways that they have articulated based on their own experiences. We begin developing these personal frameworks for science teaching on the first day of class. I invite the prospective teachers to reflect on
positive experiences they have had in learning science at some time in
their lives, inside or outside of school, and to identify factors that had fostered their learning in those instances (van Zee & Roberts, 2001). The
spring 2002 class, for example, identified the following factors: learning
by doing, hands-on, created interests, students relate to experience, self-
discovery, encouraging teacher, accomplishing goals, working in groups,
getting out of the classroom, guidance, tangible end product, observation,
brainstorming, experimenting, opportunities for things to go wrong, discussion, and discrepant event. Few raised their hands, however, to indicate that these factors had been typical of their own science instruction. I
noted that this indicated the need for reform and the factors they had
identified as fostering science learning were well aligned with national
and state recommendations for such reform. Thus, I was able to assure
them that they already had a strong knowledge base about good science
teaching on which to build.
    Throughout the semester, the prospective teachers wrote journals in
which they described science learning they observed or experienced
themselves and then reflected on factors that had fostered learning in
those instances (see Fig. 13.2). On the last day of class, they analyzed their
journals for common themes by highlighting statements about factors that
fostered science learning, cutting these out, sorting them into piles, taping
each pile of similar statements to a sheet of paper, and writing a claim at
the top about a factor that fostered science learning, based on the evidence
assembled from the journals (van Zee, 1998b). For the final examination,
the prospective teachers first presented their data (a set of journals) and
analyses (claims about factors that fostered science learning). Next they
wrote recommendations for science teaching based on their claims, as a
statement of their personal framework for science teaching. Then they de-

scribed how they would meet these recommendations while teaching a lesson of their choice, drawn from any resource. They also discussed ways in which they would meet the needs of diverse learners in teaching this lesson, described ways they would integrate across the disciplines, and made connections to issues discussed in the readings as well as to state and national standards. Finally, they wrote abstracts articulating questions about science learning and teaching that they would like to explore while student teaching.

Questions posed by the spring 2002 class were similar to those formulated for the research projects, with a focus on gender effects, conceptual development, and issues related to teaching science through inquiry. The prospective teachers modeled their abstracts on those written by the SING teachers for the Research Festival flyer. Such an abstract begins with a statement about an issue of interest and then presents a plan for exploration. For example, one prospective teacher proposed the following study:

Third Graders' Understanding of How a Seed Turns into a Plant

The purpose of this study is to better understand how third graders foresee how a seed turns into a plant. I plan to allow the children to discover on their own the inside of a seed and the factors that are needed for a seed to grow into a plant. I also plan to ask students to raise questions about things they observe on their nature walk and experiments. I am interested in seeing if they can interpret the data they collect from their nature walk, experiments, and observations and make a logical conclusion of how seeds turn into plants. I will collect evidence from class recordings, videoclips, journal writings, posters, observations, and notes from our class science talks. (prospective teacher, final, spring 2002)

The prospective teachers leave my course with a focus for exploring pedagogical issues in the context of science instruction while student teaching.

Level 4: Instructor and Prospective Teachers Inquire into the Instructor's Teaching Practices

At several points throughout the course, I ask the prospective teachers to join me in reflecting on what is happening by providing feedback. Some questions are open-ended, such as "What did you learn from the group investigation?" Other questions ask for ratings on a Likert-like scale and a brief comment. I e-mail the class a summary of responses on the anonymous questionnaires, both positive and negative, with commentary and explanations. Negative comments such as "When are you going to *teach* us how to teach science?!" have prompted me to be more explicit about what I am trying to do and why. I use Fig. 13.1, for example, to talk about different ways I might teach the course. Papers about the course provide examples of student work and elaborate on my rationale for designing such assignments.

On an anonymous questionnaire at the end of the spring 2002 course, I asked "How likely is it that you would teach science through inquiry if you had not had this course?" and also "How likely is it that you will teach science through inquiry now that you have had this course?" The difference in means was significant (paired t test, $p < .001$), with means of 2.1 and 4.5, respectively, on a scale of 1 (not likely) to 5 (very likely). All but one indicated an increase in likelihood of teaching through inquiry, with more than one half indicating a change of three units, a few even from 1 to 5. The one who marked the same rating, a 5 for both, commented, "No change—I already buy into the theory that students construct meaning best when they construct it themselves." These data indicate a change in orientation toward methods of teaching. I do not know, however, to what extent graduates will actually enact inquiry approaches as practicing teachers.

In addition, I asked the prospective teachers to rate each component of the course on a scale from 1 (not good use of my time) to 5 (good use of my time). There were many 4s and 5s, with means in the high 3s or 4s. However, some in the group marked 1s or 2s. The difficulties some students experience in my course have made me curious about epistemological differences among my students. Understanding these better may help me become more effective in assisting all in learning to teach science through inquiry.

REFLECTION

Reflecting on one's own teaching practices can be an intense but useful process (Loughran, 2002). A variety of rubrics exist for assessing inquiry instruction (e.g., Chinn & Malhotra, 2002; Luft, 1999; NRC, 2000). How well do the activities and assignments in my course align with such recommendations for inquiry teaching?

The continuum in Fig. 13.1 is similar to the table of variations in learner self-direction for essential features of classroom inquiry in *Inquiry and the National Science Education Standards* (NRC, 2000, p. 29; http://www.books.nap.edu/books/0309065767/html/20.html). This table can serve as an observational tool for supervisors and researchers interested in assessing the extent to which inquiry is instantiated in science classrooms (Windschitl, personal communication, June 14, 2002). I adapted this framework to reflect on my efforts to teach both science content and science pedagogy through inquiry:

- To what extent are learners engaged in scientifically oriented questions they pose themselves?
- To what extent do the learners determine what constitutes evidence and collect it? To what extent do the learners formulate explanations after summarizing evidence?
- To what extent do learners connect explanations to scientific knowledge independently?
- To what extent do learners form arguments to communicate and justify explanations?

With respect to science content, I have chosen "What causes the changing phases of the moon?" to be the driving question (Krajcik, Czerniak, & Berger, 1999) for the prospective teachers to explore. Within this context, however, the small groups design investigations of specific questions they generate from their own observations. Like Abell et al. (2002), I demonstrate how to collect some kinds of moon data, but each group decides when and where to look for the moon and what to record to answer the questions they have generated within this context. Like Duckworth (1987), I provide plenty of "time for confusion" as the prospective teachers work to make sense together of their moon observations. However, I also structure some experiences that lead them through development of an explanatory model. The activities and assignments require making connections to state and national standards, as well as to issues raised in the readings. The culmination of the project is writing a paper in which the prospective teachers articulate in their own words arguments and explanations for the moon's changing phases.

With respect to science pedagogy, the driving question for the course is "What factors foster science learning?" Within this context, the prospective teachers formulate and explore their own questions about science learning and teaching during the group investigation, individual research projects, and final examination. As with many beginning teacher researchers (Hubbard & Power, 1999), the prospective teachers initially think of collecting data as obtaining numbers to be manipulated in some way. Therefore, during the group investigation, the SING teachers and I provide scaffolding for videotaping instruction and making copies of student work to use as evidence in developing claims relevant to educational issues of interest. This prepares the prospective teachers for the individual research projects, for which they decide for themselves what data to gather while teaching and researching in their placement classrooms. Along with examples of teacher research (Doris, 1991; Saul et al., 1993), I also provide substantial guidance in developing interpretations of video recordings of instruction, copies of student work, and their own reflective journals, given that few, if any, have analyzed such data before. My course is a form of project-based learning (Krajcik et al., 1999) in which the prospective teachers communicate their findings by presenting posters and reports at the Research Festival concluding the group investigation and at a similar one concluding the research project. A few also have presented at local and national conferences.

Through these experiences I hope the prospective teachers begin to view themselves as generators of knowledge, capable of formulating questions about science and about science learning, interpreting data, and communicating findings to colleagues.

ACKNOWLEDGMENTS

Preparation of this chapter was supported in part by a Pew National Fellowship to participate in the Carnegie Academy for the Scholarship of Teaching and Learning. The opinions expressed are those of the author and do not necessarily reflect those of the funding agency.

REFERENCES

Abell, S., George, M., & Martini, M. (2002). The moon investigation: Instructional strategies for elementary science methods. *Journal of Science Teacher Education, 13,* 85–100.

Allen, S. (1997). Using scientific inquiry activities in exhibit explanations. *Science Education, 81,* 715–734.

Anderson, R. D. (2002). Reforming science teaching: What research says about inquiry. *Journal of Science Teacher Education, 13,* 1–12.

Appleton, K., & Kindt, I. (2002). Beginning elementary teachers' development as teachers of science. *Journal of Science Teacher Education, 13,* 43–61.

Barton, A. C., Hindin, T. J., Contento, I. R., Trudeau, M., Yang, K., Hagiwara, S., & Koch, P. D. (2001). Underprivileged urban mothers' perspectives on science. *Journal of Research in Science Teaching, 38,* 688–711.

Brickhouse, N., & Bodner, G. M. (1992). The beginning science teacher: Classroom narratives of convictions and constraints. *Journal of Research in Science Teaching, 29,* 471–485.

Briscoe, C., & Wells, E. (2002). Reforming primary science assessment practices: A case study of one teacher's professional development through action research. *Science Education, 86,* 417–435.

Bybee, R. (1997). *Achieving scientific literacy: From purposes to practices.* Portsmouth, NH: Heinemann.

Cennamo, K. S., & Eriksson, S. C. (2001). Supporting scientific inquiry through museum websites. *Educational Technology, 41*(3), 50–55.

Chang, C. Y., & Mao, S. L. (1999). Comparison of Taiwan science students' outcomes with inquiry-group versus traditional instruction. *Journal of Educational Research, 92,* 340–346.

Chinn, C. A., & Malhotra, B. A. (2002). Epistemologically authentic inquiry in schools: A theoretical framework for evaluating inquiry tasks. *Science Education, 86,* 175–218.

Cochran-Smith, M., & Lytle, S. L. (1993). *Inside/outside: Teacher research and knowledge*. New York: Teachers College Press.

Crawford, B. A. (1999). Is it realistic to expect a preservice teacher to create an inquiry-based classroom? *Journal of Science Teacher Education, 10*, 175–194.

Dewey, J. (1900). The school and the life of the child. In J. Dewey (Ed.), *The school and society* (Rev. ed., pp. 30–62). Chicago: University of Chicago Press.

Doris, E. (1991). *Doing what scientists do: Children learn to investigate their world*. Portsmouth, NH: Heinemann.

Driver, R., Newton, P., & Osborne, J. (2000). Establishing the norms of scientific argumentation in classrooms. *Science Education, 84*, 287–312.

Duckworth, E. (1987). *The having of wonderful ideas and other essays on teaching and learning*. New York: Teachers College Press.

Edelson, D. C. (2001). Learning-for-use: A framework for the design of technology-supported inquiry activities. *Journal of Research in Science Teaching, 38*, 355–385.

Erickson, G. L. (1991). Collaborative inquiry and the professional development of teachers. *Journal of Educational Thought, 25*, 228–245.

Flick, L. (2000). Cognitive scaffolding that fosters scientific inquiry in middle level science. *Journal of Science Teacher Education, 11*, 109–129.

Fradd, S. H., & Lee, O. (1999). Teachers' roles in promoting science inquiry with students from diverse language backgrounds. *Educational Researcher, 28*(6), 14–20.

Gallas, K. (1995). *Talking their way into science: Hearing children's questions and theories, responding with curricula*. New York: Teachers College Press.

Gess-Newsome, J. (2002). The use and impact of explicit instruction about the nature of science and science inquiry in an elementary science methods course. *Science and Education, 11*, 55–67.

Hammer, D. (1995). Student inquiry in a physics class discussion. *Cognition and Instruction, 13*, 401–430.

Herrenkohl, L., Palinscar, A. S., DeWater, L. S., & Kawasaki, K. (1999). Developing scientific communities in classrooms: A sociocognitive approach. *Journal of the Learning Sciences, 8*, 451–493.

Hogan, K. (1999). Thinking aloud together: A test of an intervention to foster students' collaborative scientific reasoning. *Journal of Research in Science Teaching, 36*, 1085–1109.

Hubbard, R. S., & Power, B. M. (1993). *The art of classroom inquiry: A handbook for teacher research*. Portsmouth, NH: Heinemann.

Hubbard, R. S., & Power, B. M. (1999). *Living the questions: A guide for teacher-researchers*. York, ME: Stenhouse Publishers.

Huber, R. A., & Moore, C. J. (2001). A model for extending hands-on science to be inquiry based. *School Science and Mathematics, 10*, 32–41.

Iwasyk, M. (1997). Kids questioning kids: "Experts" sharing. *Science and Children, 35*(1), 42–46.

Keys, C. W., & Bryan, L. A. (2001). Co-constructing inquiry-based science with teachers: Essential research for lasting reform. *Journal of Research in Science Teaching, 38*, 631–645.

Keys, C. W., & Kennedy, V. (1999). Understanding inquiry science teaching in context: A case study of an elementary teacher. *Journal of Science Teacher Education, 10*, 315–333.

Krajcik, J., Czerniak, C. M., & Berger, C. F. (1999). *Teaching children science: A project-based approach*. New York: McGraw-Hill.

Kuhn, D. (1993). Science as argument: Implications for teaching and learning scientific thinking. *Science Education, 77*, 319–337.

Kurose, A. (2000). Eyes on science: Asking questions about the moon on the playground, in class, and at home. In J. Minstrell & E. H. van Zee (Eds.), *Inquiring into inquiry learning and teaching in science* (pp. 139–147). Washington, DC: American Association for Advancement of Science.

Layman, J., Ochoa, G., & Heikkinen, H. (1996). *Inquiry and learning: Realizing science standards in the classroom.* New York: The College Board.

Lederman, N. G., & Niess, M. L. (1998). Survival of the fittest. *School Science and Mathematics, 98,* 169–172.

Loughran, J. J. (2002). Effective reflective practice: In search of meaning in learning about teaching. *Journal of Teacher Education, 53,* 33–43.

Lowery, L., Schoenfeld, A., & White, B. (1990). *Master's and credential in science and mathematics education (MACSME) program* (National Science Foundation TPE91–50028). Berkeley, CA: University of California, Berkeley.

Luft, J. A. (1999). Assessing science teachers as they implement inquiry lessons: The extended inquiry observational rubric. *Science Educator, 8*(1), 9–18.

Luft, J. A. (2001). Changing inquiry practices and beliefs: The impact of an inquiry-based professional development programme on beginning and experienced secondary science teachers. *International Journal of Science Education, 23,* 517–534.

Mastropieri, M. A., Scruggs, T. E., & Boon, R. (2001). Correlates of inquiry learning in science: Constructing concepts of density and buoyancy. *Remedial and Special Education, 22,* 130–137.

McDermott, L. C. (1990). A perspective on teacher preparation in physics and other sciences: The need for special science courses for teachers. *American Journal of Physics, 58,* 734–742.

McDermott, L. C. (1996). *Physics by inquiry.* New York: Wiley.

McGilly, K. (Ed.). (1994). *Classroom lessons: Integrating cognitive theory and classroom practice.* Cambridge, MA: MIT Press.

McGonigal, J. A. (1999). How learning to become a teacher researcher prepared an educator to do science inquiry with elementary grade students. *Research in Science Education, 29,* 5–23.

Minstrell, J. (1989). Teaching science for understanding. In L. Resnick & L. Klopfer (Eds.), *ASCD 1989 Yearbook: Toward the thinking curriculum: Current cognitive research* (pp. 131–149). Alexandria, VA: Association for Supervision and Curriculum Development.

Minstrell, J., & van Zee, E. H. (Eds.). (2000). *Inquiring into inquiry learning and teaching in science.* Washington, DC: American Association for the Advancement of Science.

National Research Council. (1996). *National science education standards.* Washington, DC: National Academy Press. Retrieved September 1, 2004, from http://www.nap.edu/catalog/4962.html

National Research Council. (2000). *Inquiry and the national science education standards.* Washington, DC: National Academy Press. Retrieved September 1, 2004, from http://www.nap.edu/catalog/9596.html

Pearce, C. (1999). *Nurturing inquiry.* Portsmouth, NH: Heinemann.

Polman, J. L., & Pea, R. D. (2001). Transformative communication as a cultural tool for guiding inquiry science. *Science Education, 85,* 223–238.

Roberts, D. (1999). The sky's the limit: Parents and first-grade students observe the sky. *Science and Children, 37,* 33–37.

Roth, W. M., & Bowen, M. (1995). Knowing and interacting: A study of culture, practices, and resources in a grade 8 open-inquiry science classroom guided by a cognitive apprenticeship metaphor. *Cognition and Instruction, 13,* 73–128.

Saul, W., & Reardon, J. (1996). *Beyond the science kit: Inquiry in action.* Portsmouth, NH: Heinemann.

Saul, W., Reardon, J., Schmidt, A., Pearce, C., Blackwood, D., & Bird, M. (1993). *Science workshop.* Portsmouth, NH: Heinemann.

Schwab, J. J. (1962). The teaching of science as enquiry. In J. J. Schwab & P. F. Brandwein (Eds.), *The teaching of science* (pp. 1–103). Cambridge, MA: Harvard University Press.

Scott, C. (1994). Project-based science: Reflections of a middle school teacher. *The Elementary School Journal, 95,* 75–94.

Shulman, L. (1998). Course anatomy: The dissection and analysis of knowledge through teaching. In P. Hutchings (Ed.), *The course portfolio: How faculty can examine their teaching to advance practice and improve student learning* (pp. 5–12). Washington, DC: American Association for Higher Education.

Simpson, D. (1997). Collaborative conversations: Strategies for engaging students in productive dialogues. *The Science Teacher, 64*(8), 40–43.

Smith, D., & Anderson, C. W. (1999). Appropriating scientific practices and discourses with future elementary teachers. *Journal of Research in Science Teaching, 36,* 755–76.

Songer, N. B., Lee, H. S., & Kam, R. (2002). Technology-rich inquiry science in urban classrooms: What are the barriers to inquiry pedagogy? *Journal of Research in Science Teaching, 39,* 128–150.

Tippins, D. J., Koballa, T. R., & Payne, B. D. (Eds.). (2002). *Learning from cases: Unraveling the complexities of elementary science teaching.* Boston: Allyn & Bacon.

Tobin, K. G., & McRobbie, C. J. (1996). Cultural myths as constraints to the enacted science curriculum. *Science Education, 80,* 223–241.

van Zee, E. H. (1998a). Fostering elementary teachers' research on their science teaching practices. *Journal of Teacher Education, 49,* 245–254.

van Zee, E. H. (1998b). Preparing teachers as researchers in courses on methods of teaching science. *Journal of Research in Science Teaching, 35,* 791–809.

van Zee, E. H. (2000). Analysis of a student-generated inquiry discussion. *International Journal of Science Education, 22,* 115–142.

van Zee, E. H., Iwasyk, M., Kurose, A., Simpson, D., & Wild, J. (2001). Student and teacher questioning during conversations about science. *Journal of Research in Science Teaching, 38,* 159–190.

van Zee, E. H., Lay, D., & Roberts, D. (2003). Fostering collaborative inquiries by prospective and practicing elementary and middle school teachers. *Science Education, 87,* 588–612.

van Zee, E. H., & Minstrell, J. (1997a). Reflective discourse: Developing shared understandings in a high school physics classroom. *International Journal of Science Education, 19,* 209–228.

van Zee, E. H., & Minstrell, J. (1997b). Using questioning to guide student thinking. *Journal of the Learning Sciences, 6,* 229–271.

van Zee, E. H., & Roberts, D. (2001). Using pedagogical inquiries for the basis for learning: Analysis of prospective teachers perceptions of positive science learning experiences. *Science Education, 85,* 733–757.

Wells, G. (1993). *Changing schools from within: Creating communities of inquiry.* Portsmouth, NH: Heinemann.

Wells, G. (2001). *Action, talk & text: Learning and teaching through inquiry.* New York: Teachers College Press.

White, B., & Frederiksen, J. (1998). Inquiry, modeling, and metacognition: Making science accessible to all students. *Cognition & Instruction, 16,* 3–118.

White, R., & Gunstone, R. (1992). *Probing understanding.* London: Falmer.

Whitin, P., & Whitin, D. J. (1996). *Inquiry at the window: Pursuing the wonders of learners.* Portsmouth, NH: Heinemann.

Windschitl, M. (2003). Inquiry projects in science teacher education: What can investigative experiences reveal about teacher thinking and eventual classroom practice? *Science Education, 87,* 112–143.

Yerrick, R. K. (2000). Lower track science students' argumentation and open inquiry instruction. *Journal of Research in Science Teaching, 37,* 807–838.

14

The Elementary Science
Teacher as Researcher

Valarie L. Akerson
Indiana University

Amy Roth McDuffie
Washington State University

Elementary teachers are usually generalists, without specialty or special preparation in either science content or pedagogy. It can reasonably be argued that their primary role is to prepare their students to be literate adults, and thus, many are literacy specialists. Oftentimes, elementary teachers lack confidence in teaching science (Cox & Carpenter, 1989; Perkes, 1975; Tilgner, 1990) and so avoid science because it is not their specialty (Atwater, Gardener, & Kight, 1991; Schoeneberger & Russell, 1986). Most elementary teachers have never been engaged in scientific inquiry, yet they are being asked to teach science as inquiry (Kielborn & Gilmer, 1999). Even elementary teachers who are confident in their science backgrounds and teaching approaches could benefit from conducting an inquiry project, and could improve their teaching practice with systematic study. Although a teacher research project is not the same as a scientific inquiry, it can still provide a way that helps teachers experience some aspects of inquiry. The purpose of a teacher research project, as defined in this chapter, is to conduct classroom-based inquiry on teaching practice. Thus, there are similarities in that teachers are using an inquiry in their teacher research projects (see also chap. 13). Teacher research allows elementary teachers to raise a researchable question, design a study, and analyze and interpret results. Whereas the context is different from scientific inquiry, the process gives them some experience with a social science inquiry. Therefore, an appropriate strategy for fulfilling both a need to engage in inquiry, and a need for professional development in science teaching, is to prepare teachers to use action, or teacher research, in their teaching practice.

INQUIRY

The *National Science Education Standards* (NRC, 1996, 2000) recommended that all science teachers continue to develop their pedagogy and content knowledge through inquiry. Inquiry is defined as raising an investigable question, developing methods to answer that question, carrying out those methods, analyzing the data, drawing conclusions, and reporting findings. It has been traditionally thought difficult to prepare elementary teachers to use inquiry methods to teach science, partially because they may have limited science backgrounds, and are not likely to have had experience in conducting scientific inquiry (Kielborn & Gilmer, 1999).

Giving K–8 teachers experiences with scientific inquiry has been shown to improve their understandings of inquiry, which they hopefully then relate to their ability to teach about inquiry and use in their classrooms (Kielborn & Gilmer, 1999). Learning to use inquiry in their own classrooms is important because teachers need to translate what they know (head knowledge) into the context of their classroom practice (knowledge in practice). Situating learning in a meaningful context is important (Putnam & Borko, 2000; Saxe, 1988), so researching one's own teaching can provide a context for this form of inquiry. Teacher research provides a meaningful context for inquiry into teaching, can help develop elementary teachers' science teaching skills in the context of their classrooms, and give them a sense of inquiry in their practice.

TEACHER DEVELOPMENT

There have been recommendations to provide both pedagogy and content support for elementary teachers during professional development in science teaching (National Commission on Science and Mathematics Teaching for the 21st Century, 2000; NRC, 1996). Teachers often receive materials or textbooks to use for science instruction, but no guidance for their effective use. Just getting materials does not guarantee an improvement in teaching. Rather, it is the professional development that helps teachers use the materials effectively, to improve their teaching. Further, not all curricula, materials, or strategies are equally effective for all teachers, grade levels, and student groups. It follows that an important question to ask of professional development for elementary teachers would be, "What can teachers do to improve their own science teaching in their teaching setting?" One appropriate strategy is for teachers to conduct research projects to actually explore a teaching strategy, some materials, or a curriculum with their students, and track the effectiveness of the strategy, materials, or curriculum. Such a teacher research project allows teachers to note under which circumstances and with which students a new strategy, materials, or curriculum is most effective. It enables teachers to formulate data-supported reasons for using particular strategies, materials, or curricula, and it shows the teachers—through the data and evidence collected—how student learning in their classroom is affected by trials of the strategy, materials, or curriculum. Teachers can make deliberate changes in their own teaching, and use their analysis and interpretation of data to support (or otherwise) the implementation of those changes.

TEACHER RESEARCH

What is teacher, or action, research? Simply put, it is when the classroom teacher conducts research on her own teaching or teaching situation. Feldman and Minstrell (2000) described teacher research as teachers inquiring into their teaching in their classrooms. The teacher systematically designs a study, collects data, analyzes the data, and interprets and reports the results, in a process that parallels scientific inquiry. Teacher research is often an iterative process—containing cycles of action followed by data collection, data assessment, reflection, and further action (Hopkins, 1993). Such research can be used to inform teaching practice, and contributes to the development of a reflective practitioner (Hubbard & Power, 1993).

Use of teacher research with preservice elementary teachers has, however, some limitations arising from their restricted (or no) teaching experience. For instance, preservice teachers' develop their research questions from their experiences as students in university courses and their early field experiences in an attempt to make meaning for the theory, research, and practice that they experience in their teacher education program. Nor are they easily able to follow the iterative "action cycle" seen as a common component of action research (e.g., Hopkins, 1993; Hubbard & Powers, 1993), because they have limited time in which to complete their program. However, it is possible for preservice teachers to develop conceptions of teacher research as a way to become more reflective about their teaching, and as a means to improve their teaching. One of our preservice teachers aptly summarized her understanding of teacher research based on her experiences:

You are doing something [in the classroom] and then you are asking your-self, "Does this really work?" And you are not relying on intuition to say, "Well, it felt like it kind of worked." You're actually looking for evidence to say, "Does this work?" . . . So, [in teacher research] you are going a step fur-ther than just a visual kind of thing, an emotional kind of thing, you are looking for evidence.

Reflective Practice

Schön (1983) recommended that practitioners in any field become reflec-tive to be aware of, and improve, their practice. With the emergence of teacher education programs based on a constructivist perspective for learning, a central goal of many programs is to develop reflective practi-tioners (Christensen, 1996; McIntyre, Byrd, & Foxx, 1996). Teachers have the opportunity through reflection to build their own knowledge about their own practice from their own experiences. By focusing on their sci-ence teaching, elementary teachers can become more reflective about what and how they teach, and make deliberate instructional decisions based on data (Roth McDuffie, 2001). It has also been shown that class-room-based research promotes reflection in preservice teachers (Valli, 2000).

Some might suggest that prospective and practicing elementary science teachers could get the same benefits in developing their teaching practice from reading others' research reports (Scott & Driver, 1997). Reading oth-ers' research is beneficial, but not solely helpful in identifying practices that would work best for individual teachers. Scott and Driver (1997) found that whereas researchers may be able to conduct research in some-one else's classroom, it is difficult to interpret the results and make recom-mendations about teaching strategies, because the researcher does not know the students as well as the teacher. However, by using a teacher-as-researcher approach, the teacher is able to decide which approaches are best for her students. Other elementary teachers and elementary teacher educators have made similar improvements in their science teaching by using reflective teacher research (e.g., Akerson, Abd-El-Khalick, & Leder-man, 2000; Dickinson, Burns, Hagen, & Locker, 1997). Indeed, several studies have pointed to the importance of action, or teacher research, in developing preservice teacher abilities to reflect on, and improve their own teaching, with the support of a university researcher—particularly in the field of science teaching (Chandler, 1999; Fueyo & Neves, 1995; Scott, 1994; Stanulis & Jeffers, 1995; van Zee, 1998; Winograd & Evans, 1995). For instance, Feldman and Minstrell (2000) described a lengthy process through which one teacher developed a teacher research agenda, and the ability to conduct research to improve his teaching of science. The teacher claimed that teacher research became a natural part of his teaching over time, allowing him to track his effectiveness and affects on students' learning while he is teaching. Van Zee (1998; van Zee et al., 2001; van Zee, Lay, & Roberts, 2003; see also chap. 13) described successful programs for preparing teachers as researchers in science methods courses.

To summarize, there is evidence that elementary teachers need experience in inquiry (Kielborn & Gilmer, 1999), and in professional development for teaching science (i.e., Atwater, Gardener, & Kight, 1991). Teacher research promises to give teachers an authentic experience in inquiry on their own science teaching as a professional development tool. Thus, teachers should learn to use teacher research as both an approach to inquiry and as a tool for professional development in science teaching.

METHODS FOR PREPARING ELEMENTARY TEACHERS TO USE TEACHER RESEARCH

Our students completed a teacher research project as part of a Master in Teaching (MIT) program. This 2-year master's degree program served preservice teachers who already held a baccalaureate degree in a field other than education and desired to become teachers. Two primary objectives of this program were: "(1) To educate teachers to become effective practitioners . . . by bringing the inquiry method of a research university to bear on the entire educational process . . . [and] (2) To empower teachers as reflective practitioners by helping them develop the multiple and critical decision making skills essential for today's classrooms" (University program description document). This research-based approach to developing reflective practitioners was evident in the design of the student teaching internship. Requirements of the internship included: 12 weeks in a K–8 school placement, solo teaching for at least 4 weeks; writing in a reflective journal at least once each week; completing a goals sheet at least once each week (identifying a goal for their teaching and reflecting on their success in meeting that goal); writing lesson plans for all lessons taught; developing a unit plan; completing at least four focused observations of teachers' teaching and writing a report on each observation; and completing a classroom-based teacher research project on their teaching.

Learning About Action Research

Regarding the teacher research project, the preservice teachers designed their studies during the fall semester as part of a course titled "Classroom Focused Research," taught by the first author. Using two texts as a framework for study (Hubbard & Power, 1993; McNiff, Lomax, & Whitehead, 1996), the preservice teachers studied methods of designing and conducting teacher research, and planned original classroom-based research projects as part of this course. The teacher research project focused on investigating a specific student-selected teaching strategy or approach. Preservice teachers were encouraged to select a teaching strategy and content area about which they felt least secure, and in which they wanted to improve their teaching. The approach to apply a preconceived research plan within a future teaching situation resembles scholarly educational research. Many would agree that teacher research should be responsive to the specific needs of specific teachers in specific teaching situations: That is, teacher research is undertaken by "teachers" who are already engaged in teaching (or action), so that this teaching provides the context from which questions are derived and within which the investigations are undertaken. However, preservice teachers do not have a teaching context from which questions arise. Instead, they have a plethora of teaching strategies to which they have been exposed in methods courses, they wish to know more about, and they would like to explore with students. The merit of the present approach is that the preservice teachers choose strategies and content areas in which "they felt least secure," so the project is responsive to the current needs of the preservice teachers, although it might not be responsive to their needs in the future situations in which they might find themselves (i.e., student teaching and classroom teaching).

Each preservice teacher worked with a faculty committee consisting of a chair (with expertise in the preservice teacher's selected area for research) and two additional faculty members from the Department of Teaching and Learning. The preservice teachers wrote literature reviews in their areas of study as part of a full study proposal. These proposals were submitted to the preservice teachers' chairs for feedback and were reviewed three times during the semester before preservice teachers presented a final version at the end of the semester.

Doing Action Research

The preservice teachers implemented their action research project the following semester during student teaching, modifying their research as a result of interacting with their mentor teacher and students during their placement. In the month after their student teaching internship, the preservice teachers analyzed their data, and wrote and presented oral and written reports of their studies to their faculty committee. Figure 14.1 shows a timeline of the research project activities.

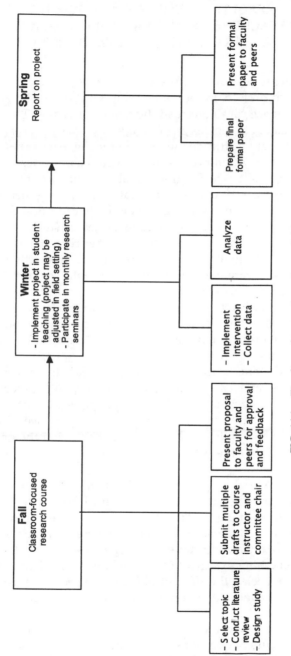

FIG. 14.1. Timeline of research project activities.

RESULTS OF ELEMENTARY SCIENCE
TEACHERS USING TEACHER RESEARCH
AS PROFESSIONAL DEVELOPMENT

After preparing four groups of preservice teachers to conduct their own teacher research projects in their internship settings, we have found that students experience common frustrations and successes. The frustrations tended to be related to the design of their study: Many preservice teachers began with a negative attitude toward conducting teacher research, in the same way that they often bring similar negative attitudes to the science methods classroom. Some frustrations also occurred while in the field conducting the research, analyzing the data, and writing their research reports. However, the preservice teachers initially did not see any point to all the work they were doing in designing a research study. Most felt quite overwhelmed by the idea of conducting teacher research as well as dealing with the already challenging activities of student teaching. For example, one student summarized her feelings of both seeing the benefit of teacher research and also feeling a bit anxious about it: "I know that it's beneficial because it's really going to force us to plan what we are doing. And to look at a specific area of interest to us. And to work on developing it . . . , but it is daunting, definitely! It's hard to know how data collection will fit in with normal teaching."

Reassuring the preservice teachers that they can do both concurrently, and that their research can support their development as a teacher, is crucial. One useful suggestion is to invite a previous student, now a classroom teacher, to share research findings and experience in conducting that research during an internship. It is inevitable that the graduate will share the feeling that the work was difficult, but worthwhile to professional development.

Our preservice teachers generally had difficulty thinking of a researchable question, partly because it was not derived from within a context in which they were teaching, and partly because they were inexperienced in research. They tended toward questions that were too broad, such as comparisons of several teaching strategies over a 4-week period, or that were focused on something external to the development of their own teaching practice, such as seeing the effect of playing background music while students work. However, with support from the course instructor and each student's individual discipline chair, feasible designs focusing on teaching strategies were developed, and then implemented, during their internships. Examples of questions that were successfully researched by the preservice teachers include: (a) studying fourth graders' conceptions of science, social studies, and language arts within an interdisciplinary program; (b) studying the effect of a conceptual change approach on fifth graders' conceptions of chemistry; (c) exploring strategies for encouraging fourth-grade girls in science; (d) exploring ways to make mathematics meaningful to third graders by using science and engineering concepts as contexts; (e) investigating strategies for including (information) technology in a middle school classroom; (f) exploring ways to encourage middle school students to participate better in their science classes; and (g) investigating influences of language arts on science, and science on language arts skills.

When the preservice teachers completed their final, oral and written research reports, their successes became vividly apparent. They were excited to share their newfound, evidence-based knowledge. It was evident from their animated presentations that they were excited about their results, and were anxious to share their information with others. Many chose to also present their work at a university-wide research symposium consisting of reports from all disciplines. In 2 of the last 3 years of the symposium, top prizes were awarded to education students' projects, which was a wonderful accomplishment given that the judges were multidisciplinary, and projects were from the sciences, humanities, and social sciences. Many of the teacher research projects have also been presented at national conferences (specifically, Akins & Akerson, 2000; Baker & Roth McDuffie, 2001; Bohrmann & Akerson, 2001; Burke & Akerson, 2002; Dickinson & Reinkens, 1997; Jardine & Roth McDuffie, 2001; Kelso & Akerson, 2000; Liu & Akerson, 2001; Nguyen & Roth McDuffie, 2001; Nixon & Akerson, 2002; Pringle & Dickinson, 1999; Stine & Akerson, 2001; Wright & Dickinson, 1999). Further, five preservice teachers' papers have been published in peer-reviewed journals (e.g., Akins & Akerson, 2002; Akerson & Reinkens, 2002; Bohrmann & Akerson, 2001; Liu & Akerson, 2002). Undergoing the extra work required to present a paper at a national level, and to submit a paper in a publishable form to a peer-reviewed journal, demonstrates the value these preservice teachers now placed on their work. Nonetheless, the preservice teachers needed support in these endeavors, and it is unlikely that any would have pursued disseminating their work to a wider audience were it not for support from a university researcher. Also, these preservice teachers would have been unlikely to engage in teacher research into investigating different approaches to teaching and learning in science if they had not been required to do so. A student commented on this issue:

> [Another preservice teacher] and I were talking on the phone the other day, and she said "Wouldn't it be easier if we didn't have to do the research projects?" And I said, "Yeah, you know, I had thought about that too. It would have been a lot easier." And then . . . I realized that it pushed me out of that comfort zone, at least in [the one area I was researching]. Where if I didn't have that requirement I would not have worked at incorporating new ideas

in teaching. I asked her, "Do you think you would have done what you did in [innovative teaching] if you hadn't done the research project?" And she said, "No!" So if nothing else, it pushes us out, at least in one content area, out of our comfort zone [to try something different].

Another preservice teacher stated, "Including teacher research is the difference between just working and being a professional." Another commented, "I hate to admit it, but doing the teacher research project forced me to test teaching methods I may not have otherwise tried. And it made me think about what I was doing."

Since commencing the research component of the preservice program, we have found consistent anecdotal evidence that teacher research has helped with the professional development in science teaching of preservice teachers. It has also given them an authentic, meaningful, contextualized inquiry experience. The anecdotal evidence points to preservice teachers improving in their science teaching by trying different approaches and reflecting on the effectiveness of those approaches. It is remains to be determined what role, if any, carrying out teacher research has played in developing a view of inquiry. That question is important for further study—to see how teacher research may partially fill the gap in elementary teachers' experience with scientific inquiry. However, the similar processes between teacher research and inquiry suggest that, at a minimum, preservice teachers benefit by completing a full cycle of inquiry, and thus having this experience as a foundation for further learning.

RECOMMENDATIONS FOR INCLUDING TEACHER RESEARCH IN ELEMENTARY SCIENCE TEACHER DEVELOPMENT

Using teacher research for elementary science teacher development has been successful. The teachers with whom we have worked have received professional development opportunities as they research, in their own classrooms, how strategies for teaching science work with their students. Additionally, these teachers have experienced an authentic inquiry project. Although not the same as a scientific inquiry, the process parallels what scientists do, particularly social scientists, and gives them a model of inquiry they may choose to have their students use.

From our experience in using teacher research to help preservice elementary science teachers both improve their teaching of science and undertake an authentic inquiry experience, we have six recommendations:

1. Emphasize that preservice teachers choose a meaningful, researchable question that focuses on their teaching practice.

2. Encourage preservice teachers to select areas for research about which they are least familiar.
3. Provide university support for the preservice teachers throughout all phases of the project.
4. Focus preservice teachers on a stringent but responsive research design.
5. Encourage students to realize they can conduct the research project.
6. Encourage preservice teachers to disseminate the results of their studies.

First, preservice teachers should select a research question that is meaningful to them, and that focuses on their teaching practice. If the requirement to focus on teaching practice is not there, then the preservice teachers may choose a research question that is not conducive to professional development. For instance, preservice teachers could select a project that studies the effects of natural light on student science performance. This could, in theory, be argued to be a valuable study, but it would not lend itself to professional development of science teachers, given that the focus of the topic is not closely related to the actions, decisions, methods, and/or thinking of teaching. Preservice teachers should therefore design studies that focus on the development of their science teaching, such as using conceptual change teaching strategies to promote student learning, exploring interdisciplinary approaches to teaching science, or using an inquiry-based teaching approach.

Second, in our experience, if preservice teachers have the freedom to choose to study any teaching strategy or content area they wish, they often select a literacy focus. Yet, their most urgent needs in terms of professional development are areas they tend not to choose, such as science. For this reason, preservice teachers should be encouraged to design studies that can help them improve their teaching of subjects for which they feel the least confident. Once they implement teaching strategies, and collect and analyze data related to the effectiveness of these strategies, they may feel more comfortable about using them, and in teaching that content area. They will, at the very least, have more experience in teaching that content area than they would if they had conducted a literacy study.

Third, university faculty should work closely with preservice teachers throughout the entire process of designing the studies, data collection, data analysis, and writing. Regular feedback during each phase is essential for students new to research. Helping preservice teachers design viable, meaningful studies, as well as collect and analyze the data, is critical in helping them get past the frustrations associated with working in a new field. As part of this process, university faculty need to encourage students to think carefully about the implications of their findings. Students often report their results and end their research report without exploring the implications of the findings for their own or others' practice. For example, Nixon and Akerson (2002) reported how a preservice teacher investigating how science can influence language arts skills initially concluded her paper with the result that her elementary students' interpretations of their own science investigations became more superficial when constrained by various writing forms. When asked to think about the implications of this result, she realized that whereas science and language arts can be thought of as interdisciplinary at times, there are still times where disciplinary instruction is most appropriate in each. Appropriate disciplinary instruction allows for appropriate development in each discipline, and allows teachers to help students meet each discipline's objectives. Without the prompting from her university mentor, she may have missed interpreting this finding, and more generally, she may not have thought beyond the data.

Fourth, focus preservice teachers on a stringent research design. They will learn little about inquiry without a robust design, and will gain valuable insight into both inquiry and educational research with a good design (Lederman & Niess, 1997). Again, preservice elementary teachers have had little, if any, experience in conducting inquiries, thus they will require support. Preservice teachers should conduct a fairly thorough literature review while designing their studies, and prior to data collection. Through this process they gain an appreciation of educational research from reading others' work (clarifying the difference between systematic research and simply reflecting on practice); clarify their own research questions/problems; and learn what we already know/have established in the field.

Although most of our work has been with preservice teachers, one inservice teacher who took a "Teacher as Researcher" methods course stated, "Just reading about all the research related to my study helps me see how my teaching might change." Thus, even the act of reading related research can help teachers see a need and process for change. In our program, the review of literature took place in the research course semester, and required preservice teachers to review at least five empirical research reports as background for their own study. As their work progressed, even through data collection and analysis, most preservice teachers continued to read related research, and modify their literature review. Consequently, they spent almost an entire school year reviewing related research, and their final literature reviews were much longer than the original five required.

As part of a stringent research design, preservice teachers should carefully develop a plan for data collection and analysis. This plan may include a timeline for these activities. Even if the students deviate from this plan during the study, having a structure in place helps them to stay focused on their research when the demands of teaching might pull them away. This plan for investigation will help them see the nature of scientific inquiry because it can deviate, if necessary, as the investigation is conducted.

Fifth, preservice teachers need encouragement that they can actually conduct a meaningful inquiry into their science teaching. They can be quite intimidated about the project, especially in the early stages of the design of the study, but need encouragement throughout the study. After the course in which the preservice teachers design their studies, we advocate monthly seminars to which they bring questions, data, problems, or other matters for discussion. These monthly seminars have been approximately one and a half hours in length. The focus is on the preservice teachers' inquiries. The format is informal, allowing the preservice teachers to raise questions regarding data collection, analysis, and interpretation, and to receive feedback from both their peers and a university researcher. Additionally, the preservice teachers should be encouraged to maintain contact with their university chairs during the entire implementation of their plans.

Finally, we recommend encouraging preservice teachers to disseminate the results of their research. When the preservice teachers recognize that their research can reach a wider audience, they determine to design a more stringent plan and more thoroughly examine the implications of their findings. They realize that the results of their research can benefit not only them and their own teaching, but also other teachers and teacher educators. This makes the teacher research a valuable addition to their development as elementary science teachers. It gives them the knowledge that their work is important, and, given the fact that other teachers and teacher educators will read their work, could boost their confidence in teaching science.

REFERENCES

Akerson, V. L., Abd-El-Khalick, F. S., & Lederman, N. G. (2000). Influence of a reflective explicit activity-based approach on elementary teachers' conceptions of nature of science. *Journal of Research in Science Teaching, 37,* 295–317.

Akerson, V. L., & Reinkens, K. A. (2002). Preparing preservice elementary teachers to teach for conceptual change: A case study. *Journal of Elementary Science Education, 14,* 29–46.

Akins, A., & Akerson, V. L. (2000). Connecting science, social studies, and language arts: An interdisciplinary approach. *Educational Action Research, 10,* 479–499.

Atwater, M. M., Gardener, C., & Kight, C. R. (1991). Beliefs and attitudes of urban primary teachers toward physical science and teaching physical science. *Journal of Elementary Science Teaching, 3*(1), 3–11.

Baker, A., & Roth McDuffie, A. (2001, October). Equivalence: Concept building in a fifth grade classroom. In R. Speiser, C. Maher, & C. Walter (Eds.), *Proceedings of the 23rd annual meeting of the North American Chapter of the International Group for the Psychology of Mathematics Education* (pp. 389–390). Snowbird, UT: ERIC Clearinghouse.

Bohrmann, S., & Akerson, V. L., (2001, January). *Improving girls' self-efficacy toward science.* Paper presented at the annual meeting of the Association for the Education of Teachers in Science, Costa Mesa, CA.

Bohrmann, M. L., & Akerson, V. L., (2001). A teacher's reflections on her actions to improve her female students' self-efficacy toward science. *Journal of Elementary Science Education, 13*(2), 41–55.

Burke, D. S., & Akerson, V. L. (2002, January). *Helping middle school science students relate to new concepts through physical modeling: A bodily-kinesthetic approach.* Paper presented at the annual meeting of the Association for the Education of Teachers in Science, Charlotte, NC.

Chandler, K. (1999). Working in her own context: A case study of one teacher researcher. *Language Arts, 77*(1), 27–33.

Christensen, D. (1996). The professional knowledge-research base for teacher education. In J. Sikula, T. Buttery, & E. Guyton (Eds.), *Handbook of research on teacher education* (2nd ed., pp. 38–52). New York: Macmillan.

Cox, C. A., & Carpenter, J. R. (1989). Improving attitudes toward teaching science and reducing science anxiety through increasing confidence in science ability in inservice elementary school teachers. *Journal of Elementary Science Education, 1*(2), 14–34.

Dickinson, V. L., Burns, J., Hagen, E. R., Locker, K. M. (1997). Becoming better primary science teachers: A description of our journey. *Journal of Elementary Science Education, 8,* 295–311.

Dickinson, V. L., & Reinkens, K. A. (1997, January). *Mr. Reinkens' neighborhood: Can you say "conceptual change"?* Paper presented at the annual international meeting of the Association for the Education of Teachers in Science, Cincinnati, OH.

Feldman, A., & Minstrell, J. (2000). Action research as a research methodology for the study of teaching and learning of science. In A. E. Kelly & R. A. Lesh (Eds.), *Handbook of research design in mathematics and science education* (pp. 429–456). Hillsdale, NJ: Lawrence Erlbaum Associates.

Fueyo, V., & Neves, A. (1995). Preservice teacher as researcher: A research context for change in the heterogeneous classroom. *Action in Teacher Education, 16*(4), 39–49.

Hopkins, D. (1993). *A teacher's guide to classroom research.* Philadelphia: Open University Press.

Hubbard, R., & Power, B. (1993). *The art of classroom inquiry: A handbook for teacher researchers.* Portsmouth, NH: Heinemann.

Jardine, T., & Roth McDuffie, A. (2001, October). Cooperative learning in a fifth grade English as a second language mathematics class. In R. Speiser, C. Maher, & C. Walter (Eds.), *Proceedings of the 23rd annual meeting of the North American Chapter of the International Group for the Psychology of Mathematics Education* (pp. 675–676)). Snowbird, UT: ERIC Clearinghouse.

Kelso, R., & Akerson, V. L. (2000, January). *Math connections: Science and engineering applications in an elementary classroom.* Paper presented at the annual international meeting of the Association for the Education of Teachers in Science, Akron, OH.

Kielborn, T. L., & Gilmer, P. J. (Eds.). (1999). *Meaningful science: Teachers doing inquiry + teaching science.* Tallahassee, FL: SERVE.

Lederman, N. G., & Niess, M. L. (1997). Action research: Our actions may speak louder than our words. *School Science and Mathematics, 97,* 397–399.

Liu, Z., & Akerson, V. L. (2001, January). *Science and language links.* Paper presented at the annual meeting of the Association for the Education of Teachers in Science, Costa Mesa, CA.

Liu, Z., & Akerson, V. L. (2002, May 31). Science and language links: A fourth grade intern's attempt to increase language skills through science. *Electronic Journal of Literacy Through Science, 1*, article 4. Retrieved May 31, 2002, from http://sweeneyhall.sjsu.edu/ejlts/vol1-2.htm

McIntyre, D., Byrd, D., & Foxx, S. (1996). Field and laboratory experiences. In J. Sikula, T. Buttery, & E. Guyton (Eds.), *Handbook of research on teacher education* (pp. 171–193). New York: Macmillan.

McNiff, J., Lomax, P., & Whitehead, J. (1996). *You and your action research project*. New York: Routledge.

National Commission on Mathematics and Science Teaching. (2000). *Before it's too late* [The Glenn Commission Report]. Washington, DC: U.S. Department of Education.

National Research Council (NRC). (1996). *National science education standards*. Washington, DC: National Academy Press.

National Research Council (NRC). (2000). *Inquiry and the national science education standards: A guide for teaching and learning*. Washington, DC: National Academy Press.

Nixon, D., & Akerson, V. L. (2002, January). *Building bridges: Using science as a tool to teach reading and writing*. Paper presented at the annual international meeting of the Association for the Education of Teachers in Science, Charleston, NC.

Nguyen, L., & Roth McDuffie, A. (2001, October). Problem solving in mathematics: Barriers to problem-centered learning. In R. Speiser, C. Maher, & C. Walter (Eds.), *Proceedings of the 23rd annual meeting of the North American Chapter of the International Group for the Psychology of Mathematics Education* (pp. 571–572). Snowbird, UT: ERIC Clearinghouse.

Perkes, V. A. (1975). Relationships between a teacher's background and sensed adequacy to teach elementary science. *Journal of Research in Science Teaching, 12*, 85–88.

Pringle, R. L., & Dickinson, V. L. (1999, January). *Classroom learning activities that generate the most participation in middle school science*. Paper presented at the annual meeting of the Association of the Education of Teachers in Science, Austin, TX.

Putnam, R., & Borko, H. (2000, January–February). What do new views of knowledge and thinking have to say about research on teacher learning? *Educational Researcher, 29*, 4–15.

Roth McDuffie, A. (2001). *Fostering the process of becoming a deliberate practitioner: An investigation of preservice teachers during student teaching*. Paper presented at the annual meeting of the American Educational Research Association, Seattle, WA. (ERIC Document Reproduction Service No. ED346082)

Saxe, G. B. (1988). Candy selling and math learning. *Educational Researcher, 17*, 14–21.

Schoeneberger, M., & Russell, T. (1986). Elementary science as a little added frill: A report of two case studies. *Science Education, 70*, 519–538.

Schön, D. (1983). *The reflective practitioner*. San Francisco, CA: Jossey-Bass.

Scott, C. A. (1994). Project-based science: Reflections of a middle school teacher. *The Elementary School Journal, 95*, 75–94.

Scott, P. H., & Driver, R. H. (1997, March). *Learning about science teaching: Perspectives from an action research project*. Paper presented at the annual meeting of National Association of Research in Science Teaching, Chicago, IL.

Stanulis, R. N., & Jeffers, L. (1995). Action research as a way of learning about teaching in a mentor/student teacher relationship. *Action in Teacher Education, 16*(4), 14–24.

Stine, E. O., & Akerson, V. L. (2001, January). *Determining how to use graphic organizers in a sixth grade science classroom.* Paper presented at the annual meeting of the Association for the Education of Teachers in Science, Costa Mesa, CA.

Tilgner, P. J. (1990). Avoiding science in the elementary school. *Science Education, 74*, 421–431.

Valli, L. (2000). Connecting teacher development and school improvement: Ironic consequences of a preservice action research course. *Teaching and Teacher Education, 16*, 715–730.

van Zee, E. H. (1998). Preparing teachers as researchers in courses on methods of teaching science. *Journal of Research in Science Teaching, 35,* 791–809.

van Zee, E. H., Iwazyk, M., Kurose, A., Simpson, D., & Wild, J. (2001). Student and teacher questioning during conversations about science. *Journal of Research in Science Teaching, 38,* 159–190.

van Zee, E. H., Lay, D., & Roberts, D. (2003). Fostering collaborative inquiries by prospective and practicing elementary and middle school teachers. *Science Education, 87,* 588–612.

Winograd, K., & Evans, T. (1995). Preservice elementary teachers' perceptions of an action research assignment. *Action in Teacher Education, 17*(3), 13–22.

Wright, A. F., & Dickinson, V. L. (1999, January). *Integrating technology into the science classroom.* Paper presented at the annual meeting of the Association of the Education of Teachers in Science, Austin, TX.

15

Cultural Considerations

J. Randy McGinnis
University of Maryland, College Park

How new, reform-oriented elementary teachers teach science in schools is of great interest to the science education research community, as well as to the larger community concerned with continuous teacher education. This chapter first delimits and summarizes research directions in the cultural perspective. Second, it presents a sample of research that takes a cultural perspective to investigate what happened to a new elementary science and mathematics specialist teacher in her first 2 years of full-time classroom teaching. The chapter concludes by calling for increased use of the school culture theoretical perspective in science education research. The school culture perspective has the potential to help foster a better understanding of new, reform-oriented elementary teachers' science teaching experiences both in the workplace and in their career decisions.

CULTURE AND SCIENCE EDUCATION RESEARCH

An emerging field within education is devoted to conceptualizing the role of cultural considerations and its impact on teachers' professional lives. Charron (1991) first called for attention to be placed on science education research that depicted with complexity the social contexts in which all science teachers taught. In the years since Charron's call, various science education theorists have developed distinct research programs within a broadly defined cultural considerations research area. What links the

studies together is that the theorists hold the culture concept as para-
mount. For heuristic purposes, I delimit the major research directions
taken in this research area as a result of the various cultures thought to impact science teachers professional lives: the culture of science, the culture
of the students and teachers, and the culture of the workplace (referred to
as school culture by leading theorists). Occasionally, theorists have blurred
the boundaries among the cultural consideration research directions (as
exemplified by Aikenhead and his associates), but for purposes of discussion the proposed divisions I make in the cultural considerations research
area have merit.

The Culture of Science

In this research direction theorists are guided by their consideration primarily of the culture of science, oftentimes seen as conflicting with the
home cultures that science teachers and their students bring to the classroom (Spector & Strong, 2001). Science as a discipline is defined as a culture that has its own history, philosophy, practices, norms, and community. One aspect of this cultural research direction places focus on
describing the nature of science and how a scientifically knowledgeable
citizenry might impact beneficially on the larger societal culture in which
science teachers and students live (Bevilacqua, Giannetto, & Matthews,
2001). Another aspect of this cultural research direction focuses on a cross-
cultural perspective to examine posited cultural clashes between the culture of science as presented in the classroom, and the teachers' and students' home cultures (Aikenhead, 1996, 1998, 2001; Aikenhead & Otsuji,
2000).

The Culture of the Students and Teachers

In this research direction theorists are guided primarily by their consideration of the various cultural backgrounds of the students and their
teachers, and how those cultures play out in science teaching and learning contexts. A key interest to theorists in this cultural research area is determining how an understanding of the teachers' and their students' cultural knowledge can be used to promote equity by instituting a more
culturally relevant science education pedagogy (Atwater, 1989, 1993;
Cabello & Burstein, 1995; Fraser-Abder, 1992; Lee, 2001; McGinnis, 1995).

The Culture of the Workplace (School Culture)

In this research direction theorists are guided by the notion of a teacher
practicing in a workplace culture commonly referred to as school culture.
Theorists, such as Costa (1997) and Page (1987, 1990), argued that an anal-

ysis of schooling should be "refracted through a schools' culture or mean-ing system" (Page, 1991a, p. 42). As Page explained (R. Page, personal communication, May 15, 2002), a focus on the school culture construct re-directs attention:

> away from the view of schooling as cultural transmission/ determinism and points us instead to the idea that individual schools (1) have distinctive meaning-systems despite their structural similarities, (2) that even though we (and policymakers) talk about THE American High [or elementary] School or THE Science Curriculum as though they are everywhere the same, they vary and the variations are consequential, so that what science (teach-ing, studenting, etc) means in one school is not necessarily what it means in another . . . and (3) schools have some limited autonomy vis-à-vis the wider culture—they do not transmit the wider culture nor can they readily trans-form it, rather they *translate* it.

Therefore, schools are not viewed as microcosms of the society in which they operate. And counter to what some theorists argue (Tesconi & Hurwitz, 1974), even given a certain time lag, they do not necessarily incorporate a community's values, trends, social conflicts, and inequali-ties—all the social and cultural proceedings of the society-at-large. In-stead, the school's meaning system is shaped by what institutional partici-pants (particularly teachers) make of the structural dimensions of schools (i.e., size and location) and of the microprocesses (i.e., curriculum, teach-ing, and learning).

Page (1991b) stressed that what teachers and other school personnel do with the information they collect about their schools is to bend them to "fit their particular circumstances and purposes" (p. 18). Theorists hold di-verse definitions of "school culture" (see Page, 1991a, for a comprehensive review of the literature). However, their definitions range primarily from culture being an inexplicit dimension of all schools (e.g., a relatively con-servative social institution) to a signal to distinguish between schools (e.g., school different personalities or essences).

THE SCHOOL CULTURE PERSPECTIVE
AS A PROMISING DIRECTION OF RESEARCH

The cultural view taken in the remainder of this chapter is informed by the notion of school culture. The focus is on this least explored direction of the three cultural consideration areas of research while acknowledging the tremendous benefit to science education of the other two directions of cultural consideration research. The school culture view forces attention to be placed on the workplace culture in which teachers at all levels and their students interact. The benefit of taking this cultural research direction is a more complete understanding of how culture and its diverse manifestations come to impact the professional lives (including their teaching practices and their career decisions) of science teachers at all levels in specific work environments.

I concur with Page that the strength of the school culture concept is that it supports a view of schools as sociocultural systems "which translate between the classroom and community contexts" (Page, 1991a, p. 42) and that it supports an emphasis on documenting and interpreting teachers' perceptions within those systems. I also believe that "specifying a school's culture is an intuitive, interpretative endeavor . . . that incorporates the domains in which the participants (and ethnographers speaking to readers) describe" (Page, 1991a, p. 47).

The goal of research using the school culture perspective is both theoretical and pragmatic. In particular, I believe the practical use of a more nuanced theoretical understanding of elementary teachers' cultural considerations to be one of the more new fruitful directions that may be taken in continuous teacher education (preservice, induction, and lifelong) to support reform-based science teaching practices and improved teacher retention.

THE CULTURE CONSTRUCT AND ANTHROPOLOGY

Those who use the culture construct as a way to understand cultural considerations in science teacher education are using a construct that comes from cultural anthropology. Interestingly, although the culture construct is the central focus of cultural anthropology, there are important differences in how members of the field define culture. Therefore, it is worthwhile to conduct a brief examination of the development of the varied views of the culture concept within the field of cultural anthropology. This summary is drawn from a more comprehensive depiction (Levinson & Ember, 1996).

The Historical Development of the Culture Construct

A historical examination of the culture construct indicates much change over time in how it has been defined and used. Nineteenth-century theorists such as Tylor, Morgan, and Spenser defined culture as equivalent to civilization. The assumption made was that culture was a quality linked to the notion of describing civilizations as more or less advanced. Culture was measured by inspection of a broad array of people's capabilities acquired by membership in a society, such as knowledge, belief, art, morals, law, and customs. This view of culture prompted researchers to conduct ethnographies of various societies so that the differing societies could be ordered on a scale of increasing complexity that signified advancement. In this view of culture, all societies followed the same linear evolutionary development model. The only differing variable was in the speed at which a society developed culture. The key question researchers sought to answer was what produced differences in the developmental speed among societies?

Early 20th-century theorists, led by Boaz, began to define the culture concept differently than their predecessors. Whereas culture was still considered a quality consisting of beliefs, customs, and social institutions, it was not linked to an evolutionary model. Instead, differences among cultures were attributed to differences in the environmental conditions in which societies lived. Ways of life or culture fit the environment in which they formed but could be influenced by intersocietial communication. Societies whose cultures were the most complex benefited from ample resources and the transmission of technological developments. A fundamental assumption made was that the content of culture was held by the society, not the individual living within the society. In addition, culture was learned—not genetically transmitted. Society's culture defined and determined the behaviors of all members of the society.

Later 20th-century theorists questioned these assumptions. Initially, theorists such as Sapir (1924) began to question whether a society's culture shaped each member within it in a similar manner or that a society was characterized by a distinct culture. Eventually, Goodenough (1981) and others took their critical appraisal further. These later theorists argued that members of a society do not all have the same knowledge of customary practices or place an equal amount of significance on them. In fact, depending on the subject matter, considerable variability existed within societies as to how activities were conducted. The primary variable these theorists identified that promoted shared understandings on subject matters, was the frequency (and depth) of interaction individuals had concerning such subjects.

The result of these later lines of cultural concept theorizing was the recognition that culture consisted of two distinct but related orders: a phenomenal order and an ideational order. The phenomenal order referred to the patterns an outside observer detected in the conduct of society. These patterns led to descriptions of peoples' way of life and were used to distinguish societies. Culture, in this sense, is viewed as a societal property. The ideational order, in contrast, referred to what an individual learned in a society that enabled the individual to accomplish desired functions. In this sense, culture was not necessarily something shared by all members of a society. The notion of a common culture or shared understanding was an illusion.

Although advocates for the two different orders of culture (phenomenal and ideational) defended their interpretation of culture and attacked the merits of the other's interpretation, Malinowski's work (1944) contributed to a deeper insight into how the content of culture benefited from consideration of both orders. Malinowski's ethnographic work made evident that there was a complicated interplay between the two orders of culture. In the context of activities, social organizations, beliefs, values, and the idiosyncratic purposes of people came together. Later anthropologists, such as Geertz (1973), extended the thrust of Malinowski's theoretical position through the examination of shared public symbols. Presently, discussion within cultural anthropology continues on the issue of the two orders of culture and their possible connection.

Use of the Culture Concept in Educational Anthropology

In the early 20th century, Boas (1923) and Malinowski (1944) devoted their attention to refuting a commonly held—yet false—theory about the reputed learning disabilities of immigrant, minority, and lower class students in public schools. As influenced primarily by Spenser's (1896) views on sociology, the canonical view of culture at that time was that certain groups of students came from *lower* cultures in contrast to the other students who came from the *higher* Anglo-American culture. Boas and Malinowski disagreed with that explanation for student differences. Boas argued that student differences resulted from students' differing socioeconomic backgrounds, not as a result of their coming from cultures more or less advanced. Malinowski argued that students who came from cultures differing from the dominant culture portrayed prominently in schools faced by academic problems as a result of cultural discontinuities. Two solutions to the problem of underachievement in the schools were proposed: a cultural relativist approach and multicultural education. Work by Spindler (1955) continued this line of research. However, up until the 1960s, a common misuse of the culture concept described students of color and others who differed in certain traits from their White, middle-class peers as "culturally deprived."

Contemporary educational anthropology is united behind the assumption that school-based learning is a form of cultural transmission. Wolcott (1991) broadened the view of cultural transmission in schools to include not only a focus on the role of adults, but also a focus on children as active participants. The consensus among educational anthropologists is that the notion of cultural discontinuities is the primary explanation for the learning difficulties of students coming from home cultures that differ from the dominant culture expressed in schooling (Erickson, 1993). Of key interest is to understand why cultural discontinuities affect certain groups differentially. Ogbu (1992) suggested that a solution to this problem came from distinguishing between the types of cultural differences that existed for certain groups in school settings. He proposed that groups (e.g., Asian Americans) who believed they could overcome cultural barriers by learning the behavior and language of the members from the mainstream school culture while maintaining their own cultures were less affected. Alternatively, he proposed that those groups (e.g., African Americans) who believed they must give up traditional ways stigmatized by the mainstream culture before they could learn the behaviors and language of the dominant culture expressed in school settings fare less well.

A SAMPLE OF RESEARCH

The following sample of research uses the school culture perspective to assist in the interpretation of what happened to a new elementary science and mathematics teacher who graduated from what was designed to be a reform-based undergraduate teacher preparation program. This sample of research is drawn from the work of McGinnis, Parker, and Graeber (2000, 2004). In the larger study examining the professional lives of five new elementary and middle level science and mathematics specialist teachers, my associates and I used an analytical framework proposed by Veenman (1984) as a way to understand the role of the teachers' perspective of school culture. This analytical framework consisted of two components, a focus on the individual teacher (i.e., intentions, needs, and capabilities) and a focus on the individual's perception of the school culture as impacting the enactment of reform-minded practices. Of particular interest to examine were the impact of mandated state and district demands, and the school level supports and constraints, on the new teachers' perspectives. Data collected included the following: the perspectives of the five new elementary teachers of science and the perspectives of those who exerted maximum power in the new teachers' schools, that is, the principals.

Context of the Study

The Maryland Collaborative for Teacher Preparation (MCTP) is a National Science Foundation funded, statewide undergraduate program for students who plan to become specialist mathematics and science upper elementary or middle level teachers. Teacher candidates selected to participate in the MCTP program were in many ways representative of typical teacher candidates in elementary teacher preparation programs. They were distinctive by agreeing to participate in a program that consisted of an extensive array of mathematics and science experiences (formal and informal) that make connections between the two disciplines.

The goal of the MCTP is to promote the development of professional teachers who are confident teaching mathematics and science using information technology, who can make connections between and among the disciplines, and who can provide an exciting and challenging learning environment for students of diverse backgrounds (University of Maryland System, 1993). This goal is in accord with the educational practice reforms advocated by the major professional mathematics and science education communities (American Association for the Advancement of Science [AAAS], 1989, 1993; National Council of Teachers of Mathematics [NCTM], 1989, 1991; National Research Council [NRC], 1996). The MCTP is designed around these salient recommendations that are intended to promote reform:

- new content and pedagogy courses that model inquiry-based, interdisciplinary approaches combined with regular opportunities for teacher candidate reflection;
- the participation of faculty in mathematics, science, and methods committed to modeling best teaching practices (especially by diminishing lecture and emphasizing problem solving);
- the development of field experiences in community schools with exemplary teachers trained to serve as mentors;
- the availability of summer internships in contexts rich in mathematics and science; and
- the support of new teachers by university and school personnel during their first years of teaching.

Theoretical Assumption and Research Questions

A fundamental assumption of the MCTP is that changes in presecondary level mathematics and science educational practices require reform within the undergraduate mathematics and science subject matter and education classes that teacher candidates take throughout their teacher preparation programs (National Science Foundation, 1993). In earlier research, my associates and I have reported that the MCTP undergraduates indicated that they held notions about the nature of mathematics and science, and about the best ways to teach mathematics and science, that are compatible with the MCTP program's reform-minded goals (McGinnis, Kramer, Graeber, & Parker, 2001; McGinnis et al., 2002). That is, the MCTP interns intended to use constructivist instructional strategies, to emphasize connections between mathematics and science, to use information technology when teaching mathematics and science, and to encourage students from diverse backgrounds to participate in challenging and meaningful learning.

This study intended to determine what happened to the new MCTP teachers' vision of teaching as reflected in their classroom practices and their discourse during their first 2 years of full-time teaching. In particular, we wondered if their vision would remain stable or would it change (and, if so, how and why?). We were aware that schools are characterized by "dynamic conservatism" (Schön, 1987) in that the dynamic pulls teachers back to a status quo, which oftentimes remains unchallenged. We anticipated that the new MCTP teachers' reform-oriented beliefs and actions would run into conflict at times with the status quo and were curious to document and to learn from these as yet unknown situations.

Two research questions guided the study:

1. As they proceed through their induction years, how do new specialist teachers of mathematics and science who graduate from an inquiry-based, standards-guided innovative undergraduate teacher preparation: (a) enact their roles as teachers; and (b) think about what they do when teaching science and mathematics with upper elementary/middle level students?

2. What supports/constraints impact the introduction of new practices (reform-based) by new specialist teachers of mathematics and science who graduate from an inquiry-based, standards-guided innovative undergraduate teacher preparation?

Mode of Inquiry

A symbolic interactionist theoretical stance was taken in this study (Blumer, 1969; Denzin, 1978). The symbolic interaction theoretical stance makes the assumption that social reality is a social production (Woods, 1992). People construct meanings through interactions; meanings are not inherently linked to inanimate objects or events. A central premise is that inquiry must be grounded in the empirical environment under study (van Sickle & Spector, 1996). This theoretical position also requires the researchers to commit to a significant period of time working in the context of the study. It places emphasis on the social construction of meaning in a culture through viewing the process of how individuals define and interpret each other's acts. By carefully examining people's interpretations of each other's acts, assertions are made as to how these interpretations sustain or transform the way they view their culture, which guides they way they act and interact.

In this study, the symbolic interaction theory provided guidance for the roles of the researchers and the interpretative domain of the study (Le-Compte, Millroy, & Preissle, 1992). The researchers held the belief that their research was a social production symbolically negotiated between the researcher and participant. Thus, explicitly revealing the purpose of the research to the participants and maintaining an openness of mind regarding interpretations of the participants' beliefs and actions complemented the research methodological approach. Communication between the researchers and with the participants regarding subjective viewpoints became imperative to conduct within a group. As a group, we researchers made sense of our experiences by revealing our different perspectives. Qualitative research assumes that there are multiple realities constructed as a function of personal interaction and perception (Merriam, 1988, p. 17).

Study Participant and Research Site

In this chapter (in which the aim is to present a sample of research) only one of the five study participants is examined. She was selected to illustrate that, even in the most supportive of contexts, the school culture concept has considerable interpretative power. Interested readers are directed to the McGinnis et al. (2000, 2004) studies for additional cases, most of which do not illustrate such a supportive context. What follows is a brief description of the participant and her school context (pseudonyms used in all cases).

Ms. Susan Lee was an Asian American woman. She was a traditional college student who finished her undergraduate elementary teacher education program in 4 years. She completed a summer MCTP research internship at a space and aeronautical lab. Ms. Lee taught fourth grade at Overlook Elementary School. Overlook Elementary School had not met any of the local standards on the district's criterion referenced tests. The school was attended by 42.2% Hispanics, 28.5% African Americans, 21.2% Whites, and 7.6% Asians. Many students (62.2%) received free or reduced meals. The school's mobility rate was 24.1%.

Findings

The findings are presented as inside and outside perspectives in response to the two research questions. The inside perspective consists of summaries of participant's comments. The summaries are supported by exemplar participant comments. We believe that the use of the participant's inside perspective (emic voice) demonstrates respect for our participant (allowing her voice to be preeminent), and enhances the study's credibility (Van Maanen, 1995). However, it should not be presumed that my colleagues and I necessarily agree with or support the participant's perspective on all matters. Furthermore, I anticipate that my readers may not agree with my interpretation of her words—that is a welcomed acknowledgment that supports the use of reported extended examples of raw qualitative data, particularly when depicting actual teaching practice. The outside perspective (etic voice) in this analysis is composed of a lesson vignette taught by the teacher, her students' comments on the lessons, her principal's voice, and the researchers' theoretical framework on the data. We believe that these two data representation strategies provide the reader with a rich depiction of the participant's professional life. Based on the information provided, and what information and conceptual orientations our readers bring to this study, we believe that readers have the minimum amount of information to start constructing a view of how the participant was perceived (by herself and others) in her school culture.

To answer the first research question, data collected from the individual and group audiotaped and transcribed interviews, videotaped classroom lessons, participant reflective writings, student reflections, and classroom artifacts were analyzed. As the lead researcher, I initially identified a pool of comments by each participant that most cogently reported their perspectives. The participants' comments were then read by my coresearchers, who were asked to select the most representative comment for inclusion in the narrative (or to suggest other comments that I may have overlooked). When disagreements among the researchers emerged in a limited number of instances, the decision on which comment to include was settled by appeal to the participant.

Inside Perspective

The participant's individual and focus interviews conducted over 2 school years were the data sources for the participant's view of how she enacted the role of a reform-prepared teacher, and what she reportedly thought about while teaching.

Enacting Her Role as a Teacher. As a teacher, Ms. Lee described herself consistently as holding a student-centered perspective. She enacted that perspective by using cooperative learning activities that she considered constructivist and that sought to make connections among different disciplines. She preferred to use alternative assessment strategies in her practice. In comparing how she taught with the other teachers in her school, she thought of herself as a more effective teacher. She stated,

I'm teaching an upper level math, a high group, and I'm using a lot of the activities that we did, the hands-on, constructivist activities that we learned in our courses, especially in the methods. I'm using a lot of those activities from my MCTP science methods professor because it goes along very nicely with the hands-on, constructivist deals, and the kids are doing great with it. They look forward to coming to math and science, and also I'm integrating the math and science activities in with the other subjects, too, like the writing activities, and so it actually is an integration of all subjects, but it's working out very nicely. We just finished a unit with a lot of hands-on activities, and actually I had the students work in groups. . . . I notice that my teaching is a lot different from the ones at the school. Most of my strategies are better. They have one other new teacher, but everyone else has been teaching more than five, ten, fifteen, maybe even thirty years, and I guess it's different in that for all of them there is more direct instruction, and more of, like, a lecture type, and there is a little bit of hands-on. (fall 1998, interview)

Thinking About What She Does When Teaching Science and Mathematics with Elementary Students.   Ms. Lee reported that when she was teaching mathematics or science, a central thought she held was a commitment to make connections between the disciplines. She kept in mind a model of her MCTP instructors' teaching method to guide her actions. Over time, she increasingly began to compare the manner in which she taught to the other teachers' method she observed in her school. As she stated,

I model everything I do after what I learned in the MCTP. When I'm teaching a science unit I tie in the science with the mathematics. And when my students do science, they know it's not just science, we're also doing mathematics. When we are doing mathematics, we do have reading, writing and science, too. (spring 1998, interview)

The way that I teach is more that they're [her students] building the knowledge, so it's not just memorizing facts [like the other teachers do]. Everything is connected, and so they never question why they're learning something. . . . And so in terms of understanding, I think that this approach is just so much better. I'm very pleased with the results that I am getting in the classroom. (fall 1999, interview)

Outside Perspective

The primary data sources that support the development of an outsider perspective of how Ms. Lee enacted her role as a teacher were videotaped observations of her teaching lessons accompanied by anonymous student comments and an audiotaped and transcribed interview with her principal. From these sets of data, the reader has access to a depiction of a complete lesson taught by the teacher that is uninterrupted by researcher analysis, and the thoughts of the teacher's principal.

What follows is a vignette of classroom teaching performed by Ms. Lee. In this vignette, the researchers can identify in Ms. Lee's teaching practices, aspects aligned with some of the recommendations contained in standards-related documents produced by AAAS, NRC, and NCTM. Specifically, we argue that as a new teacher she evidenced attempts to teach for understanding, to make connections across the curricula, to use a nonroutine and relevant problem-based learning activity, and to use alternative assessment:

Fall 1998 Vignette.

Ms. Lee introduces the lesson to her diverse students by stating that there is something special going on the entire month. Student's hands go up, and a student offers, "It's Asian Pacific Heritage Month." Ms. Lee responds, "That's right, Asian Pacific Heritage month. I am very excited to bring something from my culture, from China, to the classroom during our science time. We have been studying ecosystems. Can someone remember what you need to grow a plant?" The class begins listing things that are needed to grow a plant: soil, water, air, the right temperature, and light. Ms. Lee writes the students' suggestions on the chalkboard. The students appear eager to give answers; they almost all have their hands up. Ms. Lee states, "Today, you are going to plant something special, it is called the Dow Gawk. Can everyone say that?" Ms. Lee writes the words on the board and states, "This is also called the Chinese Long Bean. Let me tell you a little about the Chinese Long Bean before everyone is going to plant her or his own Dow Gawk plant. There is something special about this Dow Gawk which is similar to the string beans that we eat. What does the string bean look like?" Students share responses with each other in pairs. They comment on such things as "It is long," "It is nasty," "It looks like a pod." She tells them how large the bean grows, to a size "most people would not expect." She tells the students that she won't tell them how large their plants will grow, but that the plant will grow out of the cups in which they will plant the seed. Later they will have to plant the plant into a larger pot. "Eventually you'll have to grow it hanging because it won't grow straight."

Ms. Lee passes out the seeds and asks the students (who are grouped in cooperative groups of four) to observe their individual beans and to write down their observations on an index card. She asks for a student volunteer to pass out the index cards. While Ms. Lee places a bean seed on each student's desk, she says, "There is something else interesting about this bean seed, so I want you to look at it and think about ways to describe it." Students attentively study the passed out bean seeds. Ms. Lee states, "Remember, you should be describing your bean seed on the index card and be writing a sentence or two describing the bean." Ms. Lee asks for volunteers to share their observations.

She selects equitably both girl and boy volunteers and students of color and limited English speakers to respond. Students share their observations such as the size of the bean, the color, and the texture.

She continues her lesson by stating, "We will be planting these Dow Gawk seeds today and charting their growth until the last day of school. You are welcome to continue the growth measurements in the summertime, also." Ms. Lee shares with the class that her mother plants them in her vegetable garden. She tells the students that if they take care of their plants, the plants will grow very large, larger than they expect probably. The students begin to ask how large the beans will get. Ms. Lee replies, "You will find out."

Ms. Lee shows the class some graph paper. She then tells the class that every other day in class they will be charting and graphing the beans' growth. Next, Ms. Lee points out to the class that the beans are wet and asks the class why she has made them wet. One student suggests that by soaking the bean seeds it speeds up their germination. Ms. Lee praises the student for his answer and agrees with his thinking.

Ms. Lee asks for a volunteer from each table to come and retrieve supplies. Hands shoot up. A volunteer from each table retrieves supplies for each table. Ms. Lee asks the students to observe the cup in which the beans will be planted. One female student points out that there is a hole in the bottom. Ms. Lee asks, "Why do you think I put holes in the bottom of the cups?" A male student suggests, "So if you add too much water the cup won't flood." Ms. Lee then asks, "Does anyone want to add anything?" Another female student offers that the hole could make it easier for them to observe the plant root system. Ms. Lee shares with the class that while her intent in putting the hole in the bottom of the cup was to allow excess water to seep out she also now thinks the hole will indeed help with observations of the root structure.

Ms. Lee asks the class how far down they should plant their seeds. A male student suggests that if they plant the seeds too far down, the plant won't be able to reach the top. Ms. Lee demonstrates to what depth to plant the bean seeds. The students plant their seeds. She instructs the students to put their index cards underneath the cup of seeds, and then to place the cups in the tray at the center of the table.

Ms. Lee hands out graph paper to every student and informs the class that they will begin the graph today. She states, "You will be doing a line graph. You may remember that we have done this before in math class." She asks the class, "What are the types of things you need for a line graph?" Students raise their hands and one boy offers that the graph needs a title. Ms. Lee asks the class, "What would be a good title for this graph?" A student suggests "Dow Gawk." Another student suggests, "Planting a Dow Gawk." Ms. Lee passes out rulers. Ms. Lee states, "Share something else you need to make a line graph." A female student suggests lines. Another female student suggests numbers. A third male student suggests labels. Ms. Lee writes all of the students' suggestions on the board. She tells the students that the vertical axis should be "Measurement in centimeters." The students label their graphs. Next she asks, "On the bottom axis what should you put? We will be measuring the growth every other day. That is a huge hint." Several students' hands shoot up. "So what should we put on the horizontal axis?" A student suggests, "days." Ms. Lee agrees with that suggestion.

Ms. Lee walks around the room checking students' graphs. "A reminder—the days are on the bottom axis, the horizontal axis, the axis that is lying flat. And the measurement is on the vertical axis, the one that is going up and down."

The lesson ends with the students cleaning up their desks and putting their cups on the window sill.

In an audiotaped interview, Ms. Lee's principal made the following comments concerning her presence in his school:

I nominated her, along with the PTA and other community members and staff, for the Sally May First-Year Teacher Award. It's a very prestigious award. It's a national award. To put it in a better perspective I've never nominated a first-year teacher for an award, for the Sally May Award or those types of awards. I never felt that I had one that was of this caliber. . . . Her ability to integrate and make those connections with math and science is considerable. She does it in very wonderful ways, by portraying different characters. I can't remember the actual character names that she uses, but she dresses up for this. The kids love it. The parents love it. Yet, what I like from that is that it's another way to hook your child into a particular curriculum area and that she has this span, not just during the specified subject time, such as a math class, but she'll interweave this during reading and language arts, so that those connections can be there, as well as bringing it back up during her science time. Each unit is very well-scripted and she's elaborated on it in some wonderful ways that the kids have that have enabled them to be very successful—their writing about it, their data gathering, their ability to conduct investigations. . . . She's one of the best prepared people to take on [our school district's] public schools curriculum in those two areas that I've had the privilege of working with in the last, at least, ten years. . . . At times, some of the veteran teachers sort of wonder, "Well, she's only a first-year teacher, but look how fast she's picking all these things up." They've been amazed by that. I just have been very pleased.

On a short survey administered near the end of the school year, one of Ms. Lee's students made the following anonymous statement concerning how Ms. Lee taught science: "Ms. Lee helped us learn science by teaching us how to observe more carefully (for example, rocks). She also helped us understand the topics we studied (for example, why rocks change)."

To answer our second research question, data collected from the individual and group audiotaped and transcribed interviews (participants and principals) and participant reflective writings were analyzed.

Inside Perspective

The individual and focus interviews conducted over 2 school years were the data sources for the inside perspective on perceived supports and constraints the new teachers faced in enacting their reform-based visions of practice in their schools.

Supports. A common theme expressed by all of the five new teachers in the study was the direct impact of the particular school culture on what supports the new teachers had available. Ms. Lee taught in an elementary school that was distinguished by the high percentage of English as a Second Language (ESOL) students (12.8%) who were at variable levels of English proficiency. Ms. Lee characterized many of these students' parents as illiterate, which caused complications in her ability to explain to them her rationale for her student-centered, active teaching. However, a support she found, due to the active nature of her learning activities, was in the positive reaction of her students' parents to the students' ability to show at home what they did in her class. As she stated,

> What I've noticed, what I've found out, is that a lot of the parents are illiterate and they don't really, the kids don't really have much support at home, and so a lot of the things we do in class is more hands-on so that they can be more independent. Because at home they have to be that way, a lot of them do. The kids are actually able to tell their parents what they did. They can explain, "Well this is how we learned it." With the hands-on things that we did you don't have to read a worksheet and say, "Well, Mom, Dad we learned about electricity today. This is the worksheet that we did." The parents come in and they're just so excited. Sometimes they come in and they say, "My son is just so happy to be in your class and he's learning so much. Everyday he comes home and tells me everything that he learned throughout the day." (spring 1999)

Another type of support on which Ms. Lee relied to implement her vision of a reform-minded practice was her learning experience in her undergraduate MCTP teacher preparation program. Specifically, the way she was taught in her introductory MCTP physics class and the activities in which she participated during her MCTP science methods class, assisted her. She stated,

Our MCTP classes [content and pedagogy] were taught in a constructivist manner, so our learning was constructivist, too. They built on our prior knowledge, and those kinds of things, like, all these little facts that I learned in my MCTP Physics course. I didn't forget the knowledge, because the professor built a solid foundation of the science and the math. Those kinds of things I didn't forget, and I'm using that very nicely here in the classroom. (fall 1998)

Constraints

The constraints for reform-based practices identified by our five MCTP study participants were shared and also idiosyncratic. The list of shared constraints included: nonsolicited ideas on how to change the participant's practices; the number of mathematics objectives to meet; the shortage and availability of computer equipment; the diverse level of student abilities; the science kits' prescribed curriculum and schedule; the prescribed science and mathematics curricula; the district's ongoing student testing of instructional outcomes; the frequent instructional interruptions; the number and extent of standardized student testing; the more experienced teachers' expectation that the new teacher would become less active and less innovative over time; and the suspicion of some parents about alternative assessment ideas. As stated by Ms. Lee, "Obviously if there's more people watching, there's more people with ideas, and they're going to be, you know, giving suggestions, and it's not always going to be positive. They're gonna say, 'Well, why don't you do it this way?' I mean, that will place a constraint."

The list of idiosyncratic constraints identified by our five MCTP study participants included: student expectations of being taught in a traditional manner, standardized student testing, communication with non-English proficient student parents, availability of information technology equipment, student subject rotation from teacher to teacher, prescribed curriculum and schedule of the science kits, standardized testing of short-term instructional outcomes, diverse abilities of the students, team concept of teaching, and excessive parental involvement. As stated by Ms. Lee,

I do feel very comfortable teaching in the MCTP manner. My problem is I'm starting to have tension with my other two teammates because I'm teaching differently than they are. And, I am. I prefer to teach this way and I don't see myself changing. And although I am the younger one, I don't follow how they teach. And I've had a lot of parents talk to me and tell me that they want their child to be in my class next year—and the principal is saying all these things. I know I'm not doing anything wrong and the kids are learning so much this year and so, like everything I learned from MCTP physics, just

like the hands-on lab things. My teammates keep on pushing these textbooks in my face and they keep on saying, "Susan, you have to use this. You
have to use this. The students have to answer questions from the textbook."
We do answer questions, I mean, we do write in a daily log and we do discuss the book, but we don't do the whole section in a book chapter and then
answer the corresponding questions to it.

Outside Perspective

The primary informants of the outsider perspective were the participants' principals. As building managers and instructional leaders of their
schools, their perspectives were viewed as particularly powerful within
the new teachers' school cultures. During the summer immediately following the new teachers' first teaching year, they were asked to reflect on
both the supports and the constraints (or, as one principal renamed it, the
"challenges") the new teachers faced in enacting their reform-based visions of practice in their schools.

Supports.   The supports identified by the MCTP new teachers' school
principals were diverse. These supports included: the positive reaction to
the new teachers' practices by colleagues, students, and parents. The principals also voiced their appreciation of what the new teachers brought to
their schools in oftentimes glowing terms, referring to the MCTP study
participants' nominations for "Teacher of the Year" in several instances.
As reported in more detail earlier, Ms. Lee's principal was extremely supportive of her in his school: "She's one of the best prepared people to take
on [our school district's] public schools curriculum in those two areas
[mathematics and science] that I've had the privilege of working with in
the last, at least, ten years. . . . I nominated her, along with the PTA and
other community members and staff, for the First-Year Teacher Award."

Constraints

The constraints identified by the school principals for the five study
participants varied by individual. In Ms. Lee's case, the principal did not
identify any constraints.

DISCUSSION

This study has the potential to contribute toward understanding what
happens to reform-minded new teachers in science during their first years
of teaching practice. In this sample of research, I will limit discussion to

how the five participants (with a focus on Ms. Lee) reacted to their perceptions of their school cultures.[1]

In Ms. Lee's case, she entered a school culture in which she seemingly flourished, based on her principal's evaluation, her students' comments, and our observation of her teaching.[2] However, this was not the case for the majority of our five study participants. In Ms. Lee's case, and in the cases of all the other new elementary science and mathematics teachers we examined over a 2-year period, we found that as our new teachers became enculturated into their schools, they detected constraining structures in their school cultures. In response to these constraining structures, they implemented "social strategies." Social strategies in this study are actions individuals take in reaction to perceived coercive power in a community setting. The social strategies the new teachers developed were *resistance* and *moving on*.[3] These strategies were not mutually exclusive, but were used as the new teachers thought appropriate, in response to specific instances of perceived power in their school cultures.

Resistance

In several instances, the new teachers expressed resistance in their actions toward traditional ways of thinking about mathematics and science teaching that they detected in their school cultures. For example, when several veteran teachers reacted negatively to some of Ms. Lee's reform-minded ideas, she engaged in resistance by defending her practices. As she stated,

> We had a team meeting, and they [her grade level teammates] came up to me, and they just said, I mean, of course they were positive at first, I mean, we get along very well, but then they're starting to say things like, they think that my method is just a little bit, well, they think they *cover* [participant's

---

[1]An alternative discussion topic not selected for inclusion in this chapter due to page constraints includes a detailed cultural analysis of all the participants' school cultures and how the schools' ethos impacted all aspects of the new teachers' professional lives.

[2]However, in private conversation with the lead researcher, Ms. Lee expressed increasing dissatisfaction with the professional development opportunities at her school. Because she saw herself as a lifelong learner with a strong desire for intellectual growth, this dissatisfaction grew to dominate her career decisions. In addition, she mentioned that the unsupportive comments made by some of her teacher colleagues in her school influenced her to begin viewing her school as a less desirable workplace. Consequently, at the end of her third year of teaching, she resigned her teaching appointment to begin full-time graduate studies in elementary education at an out-of-state institution of higher learning. She expressed uncertainly as to her career plans following completion of her master's degree program.

[3]The social strategy *exit* (ending the teaching career by resignation) was an option identified by our participants that they would take if the other two social strategies, *resistance* and *moving on*, did not improve their circumstances. Ms. Lee eventually took this social strategy.

emphasis] more things. I said, "Well, you may cover more things and they may remember it short-term, like on short term to take a test, but what about later on when you're building from it?"

Moving On

In one case when the perceived instances of coercive power within a school culture became problematic, the new teacher decided to transfer out of her school to another school within the school district. She stated that if that action did not improve her professional life, her next step would be to leave teaching as a career. In another case, the new teacher decided to break completely with the broader-in-scope context in which his school culture was situated. He departed his school district to take a teaching position in a Native American school that was located in a different state.

CONCLUSIONS

My associates and I draw several implications from this study of five new, reform-oriented teachers that included Ms. Lee. First, the study suggests that a reform-based mathematics and science teacher preparation program can recruit, educate, and graduate a cadre of new, reform-oriented teachers who are employed by schools. Evidence collected indicates that new teachers from such a teacher preparation program have the capabilities and intentions to teach mathematics and science in a reform-minded manner by using high quality science and mathematics.

Second, the study suggests that the new teachers' perception of their school cultures in which they began their teaching practices is a major factor in the degree and manner in which reform-minded mathematics and science teaching is implemented. The supports and constraints an individual teacher encounters on a daily basis, particularly from individuals with power over their work lives (e.g., their principal and fellow teachers), are noticed by new teachers and influence how they perceive their school culture (Williams, 1961), and their curricular, instructional, and assessment actions.

Third, if this study's findings are supported by future research, to enact reform and to retain new, reform-prepared teachers a key implication is that the new teachers fare better when they are employed in supportive, reform-oriented school cultures rather than in other environments. We posit that if better matches are made initially between reform-prepared teachers and school cultures, then the extent and the quality of reform-based practices in mathematics and science teaching will increase in those specific school cultures, as will the retention of more newly prepared teachers within those school cultures. Alternatively, if school districts desire to support reform-prepared teachers in nonsupportive school cultures, then considerable additional thought needs to be expended on how to accomplish this goal. In effect, radical change to existing school cultural practices is required. For example, new teachers are evaluated currently by their school principals within their school cultures. In schools where the principals do not support the larger district's commitment to reform-based science teaching, the new teacher is vulnerable to local administrative censure. However, the situation would be changed dramatically if new teachers who graduated from teacher preparation programs based on reform-based recommendations were assessed instead by districtwide assessors (who support reform-based teaching). Furthermore, if supportive teacher networks for these new teachers that extend beyond individual school cultural boundaries could be developed and supported by school districts, then the teachers would gain strength in community. With the strength of a supportive community characterized by a commitment to reform, the new teachers would be in a much better position to resist the impact of critics in their school cultures.

Finally, what can reasonably be done in teacher preparation to more adequately prepare new, reform-minded teachers to enact reform-based practices in school cultures that are not initially supportive? A first step during their teacher preparation programs would be to alert them to the concept of school culture. A second step would be to work with them in collaboration with their school districts' induction program as they reflected on the impact of school culture on their practices and career decisions. At a minimum, presenting career decision options that are less climactic than the social strategy of exiting for those experiencing severe cultural discontinuity would be beneficial.

The school culture perspective offers a view of what can happen to new, reform prepared teachers in their induction years. This theoretical perspective underscores particularly the need for additional attention toward alerting new reform-minded mathematics and science teachers as to the potential consequences of accepting employment in nonsupportive school environments. A review of recent studies (Britton, Raizen, Paine, & Huntley, 2000) suggested that many schools are problematic workplaces for new teachers. Specifically, new teacher attrition is now 23% within the first 3 years of practice. Job dissatisfaction is reported—particularly by new mathematics and science teachers—as a major reason for exiting.

Consequently, increased attention needs to be directed toward investigating the school cultures in which new, reform-prepared teachers work. And although we place considerable value on documenting and interpreting the perception of school culture from the new teachers' perspectives, we believe also that documenting and interpreting simultaneously others' perception (beyond those held by the researchers) of the new teachers' school cultures can be useful. Our work suggests that school cultures, which from the new teachers' perspective, actively support and respect the reform-oriented culture, result in workplaces characterized as having contented, stable, and effective personnel. In school cultures that from the new teachers' perspective pose a culture discontinuity between their expectations and their circumstances, an undesired workplace produces an opposite result. The latter situation, unfortunately, predominated in our extended investigation of a select sample of new, reform-oriented elementary teachers with significant science teaching responsibilities. It is our hope that this situation can be improved by recognizing the power potentially of the school culture concept in the conceptualization and implementation of all aspects of elementary science teachers' continuous professional education (teacher preparation and inservice).

ACKNOWLEDGMENT

This research was supported by grant no. DUE 9814650 from the National Science Foundation's Collaboratives for Excellence in Teacher Preparation Program (CETP).

REFERENCES

Aikenhead, G. (1996). Science education: Border crossing into the subculture of science. *Studies in Science Education, 27*, 1–52.

Aikenhead, G. (1998). Teachers, teaching strategies, and culture. *Science Education International, 9*, 7–10.

Aikenhead, G. (2001). Student's ease in crossing cultural borders into school science. *Science Education, 85*, 180–188.

Aikenhead, G., & Otsuji, H. (2000). Japanese and Canadian science teachers' views on science and culture. *Journal of Science Teacher Education, 11*, 277–279.

American Association for the Advancement of Science. (1989). *Science for all Americans.* New York: Oxford University Press.

American Association for the Advancement of Science. (1993). *Benchmarks for science literacy.* New York: Oxford University Press.

Atwater, M. (1989). Including multicultural education in science education. Definitions, competencies and activities. *Journal of Science Teacher Education, 1*(1), 17–19.

Atwater, M. (1993). Multicultural science education. *The Science Teacher*, March, 33–37.

Bevilacqua, F., Giannetto, E., & Matthews, M. R. (Eds.). (2001). *Science education and culture: the contribution of history and philosophy of science.* Boston: Kluwer Academic.

Blumer, H. (1969). *Symbolic interaction [instructional outcomes].* Berkeley, CA: University of California Press.

Boas, F. (1928). *Anthropology and modern life.* New York: Dover.

Britton, E., Raizen, S., Paine, L., & Huntley, M. A. (2000, March). *More swimming, less sinking: Perspectives on teacher induction in the U.S. and abroad.* Paper presented at the meeting of the National Commission on Teaching Mathematics and Science in the 21st Century. Washington, DC.

Cabello, B., & Burstein, N. D. (1995). Examining teachers' beliefs about teaching in culturally diverse classrooms. *Journal of Teacher Education, 46,* 285–295.

Charron, E. (1991). Toward a social-contexts frame of reference for science education research. *Journal of Research in Science Teaching, 28,* 609–618.

Costa, V. (1997). Honors chemistry: High status knowledge or knowledge about high status? *Journal of Curriculum Studies, 29,* 289–313.

Denzin, N. K. (1978). *The research act: a theoretical introduction to sociological methods* (2nd ed.). New York: McGraw-Hill.

Erickson, F. (1993). Transformation and school success: The politics and culture of educational achievement. In E. Jacobs & C. Jordon (Eds.), *Minority education: Anthropological perspectives* (pp. 27–51). Norwood, NJ: Ablex.

Fraser-Abder, P. (1992). How can teacher education change the downhill trend of science teacher education? *Journal of Science Teacher Education, 3,* 21–26.

Geertz, C. (1973). *The interpretation of cultures.* New York: Basic Books.

Goodenough, W. (1981). *Culture, language, and society* (2nd ed.). Menlo Park, CA: Benjamin-Cummings.

LeCompte, M. D., Millroy, W. L., & Preissle, J. (Eds.). (1992). *The handbook of qualitative research in education.* New York: Academic Press.

Lee, O. (2001). Cultural and language in science education: What do we know and what do we need to know? *Journal of Research in Science Teaching, 38,* 499–501.

Levinson, D., & Ember, M. (1996). *Encyclopedia of cultural anthropology* (Vol. 1). New York: Holt.

Malinowski, B. (1944). *A scientific theory of culture and other essays.* Chapel Hill, NC: University of North Carolina Press.

McGinnis, J. R. (1994). Preparing to teach middle level science in the 1990s: Embedding science instruction in an awareness of diversity. *Journal of Science Teacher Education, 6,* 108–111.

McGinnis, J. R., Kramer, S., Graeber, A., & Parker, C. (2001). Measuring the impact of reform-based teacher preparation on teacher candidates' attitudes and beliefs (mathematics and science focus). In J. Rainer & E. Guyton (Eds.), *Research on the effects of teacher education on teacher performance: Teacher education yearbook IX (Association for Teacher Education)* (pp. 9–28). Dubuque, IA: Kendall/Hunt.

McGinnis, J. R., Kramer, S., Shama, G., Graeber, A., Parker, C., & Watanabe, T. (2002). Undergraduates' attitudes and beliefs of subject matter and pedagogy measured periodically in a reform-based mathematics and science teacher preparation program. *Journal of Research in Science Teaching, 39,* 713–737.

McGinnis, J. R., Parker, C., & Graeber, A. (2000, April). *An examination of the enculturation of five reform-prepared new specialist teachers of mathematics and science.* A paper presented at the American Educational Research Association, New Orleans, LA.

McGinnis, J. R., Parker, C., & Graeber, A. (2004). A cultural perspective on the induction of five reform-minded new specialist teachers of mathematics and science. *Journal of Research in Science Teaching, 41*(7), 720–747.

Merriam, S. B. (1988). *Case study research in education: A qualitative approach.* San Francisco: Jossey-Bass.

National Council of Teachers of Mathematics. (1989). *Curriculum and evaluation standards for school mathematics.* Reston, VA: Author.

National Council of Teachers of Mathematics. (1991). *Professional standards for teaching mathematics.* Reston, VA: Author.

National Research Council. (1996). *National Science Education Standards.* Washington, DC: National Academy Press.

National Science Foundation. (1993). *Proceedings of the National Science Foundation workshop on the role of faculty from scientific disciplines in the undergraduate education of future science and mathematics teachers* (NSF 93-108). Washington, DC: National Science Foundation.

Ogbu, J. (1992). Understanding cultural diversity and learning. *Educational Researcher, 21*(8), 5–14, 24.

Page, R. (1987). Teachers' perceptions of students: A link between classrooms, school cultures, and the social order. *Anthropology & Education Quarterly, 18*, 77–99.

Page, R. (1991a). Kinds of schools. In K. Borman (Ed.), *Contemporary issues in U.S. education* (pp. 38–60). Norwood, NJ: Ablex.

Page, R. (1991b). *Lower-track classrooms: A curricular and cultural analysis.* New York: Teachers College Press.

Page, R. N. (1990). Cultures and curricula: Differences between and within schools. *Educational Foundations, 4*, 49–76.

Sapir, E. (1924). Culture, genuine and spurious. *American Journal of Sociology, 29*, 401–429.

Schön, D. (1987). *Educating the reflective practitioner: Toward a new design for teaching and learning in the professions.* San Francisco: Jossey-Bass.

Spector, B. S., & Strong, P. N. (2001). The culture of traditional preservice elementary science methods students compared to the culture of science: A dilemma for teacher educators. *Journal of Elementary Science Education, 13*, 1–20.

Spenser, H. (1896). *The study of sociology.* New York: D. Appleton.

Spindler, G. D. (1955). From omnibus to linkages: Cultural transmission models. In J. I. Roberts & S. Akinsanya (Eds.), *Educational patterns and cultural configurations: The anthropology of education* (pp. 177–183). New York: David McKay.

Tesconi, C. A., & Hurwitz, E. (1974). *The question of equal educational opportunity.* New York: Harper & Row.

University of Maryland System. (1993). Special teachers for elementary and middle school science and mathematics: A proposal submitted to the National Science Foundation Teacher Preparation and Enhancement Program. Unpublished manuscript, University of Maryland, Georgia.

Van Sickle, M., & Spector, B. (1996). Caring relationships in science classrooms: A symbolic interaction study. *Journal of Research in Science Teaching, 33*, 433–453.

Van Maanen, J. (1995). *Representations in ethnography.* Thousand Oaks, CA: Sage.

Veenman, S. (1984). Perceived problems of beginning teachers. *Review of Educational Research, 54*, 143–178.

Wolcott, H. F. (1991). Propriospect and the acquisition of culture. *Anthropology and Education Quarterly, 22*, 274–278.

Woods, P. (1992). Symbolic interaction [instructional outcomes]: Theory and method. In M. D. LeCompte, W. L. Millroy, & J. Preissle (Eds.), *The handbook of qualitative research in education* (pp. 337–404). New York: Academic Press.

Williams, R. (1961). *Culture and society 1780–1950.* Harmondsworth: Penguin.

16

Employing Case-Based Pedagogy Within a Reflection Orientation to Elementary Science Teacher Preparation

Lynn A. Bryan
Deborah J. Tippins
University of Georgia

In the last decade of the 20th century, the field of science education emerged from a process-product perspective on teacher thinking and learning (Clark, 1988; Clark & Peterson, 1986; Lanier & Little, 1986) to a view of teaching as a professional practice based on standards that called for teachers to have an empowered role in developing knowledge and educational change. Reform documents such as Benchmarks for Scientific Literacy: Project 2061 (American Association for the Advancement of Science [AAAS], 1993) and the National Science Education Standards (National Research Council [NRC], 1996) present a vision of science teaching and learning that is intimately connected to experience, a vision in which the professional development of science teachers is a continuous and reflective process.

The goal of science reform documents is to provide a blueprint for the transformation of science teaching and learning—in terms of teacher preparation, curricula, teaching strategies, assessment, and so on. The reality for teachers and teacher educators who wish to implement these standards and benchmarks is that there is little guidance for making sense of science teaching and learning, especially in the context of their local schools and communities (Koballa & Tippins, 2000). Classroom cases appeal to science teacher educators as one strategy for helping prospective and practicing teachers to develop, refine, and/or transform their thinking about science teaching and learning, and subsequently their practice.

EPISTEMOLOGICAL PERSPECTIVE

Classroom cases are instructional "tools" that can be used to explicate and clarify the beliefs and knowledge of teachers. Case-based pedagogy enables teacher educators to provide their students with meaningful experiences and prepare them for reflective activity once they arrive in the field. The case creates a focused experience that reduces the complexity of teaching into manageable stories situated within a specific context (Schön, 1987). Cased-based pedagogy is a means by which both prospective and practicing teachers can examine problems of practice as well as their assumptions, beliefs, and knowledge about teaching and learning.

Our perspective of case-based pedagogy as an alternative professional development model is grounded in a reflection orientation to teacher education. A reflection orientation (Abell & Bryan, 1997) is rooted in the scholarship of Dewey (1933) and Schön (1987), in which learning is viewed as a process of knowledge construction through framing and reframing experiences. A reflection orientation to teacher education is characterized by "asking students to describe their ideas, beliefs, and values about science teaching and learning and by offering experiences that help them clarify, confront, and possibly change their personal theories" (Abell & Bryan, 1997, p. 157.)

Developing reflective practice in science teacher education is in many ways similar to conceptual change learning in science. In both processes, the learners work to clarify and refine their beliefs and apply new understandings to solving practical problems. Likewise, a concomitant set of actions defines the instructor's role in both conceptual change teaching and in coaching reflection. The teacher/teacher educator must begin with an appreciation of the students' ideas and then offer opportunities for conceptual change/reflection. Parallels between these instructor actions are further delineated in Table 16.1.

Although useful in providing a structure for science teacher education instruction, this analogy is drawn loosely. In conceptual change teaching, the science teacher guides the learner in understanding the accepted scientific view of the concept. However, in the process of coaching reflection, the teacher educator is not striving to bring the beginning reflective practitioner toward one accepted view of teaching and learning. Teachers develop and refine their theories and solutions based on the community, school, and classroom contexts in which they are engaged. Hence, what works for one teacher in a particular situation may be unique and not fully applicable to another similar situation. Second, in conceptual change teaching, the teacher generally focuses the students' attention on one or a small number of scientific concepts at a time. However, in reflecting on practice, classroom situations are often so complex that prospective teach-

TABLE 16.1
Conceptual Change Learning Versus Reflective
Practice: Instructor Actions

Science Teacher[a]	Teacher Educator
Ascertains students' existing ideas about the science concept; involves students in exploration of concept.	Elicits students' beliefs; guides students in identifying and framing issues of practice.
Provides experiences that perturb students' thinking; provides opportunities for students to compare their views with other students' and experts' views; assists students in clarifying new understanding of the science concept.	Provides experiences that perturb students' beliefs; provides opportunities for students to compare their views with other students' and experts' views; helps students clarify new frames from which they can interpret practice.
Provides opportunities for students to apply new ideas to practical situations.	Provides opportunities for students to apply solutions and determine the consequences and implications of the solutions.

[a]See Cosgrove and Osborne (1985).
Note. From *A Case of Learning to Teach Elementary Science: Investigating Beliefs, Experiences, and Tensions* (p. 39), by L. A. Bryan, 1997. Unpublished doctoral dissertation, Purdue University, West Lafayette, IN. Reprinted with permission.

ers may focus on multiple issues. The issues that prospective teachers choose to individually or collaboratively frame largely depend on their beliefs, experiences, interests, and disciplinary backgrounds (Barnes, 1992).

The use of cases within a reflection orientation is based on the assumption that cases facilitate inquiry into issues about science teaching and learning and promote the social construction of knowledge. Cases embed historical elements, social relations, and practices of the communities in which the portrayed science teaching and learning is carried out (Koballa & Tippins, 2000). Such richness of the classroom cases provides the content and simulated experiences from which prospective and practicing teachers may uncover, analyze, and refine their own conceptions about teaching and learning. The use of classroom cases from a reflection orientation is contextualized, problem-based, and premised on the teacher as generator of knowledge.

Review of Relevant Works

The use of cases in education has a well-established history that dates back to the late 1920s (Kagan, 1993; McAninch, 1993). First modeled after the case-based pedagogy in law, medicine, and business, case-based pedagogy in education has been viewed from different perspectives and used for various educational goals. For some, the structure and substance of the case take prominence (L. Shulman, 1986); for others, the deliberation and discussion processes share center stage with the substantive content (e.g., Merseth, 1991).

There are a variety of approaches to developing and using classroom cases in science teacher education. Over the last 80 years, various case formats have been developed to illustrate the dilemmas and challenges of teaching and learning (Koballa & Tippins, 2000). Cases in written medium use a narrative structure in the form of critical incidents, vignettes, and simulations to describe a given situation or experience (e.g., Greenwood & Parkway, 1989; Howe & Nichols, 2001; Kowalski, Weaver, & Henson, 1990; Silverman, Welty, & Lyon, 1992; J. H. Shulman & Mesa-Bains, 1993; Wasserman, 1993). More recently, however, pedagogical approaches to developing cases are emerging in formats such as case-as-layered-commentary (Koballa & Tippins, 2000; J. Shulman, 1992; Tippins, Koballa, & Payne, 2002), video cases (e.g., Annenberg/CPB Case Studies in Science Education, 1997; Bryan, 2000; Bryan & McLaughlin, 2002; Stigler, Gonzales, Kawanaka, Knoll, & Serrano, 1999), and integrated media cases (Abell et al., 1996; Abell, Cennamo, & Campbell, 1996; Krajcik, Ladewski, Blumenfeld, Marx, & Soloway, 1995; Ladewski, 1996).

The latter "emerging case formats" are more conducive to critical reflection (Schön, 1987; Smyth, 1989; Zeichner & Tabachnick, 1991) in that they are less characterized by a technical or procedural focus that takes context and unchallenged assumptions for granted. Instead, emerging case formats allow teaching to be viewed as professional artistry, recognizing the value of the political, economic, and/or social context; prior beliefs; tacit knowledge; and intuitive understandings. Uncertainty is a central element that highlights a more realistic nature of learning to teach. Cases-as-layered-commentary, video cases, and integrated media cases introduce an element of flexibility for accessing information that shapes and informs decisions, questions, assertions, and solutions. Such flexibility enhances the pedagogical richness of cases and is resonant with the goals and purposes of a reflection orientation to teacher education. Emerging case formats support and encourage multiple perspectives (or frames, as Schön might say), thereby enhancing the sophistication of students' questions and the richness of their observations about educational situations and dilemmas of practice (Abell, Bryan, & Anderson, 1998; Abell et al., 1996; Ladewski, 1996; Lampert & Eshelman, 1995).

CASE-BASED PEDAGOGY IN ELEMENTARY
SCIENCE TEACHER EDUCATION

This section describes two emerging case formats and how they are used to fit within the goals of a reflection orientation to elementary science teacher education. Specifically, they illustrate the way in which employing case-based pedagogy from a reflection orientation is resonant with the belief that learning to teach science is a process of conceptual change (Duschl & Gitomer, 1991; Hewson & Thorley, 1989; Osborne & Wittrock, 1983). In doing so, we emphasize our belief that students should learn to teach science in the same way that we desire for them to learn science—through inquiry into issues, problems, and dilemmas rather than learning that is procedural and prescriptive (Abell & Bryan, 1997; Bryan, 1997). Furthermore, the use of case-based pedagogy within a reflection orientation to science teacher education resonates with the current vision of the professional development of science teachers as outlined in the *National Science Education Standards* (NRC, 1996, 2000). The two examples presented are derived from our own teaching and research.

Example 1: Case-as-Layered Commentary

J. Shulman (1992) initially introduced the idea of case-as-layered-commentary, a format comprised of multiple voices. A case-as-layered-commentary features narrative accounts of classroom dilemmas written by prospective and practicing teachers. Peers and other educational scholars with unique knowledge of the particular situation develop commentaries and responses that provide context-specific insights into the case. When case narratives are developed with layered commentary, they create an opportunity for shared inquiry and a link to research, transcending the experience of the individual classroom teacher. The permeability of the layered case encourages inclusion of multiple perspectives and alternative ways of framing and comprehending the dilemmas of practice.

Our example of a case-as-layered-commentary is one that, although used in our classrooms, was written by a student teacher and educational professionals from Iloilo City in the Philippines. We often use cases in our own classrooms that incorporate cultural diversity and examine issues of cultural knowledge that students may bring with them to the science classroom. The following case-as-layered-commentary entitled, *A Rare Animal*, raises significant questions for our prospective elementary teachers and ultimately challenges their notions of, "What knowledge is important in the elementary science classroom?"

A rare animal

Layer I: Encountering the Rare Animal. Kate's Story

I am Kate Hualde Miguel, a fourth year education student majoring in Elementary and Health at West Visayas State University. I just completed my semester of practice teaching at Ticud Elementary, a city school that is frequently flooded from the overflow of the Jaro River. For me, science teaching is an adventure, a noble profession. And the world of practice teaching is a training ground for becoming an effective teacher in the years to come. During my practice teaching there were various encounters and experiences—both wholesome and dreadful. But all of the experiences taught me lessons, which will help me become a better teacher and person. Many of my experiences as a practice teacher are vivid flashbacks—images that at times make me laugh at myself, or provide me a dose of lesson. Such is the case when I encountered a rare and unfamiliar animal.

One Thursday afternoon, before holding my 2:20 class period, I took a nap. I could hardly resist my sleepiness so I fell into a deep slumber. A short while later I was awakened when Ma'am Jenetillo, my cooperating teacher, entered the classroom. I was embarrassed to be caught napping, so quickly I shook the sleepiness away. At that moment Ma'am Jenetillo informed me that the next day I would temporarily take charge of the Grade 3 science class while the teacher was away attending a 3-day seminar. I was quite lucky because the teacher in charge had prepared in advance the lesson plans and visual aids for the 3-day period. Since I knew that everything would be ready in advance, I didn't bother to study or even scan through the books to prepare for the lessons. Instead, I engrossed myself in making some sociograms of the class.

The next day, immediately after the flag ceremony, I proceeded to the Grade 3 classroom. As I entered the room 47 faces greeted me with silence. Without hesitation, I introduced myself and quickly established a rapport with students. So a few minutes later, I began the lesson with a classification chart to review the characteristics of mammals. When we arrived at the discussion of rare animals, most of the students were freely raising questions. It was in this situation that I was caught unaware.

Ma'am Jenetillo had left a list of rare animals to discuss with the class. As I made my way down the list, I stopped short when I came to porcupine. I had not expected one of the rare animals on the list to be a porcupine. This animal was unfamiliar to me. Nevertheless, I proceeded with the lesson:

Ma'am Kate:	How about porcupines? Are you familiar with this animal?
Class:	No!
Jonas:	Maybe it looks like a cat!
Ma'am Kate:	It has a spine.

Since, I really didn't have any idea what a porcupine looked like, I let the students continue to guess. I used the dialogue of the children to illustrate the features of the porcupine.

Jessie:	Ma'am, can we find porcupines here?
Ma'am Kate:	We cannot find them here at school. We can find them in the mangrove forests and far away places where no one can disturb or kill them.
Jethro:	Is it small like a bee?
Ma'am Kate:	No, it's not so small but not so big.

| Thelma: | Maybe it looks like a caribou because they eat grass like in a forest. |

As the conversation continued, I felt that I was skinned alive—I didn't know how to answer. My pupils' curiosity continued to grow—they wanted to know even more about the porcupine, especially the physical appearance of the animal. Soon I had no choice but to draw a picture of this animal, even though I had no idea about the real image of a porcupine. So that my pupils would not detect my pretension, I hid the truth by drawing an animal similar to an Iguana with a spine on its back. After I drew it they commented:

Thelma:	It looks like a crocodile!
Jethro:	Oh! What an ugly creature this porcupine is!
Thelma:	I think the porcupine is a twin to the Iguana.
Jessie:	It looks like a monster.
Jethro:	Maybe the spines on the back are made of metal.

After class, I sat down and tried to remember what a porcupine looked like to no avail. I felt guilty because I had taught the students a falsehood. At the end of the three days I returned to my Grade 5 class permanent assignment. I happened to be glancing through the pages of a science dictionary when I came across a picture of a porcupine. I was shocked! I realized that I had made a big mistake by drawing a very different creature. I was quite bewildered. What if Ma'am Jenetillo does a review of the lesson I taught? What if the pupils find a real picture of a porcupine—will they form the impression that I don't know any science? Well, this lesson on porcupines will always teach me not to take the small things for granted in teaching. Like the spines on the porcupine, it will continue to prick my memory.

Layer II: Expect the Unexpected: Ma'am Jenetillo's Story

I had just returned from an exhausting but inspirational seminar on multiple intelligences in the science classroom. Upon entering my room, I was pleased to see that Kate, the student teacher, had left the room orderly and had completed the lessons I had left in her care. I wanted to give my pupils a chance to show me just how much they had learned about rare animals in my absence. I placed the names of these animals in a box—spotted deer, anteater, warty hog and others. Then I divided my Grade 3 class into groups, inviting a representative from each group to draw an animal name from the box. After passing out large sheets of yellow cartalena, I directed each group to draw a picture of its animal in the natural habitat where it lives. Later that morning we had a parade of rare animals. Each group taped their animal's picture to the wall and then took turns describing its characteristics and habitat. Suddenly, my attention was drawn to a picture of a strange lizard-like creature. I had no recognition of this animal, so I quietly listened as students began to share.

| Jethro: | As you can see in our picture, the porcupine lives in a mangrove forest. It's a little bit smaller than a caribou. |

Thelma:	It has scales similar to a crocodile and has a long back-bone.
Jessie:	Probably the porcupine is related to the iguana, only larger.

I stared with amazement at the picture this bright group of students had drawn. I wondered where I had gone wrong. Were my pupils somehow confusing mammals with reptiles? How could their drawing of a porcupine look so dramatically different than the real animal? I couldn't wait to share this drawing with Kate to see if the pupils had shared their similar ideas of a porcupine with her.

Layer III: Porcupines in the Philippines? Sir Roger's Story

As the science coordinator at Lopez-Reyes Elementary School I have many responsibilities. In addition to making frequent classroom observations, I develop model lessons to share with teachers, facilitate their action research projects, and, in general, expose them to new teaching strategies and trends in science education. This year, in particular, I have worked hard to reduce teachers' dependence on the science textbook as the only source of all knowledge. Often times the examples used in these textbooks are not relevant to the experiences and everyday lives of Filipino children. The use of the English language in these books poses an additional dilemma for children who speak the vernacular language of Hiligarynon in the home.

As I was walking down the hallway this morning, on my way to begin "ground rounds," I happened to glance in the doorway of Ma'am Jenetillo's Grade 3 classroom. I was startled to see a drawing of an animal resembling a large lizard on the chalkboard, with the word PORCUPINE written in large block letters below. I thought immediately that I was observing some kind of joke. Why would Ma'am Jenetillo include the porcupine in her science lessons? With no porcupines to be found here in the Philippines, I have emphasized to teachers the importance of using examples of local flora and fauna. And why did this strange creature look nothing like a porcupine? I was curious about the drawing, but assumed Ma'am Jenetillo had only used it as a joke or trick. In order to satisfy my curiosity, I made a note to ask her about it at our noon mealtime.

How the use of a case-as-layered-commentary is structured in the teacher preparation course will vary from instructor to instructor. However, when multiple perspectives for each case are emphasized, the commentaries enrich opportunities for reflection and analysis of teaching and learning issues. Resonant with the belief that learning to teach science is similar to learning science, we align the structure of our use of this case with Table 16.1.

The Teacher Educator Elicits Students' Beliefs and Guides Students in Identifying and Framing Issues of Practice.   In conjunction with prospective teachers' examination of the *National Science Education Standards* for teaching and for content, as well as the state-mandated *Quality Core Curriculum* (Georgia Department of Education, 1999), we conduct small group discussions in which prospective teachers share their ideas and beliefs about what is important to teach in the science classroom. Relevant questions that arise include: Who determines what is important? What characterizes the information and skills most necessary for students' learning in the science classroom? How am I going to learn all of this science before going into the classroom? In a subsequent large group discussion, prospective teachers offer their multiple perspectives and engage in critical analysis and debate about their ideas.

The Teacher Educator Provides Experiences that Perturb Students' Beliefs, Provides Opportunities for Students to Compare Their Views with Other Students' and Experts' Views, and Helps Students Clarify New Frames From Which They Can Interpret Practice.   Next, prospective teachers read the first layer of the case-as-layered-commentary. In their discussion of the first layer of this case, prospective teachers focused almost entirely on the moral nature of teaching, examining the question of what it means to be honest within the teaching profession. Because a teacher naturally assumes a position of authority of knowledge in a classroom, prospective teachers felt that they have the responsibility to know "everything." Furthermore, admitting to children a lack of knowledge about something for which the teacher is responsible puts the teacher in a vulnerable position. Prospective teachers began to question, "How does one handle such a situation? What if 'not knowing' happens too often? What am I expected to know?"

Prospective teachers then read the second layer of the case. As they began to discuss the second layer from Ma'am Jenetillo's point of view, prospective teachers' discussions expanded to examine more closely the cultural and sociopolitical aspects of teaching. In this phase of the discussion, they questioned traditional hierarchical status structures within the context of schooling and culturally embedded ideas such as the notion of "saving face."

It was not until the prospective teachers discussed the final case commentary of Sir Espisito that they began to question the legitimacy of decontextualized science teaching. This discussion was ultimately framed within a context of prospective teachers questioning a blind acceptance of outside forms of knowledge and authority with respect to science teaching and learning.

The Teacher Educator Provides Opportunities for Students to Apply Solutions and Determine the Consequences and Implications of the Solutions. In this phase of the case-as-layered-commentary lesson, prospective teachers write their responses to the case, considering what should have been done about this dilemma from the perspective of at least one of the authors (student teacher, mentor teacher, or science coordinator). Reflecting on the multiple perspective offered in *A Rare Animal* provided prospective teachers with a new view of curriculum decision making, especially in the context of learning about the *National Science Education Standards* (NRC, 1996, 2000) and state-mandated curricula. Although their final discussions fostered some skepticism and provoked them to question who determines what "counts" as knowledge, ultimately prospective teachers came away from the experience with a healthy regard and respect for reform initiatives that outline the fundamental concepts and processes of science. They realized the critical role that they will have in making curricular decisions and developed an awareness of what a community of educators means for science education. Furthermore, the prospective teachers understood through the context of the case how important relevancy is to science instruction—that one aspect of "what counts as important" for science content is its application "to situations and contexts common to everyday experiences" (NRC, 1996, p. 109).

Example 2: Integrated Media Cases

In classrooms where many events occur simultaneously, videotape can capture the richness of classrooms from several perspectives. In recent years, videotaped cases have emerged as a popular tool for teacher education (e.g., Annenberg/CPB's *Case Studies in Science Education*). However, both written and videotaped cases share a limitation as a medium for case-based pedagogy. Both media are linear in nature, contradicting the complex, nonlinear nature of reflective thinking and classroom decision making. *Integrated multimedia cases*, on the other hand, can be more effective in presenting the complexity, ambiguity, and continuity of real-world classrooms over time.

A growing number of video cases have been developed within the context of other technology such as with hypermedia and CD-ROM (Abell, Cennamo et al., 1996; Abell et al., 1996; Krajcik et al., 1995; Ladewski, 1996). Video cases integrated with other media (hence, *integrated media cases*) extend the use of the case by allowing the users to interact with the case—that is, make decisions about which parts of the case to view, what order to view parts of a case, what supporting information they wish to examine, and so on. Integrated media cases permit user access to multiple perspectives from a wealth of sources (teachers' reflections, students' perspectives, theories presented in the "resource library," peer discussions in teacher education classrooms). Learners are able to make branching decisions and juxtapose scenes for comparison and discussion. Such a rich context for decision making supports "in-depth, open-ended, student-centered explorations" of alternative ways to think about teaching and learning (Ladewski, 1996, p. 174).

The example of an integrated media case comes from Abell, Cennamo et al. (1996). We use in our elementary science methods courses one of several classroom cases that they developed featuring a fifth-grade teacher named Mrs. Clark. In the video portion of the integrated media case, Mrs. Clark carries out a lesson on levers that is based on the "Generative Learning Model (GLM)" (Cosgrove & Osborne, 1985), which promotes a conceptual change view of teaching and learning. The following is a description of the case:

> Mrs. Clark introduces levers with a game in which a carnival man can guess a person's weight and place her on a hanging teeter totter so that she will balance a dummy on the other side. Students, working in pairs with their own materials, explore how he can do this. Next the pairs solve a series of balance problems designed by Mrs. Clark. For example, if 3 washers are placed at the end of one side of the balance, where do you need to place 6 washers on the other side to make the system balance? Finally the groups solve new problems which require moving the fulcrum to balance the system. When they come together to discuss their findings, most of the fifth graders recognize a relationship between the mass of the load and its placement on the hanging teeter-totter. However, when Mrs. Clark pushes them to explain why an off-center teeter totter with a heavy load on one side and a light load on the other balances, students decide that the sides must weigh the same, thus revealing a common misconception.
>
> To challenge this idea, Mrs. Clark shows the students a broom and asks where it would balance if placed on a finger. Students find the balance point and agree that if they cut the broom at that point and weigh each piece, the pieces would weigh the same. The class then proceeds to saw and weigh the broom. When confronted with the data (the short end of the broom weighs twice as much as the long end), one student exclaims, "We need a new scale!" Mrs. Clark encourages the class to find another way to explain the discrepancy, focusing their attention on the relationship between load and distance. In the final lesson, we see the class solving seesaw problems using an 8-foot board, a half-log fulcrum, and the students themselves as loads. (Abell, 1996)

A series of reflection tasks and discussions accompany the use of the integrated media levers case. Reflection tasks and discussions require students to clarify their beliefs about teaching and learning, compare their ideas to their peers' and instructors' ideas, and examine evidence to support their ideas and beliefs about science teaching and learning. The video case of Mrs. Clark's lesson is like a window into her classroom—a means for elementary teachers to reflect on someone else's practice in the process of developing and refining their own practice (Abell & Bryan, 1997). The supporting materials provide access to Mrs. Clark's reflections on her teaching, demographic and geographic information about the school and community, curriculum materials, and examples of student work (e.g., pre-lesson assessments, data sheets, and student drawings). Once again, using the reflection cycle depicted in Table 16.1, we illustrate the structure of our use of the integrated media levers case.

The Teacher Educator Elicits Students' Beliefs and Guides Students in Identifying and Framing Issues of Practice. Prior to viewing the video of Mrs. Clark's instruction, prospective teachers are asked to write about what they expect to see. Guiding questions for their writing include: What do you expect to happen in this science lesson? What do you think the teacher will be doing? What do you think the children will be doing? A subsequent large group discussion allows the elementary science methods students to share their ideas and explain from where their ideas originate. Over the years, we have found some students to be very specific in their description of what they expect. For example, one student expected to see Mrs. Clark introduce levers by showing pictures of levers such as scissors, a can open, and a wheel barrow. She thought that Mrs. Clark then would allow children to explore with levers while working in small groups. Finally, the prospective teacher predicted that the children might be asked to write a report on the use of levers in their homes or at school. Although this student's response showed some level of specificity, other students are relatively vague about their expectations. For example, one student responded, "I expect to see children engaged in hands-on activities." Many prospective teachers base their ideas on previous observations during their education program, others base their ideas on their own science experiences (or lack of science experiences), and some develop their ideas from current research and expert opinions to which they have been introduced in education courses.

The Teacher Educator Provides Experiences that Perturb Students' Beliefs, Provides Opportunities for Students to Compare Their Views with Other Students' and Experts' Views, and Helps Students Clarify New Frames From Which They Can Interpret Practice. During the viewing of the video, prospective teachers' are asked to chart the teacher's actions and students' actions in the video. Prospective teachers view the case first in its entirety, and then may review any segments for which they need clarification or more information. Class discussion provides a forum for the prospective teachers to discuss what they have viewed in light of their own beliefs about teaching and learning. In addition, they access Mrs. Clark's reflections about her teaching to see how Mrs. Clark's views compare with their own. Guiding questions for this part of the methods lesson include: Given that one's teaching actions are in many ways a reflection of one's beliefs, what do you think Mrs. Clark's beliefs are about: (a) how students learn; (b) the role of the teacher in the science classroom; and (c) the role of students in the science classroom? What evidence supports your assertions? How do her beliefs compare to your own? What are some alternative ways of teaching this lesson? On what epistemological assumptions are your alternatives based? As these questions are addressed in small and large group settings, prospective teachers utilize the opportunity to review video, access a resource library within the integrated media materials (e.g., to find out more information about the theoretical tenets of constructivism or the *National Science Education Standards* that align with the lesson), and observe Mrs. Clark reflecting on her teaching.

The Teacher Educator Provides Opportunities for Students to Apply Solutions and Determine the Consequences and Implications of the Solutions. After the prospective teachers complete their viewing of the video case and seek other available resources as necessary, we assist them in examining their refined or confirmed beliefs by providing field experiences in which they plan and implement their own GLM science lesson in a local elementary school. For many prospective teachers, the field experience is a theory-changing step in the process of learning to teach elementary science. They recognize that their beliefs are still in question and they will have to rethink many pedagogical issues that frustrate, confuse, or perplex them as they gain experience in the profession.

CONSIDERATIONS FOR USING CASE-BASED PEDAGOGY IN ELEMENTARY SCIENCE TEACHER PREPARATION

In this chapter, we have presented our epistemological perspective and use of case-based pedagogy as an alternative professional development model that is grounded in a reflection orientation to elementary science teacher education. A crucial aspect of employing case-based pedagogy within a reflection orientation is its potential to align with the current vision of science teacher professional development as outlined in the *National Science Education Standards* (NRC, 1996, 2000). Cases, especially those of emerging case formats, are resonant with the kinds of professional development strategies suggested in current reform initiatives:

"Strategies in which teachers study their own and others' practice are especially powerful in building their knowledge. Some examples of this kind of professional development are the study of videos of classroom teaching; discussion of written cases of teaching dilemmas; and the study of curriculum materials and related student work (assignments, lab reports, assessments, etc.)" (NRC, 2000, p.105). Nonetheless, employing case-based pedagogy requires consideration of several issues. First, case-based discussions involve complex interactions that necessitate experienced facilitators. The pedagogical assumptions and communication styles of the facilitator can have a significant influence on the outcome of a discussion. For example, some facilitators emphasize the "exemplary case" from which core principles can be extracted for group discussion. Other facilitators are comfortable with viewing each case as unique, idiosyncratic, and replete with uncertainties. In our example of case-as-layered-commentary, the dilemma is multilayered and cannot simply be extracted from the classroom case context and reduced to a simplistic set of practical guidelines for handling similar situations. The situated nature of the events and relationships that surround the dilemma will influence how prospective teachers understand the contextual and cultural knowledge of the case.

Second, teacher educators who use cases that portray science instruction should be cognizant of their students' content knowledge. Few elementary teachers feel well qualified to teach science, often expressing a lack of confidence in their knowledge of content and pedagogy for science teaching (Lederman, Gess-Newsome, & Zeidler, 1993; Weiss, Banilower, McMahon, & Smith, 2001). If teachers are not comfortable with their knowledge of the science content portrayed in a case, then the efforts of a teacher educator to use the case for reflection on pedagogical content issues may be hindered. Hence, we often preface the use of a case with science instruction that reflects our advocated view of how learners construct knowledge of science via scientific inquiry.

Finally, within a reflective learning experience, the teacher educator's role of coaching reflection, like reflection itself, is a nonlinear process. There are no prescribed, specified series of steps that one takes in order to coach someone to develop reflective thinking. Learning to observe and analyze teaching; learning to isolate, frame, and reframe problems of practice; and learning to take action and interpret that action are skills that take time and practice for a teacher to develop. Hence, a teacher's interpretations are fundamental to the process, but may not always be the same as the teacher educators' interpretations. Rather than confronting students of teaching with alternative conceptions and administering prescriptions for improving their practice, teacher educators should strive to share responsibility in the learning process. They may accomplish shared responsibility by modeling reflective thinking with their students. Shared responsibility also suggests that teacher educators listen to and understand students' experiences, beliefs, and interpretations of practice. Like students of science, students of teaching construct their own understandings and meanings based on their existing conceptions about teaching and learning. Coaching reflection requires that the coach (the teacher educator) engage in close observation and discussion to enable teachers to explore any disparities between their existing conceptions/beliefs and observation of others' (or their own) practices.

The *National Science Education Standards* (NRC, 1996, 2000) define inquiry as a central component of the teaching standards. We, as teacher educators, are challenged to design experiences in which teachers learn to teach science in the same way that we desire for them to learn science. Case-based pedagogy allows teachers to learn through inquiry into issues, problems, and dilemmas; to collect evidence to address their questions and support their refined or modified beliefs; to communicate within a community of learners their explanations from the evidence; and to consider their beliefs in light of alternative explanations and others' perspectives (NRC, 2000). Cased-based pedagogy provides a vehicle for coaching students of teaching to inquire systematically and purposefully into their own practice, encouraging them to make such inquiry a habit that will become increasingly valuable throughout their careers.

REFERENCES

Abell, S. (Producer). (1996). *Levers: A fifth grade module by Mrs. Clark* [Video case study, classroom 3]. (Available online at http://www.coe.missouri.edu/~abells/roes/roes.html). Used with permission.

Abell, S. K., & Bryan, L. A. (1997). Reconceptualizing the elementary science methods course using a reflection orientation. *Journal of Science Teacher Education, 8*, 153–166.

Abell, S. K., Bryan, L. A., & Anderson, M. A. (1998). Investigating preservice elementary science teacher reflective thinking using integrated media case-based instruction in elementary science teacher preparation. *Science Education, 82*, 419–509.

Abell, S., Cennamo, K., Anderson, M., Bryan, L., Campbell, L., & Hug, J. W. (1996). Integrated media classroom cases in elementary science teacher education. *Journal of Computers in Mathematics and Science Teaching, 15*, 137–151.

Abell, S., Cennamo, K., & Campbell, L. (1996). The development of integrated video cases for use in elementary science methods courses. *Tech Trends, 41*(3), 20–23.

American Association for the Advancement of Science (AAAS). (1993). *Benchmarks for science literacy: Project 2061*. New York: Oxford University Press.

Annenberg, C. P. B. (1997). *Case studies in science education.* Smithsonian Institution Astrophysical Observatory in association with the Harvard–Smithsonian Center for Astrophysics.

Barnes, D. (1992). The significance of teachers' frames for teaching. In T. Russell & H. Munby (Eds.), *Teachers and teaching: From classroom to reflection* (pp. 9–32). New York: Falmer.

Bryan, L. A. (1997). *A case of learning to teach elementary science: Investigating beliefs, experiences, and tensions.* Unpublished doctoral dissertation, Purdue University, West Lafayette, IN.

Bryan, L. A. (2000). *Science education in the Japanese classroom: A video case.* University of Georgia, Athens, GA.

Bryan, L. A., & McLaughlin, H. J. (2002). *Everyday life in a rural Mexican escuela unitaria: A video case.* University of Georgia, Athens, GA.

Clark, C. (1988). Asking the right questions about teacher preparation: Contributions of research on teacher thinking. *Educational Researcher, 17*(2), 5–12.

Clark, C., & Peterson, P. (1986). Teachers' thought processes. In M. C. Wittrock (Ed.), *Handbook of research on teaching* (3rd ed., pp. 255–296). New York: Macmillan.

Cosgrove, M., & Osborne, R. (1985). Lesson frameworks for changing children's ideas. In R. Osborne & P. Freyberg (Eds.), *Learning in science: Implications of children's science* (pp. 101–111). Portsmouth, NH: Heinemann.

Dewey, J. (1933). *How we think.* New York: Heath.

Duschl, R. A., & Gitomer, D. H. (1991). Epistemological perspectives on conceptual change: Implications for educational practice. *Journal of Research in Science Teaching, 28,* 839–858.

Georgia Department of Education. (1999). *Georgia learning connections—Quality core curriculum standards introduction.* Retrieved April 14, 2002, from http://www.glc.k12.ga.us/qstd-int/homepg.htm

Greenwood, G., & Parkway, F. (1989). *Case studies for teacher decision-making.* New York: Random House.

Hewson, P. W., & Thorley, N. R. (1989). The conditions of conceptual change in the classroom. *International Journal of Science Education, 11,* 541–553.

Howe, A. C., & Nichols, S. E. (2001). *Case studies in elementary science: Learning from teachers.* Upper Saddle River, NJ: Merrill.

Kagan, D. (1993). Contexts for the use of classroom cases. *American Educational Research Journal, 30,* 703–723.

Koballa, T., Jr., & Tippins, D. (2000). *Cases in middle and secondary science education: The promise and dilemmas.* Upper Saddle River, NJ: Merrill.

Kowalski, T., Weaver, R., & Henson, K. (1990). *Case studies on teaching.* New York: Longman.

Krajcik, J., Ladewski, B., Blumenfeld, P., Marx, R., & Soloway, E. (1995, April). *Multiple perspectives on designing, developing, and using integrated multimedia for teacher enhancement.* Paper presented at the annual meeting of the American Educational Research Association, San Francisco, CA.

Ladewski, B. (1996). Integrated multimedia learning environments for teacher education: Comparing and contrasting four systems. *Journal of Computers in Mathematics and Science Teaching, 15,* 173–197.

Lampert, M., & Eshelman, A. (1995, April). *Using technology to prepare effective and responsible educators: The case integrated multimedia in mathematics methods courses.* Paper presented at the annual meeting of the American Educational Research Association, San Francisco, CA.

Lanier, J., & Little, J. (1986). Research on teacher education. In M. C. Wittrock (Ed.), *Handbook of research on teaching* (3rd ed., pp. 527–569). New York: Macmillan.

Lederman, N. G., Gess-Newsome, J., & Zeidler, D. L. (1993). Summary of research in science education—1991. *Science Education, 77,* 465–559.

McAninch, A. (1993). *Teacher thinking and the case method: Theory and future directions.* New York: Teachers College Press.

Merseth, K. (1991). *The case for cases in teacher education.* Washington, DC: American Association for Higher Education and American Association of Colleges for Teacher Education.

National Research Council (NRC). (1996). *National science education standards.* Washington, DC: National Academy Press.

National Research Council (NRC). (2000). *Inquiry and the national science education standards.* Washington, DC: National Academy Press.

Osborne, R. J., & Wittrock, M. C. (1983). Learning science: A generative process. *Science Education, 76,* 489–508.

Schön, D. (1987). *Educating the reflective practitioner: Toward a new design for teaching and learning in the professions.* San Francisco, CA: Jossey-Bass.

Shulman, J. (Ed.). (1992). *Case methods in teacher education.* New York: Teachers College Press.

Shulman, L. (1986). Those who understand: Knowledge growth in teaching. *Educational Researcher, 15,* 4–14.

Shulman, J. H., & Mesa-Bains, A. (Eds.). (1993). *Diversity in the classroom: A casebook for teachers and teacher educators.* Hillsdale, NJ: Lawrence Erlbaum Associates.

Silverman, R., Welty, W., & Lyon, S. (1992). *Case studies for teacher problem solving.* New York: McGraw-Hill.

Smyth, J. (1989). Developing and sustaining critical reflection in teacher education. *Journal of Teacher Education, 40*(2), 2–9.

Stigler, J. W., Gonzales, P., Kawanaka, T., Knoll, S., & Serrano, A. (1999). *The TIMSS videotape classroom study. Methods and findings from an exploratory research project on eighth-grade mathematics instruction in Germany, Japan, and the United States.* Washington, DC: U.S. Government Printing Office.

Tippins, D. J., Koballa, T. R., & Payne, B. D. (2002). *Learning from cases: Unraveling the complexities of elementary science teaching.* Boston: Allyn & Bacon.

Wasserman, S. (1993). *Getting down to cases: Learning to teach with case studies.* New York: Teachers College Press.

Weiss, I. R., Banilower, E. R., McMahon, K. C., & Smith, P. S. (2001, December). *Report of the 2000 National Survey of Science and Mathematics Education.* Retrieved September 5, 2002, from http://2000survey.horizon-research.com/reports/status/complete.pdf

Zeichner, K., & Tabachnick, R. (1991). Reflections on reflective teaching. In R. Tabachnick & K. Zeichner (Eds.), *Issues and practices in inquiry-oriented teacher education* (pp. 1–21). London: Falmer.

17

Models of Elementary Science Instruction: Roles of Science Specialist Teachers

M. Gail Jones
North Carolina State University

Julie Edmunds
University of North Carolina at Chapel Hill

Elementary science instruction plays a critical role in providing students with the fundamental educational experiences that contribute to the long-term development of sophisticated understandings of science phenomena. In these earliest years of schooling, students develop the ability and desire to question; they also begin to develop strategies to answer questions about the world around them (Tilgner, 1990).

The *National Science Education Standards* (National Research Council [NRC], 1996) recommended a shift in science instruction from an emphasis on recall of scientific facts and information to more understanding and use of scientific inquiry. Beginning with the elementary grades, students are expected to design and implement their own simple investigations. In addition to understanding the process of doing science, students should develop in-depth understandings of key scientific concepts in the physical sciences, life sciences, earth and space sciences, science and technology, as well as in the areas of personal and social perspectives of science and the history of science. To foster this shift in focus, elementary teachers are expected to present a program of scientific inquiry in the classroom, guide and facilitate student learning, use multiple methods of assessments to measure students' learning, design and manage learning environments that provide students with a place and resources to learn science, develop communities of learners that engage in scientific inquiry, and support development of a school science program (National Research Council, 1996).

Although the expectations for quality science programs in elementary schools are high, in many cases elementary teachers feel unqualified to teach science and do not enjoy teaching it (Schoenberger & Russell, 1986; Tilgner, 1990). In addition, elementary teachers have typically taken, at most, five science courses in their teacher preparation (Blank, Kim, & Smithson, 2000; Weiss, Banilower, McMahon, & Smith, 2001). Only 8% of elementary teachers report having attended a state or national science teacher association meeting over the last 3 years and half of all elementary teachers report that they have spent less than 6 hours in inservice in science in the last 3 years (Weiss et al., 2001). This lack of depth of preparation coupled with a low priority placed on science, particularly in the era of accountability for reading and mathematics (Jones, Hardin, Chapman, Yarbrough, & Davis, 1999), results in a minimal amount of time spent on science instruction at the elementary school level. On average, teachers spend a very small percentage of their time teaching science, typically less than 3 hours a week (Blank et al., 2000; Goodlad, 1984), and some may not teach it at all (Schoenberger & Russell, 1990).

The lack of elementary teacher preparation in science content and elementary teachers' lack of confidence in their ability to teach science makes providing quality science education programs a challenge. One strategy for addressing this problem has been to place science specialists in elementary schools (Abell, 1990; Hounshell, 1987; Swartz, 1987). This chapter examines the roles of the science specialist in schools and discusses the different models of science instruction. By looking closely at three different school's programs, the discussion highlights the differences in program structures and discusses how school contexts may impact the implementation of the different models of science programs. It then considers elementary teacher education programs and considers how programs can be designed to prepare science specialists to work in elementary schools.

THE SCIENCE SPECIALIST

In the United States, approximately 15% of elementary students receive science instruction from a science specialist in addition to their regular teacher and another 12% of students receive science instruction from a science specialist instead of their regular classroom teacher (Weiss et al., 2001). Ideally, specialists are individuals who have a strong background in science and can lead the science program in the school. Specialists can be more knowledgeable in science content and pedagogy and can keep up with rapid changes in science more easily than the typical classroom teacher (Williams, 1990). One of the few studies of science specialists by Schwartz, Lederman, and El-Khalick (2000) found that the views and instructional practices of science specialists are likely to be more consistent with science reform efforts than those of elementary teachers. Their study also provided support for the idea that science specialists may be more "effective" at implementing the vision of science education espoused by recent science reforms.

Although there has been ongoing discussion related to the issue of science specialists on the philosophical level (Abell, 1990; Hounshell, 1987; Swartz, 1987; Williams, 1990), there is little research on the actual implementation of science specialists in the classroom. A recent study by Schwartz et al. (2000) is one of the first to begin to address this dearth of research with an evaluation of the specialist model in elementary schools.

Understanding how science specialists function in the elementary school can shed light on how general elementary teachers as well as science specialists should be prepared. Teacher preparation educators can build programs that provide teachers with the skills and competencies they need to teach science, as well as the collaborative skills needed to implement a specialist model of elementary science instruction within the larger school context.

A Closer Look at Specialist Program Models

The three schools described in the following sections were part of a study of the science programs in 11 elementary schools in an urban school system. A close look at these schools shows the interplay between context, presence of a science specialist in the school, and science program implementation. The schools: Peach Street, Harper, and Bismuth (all names are pseudonyms) illustrate the different models of elementary science programs. One of the schools did not use a science specialist and represents the *traditional model* of the classroom teacher as the teacher of science. The second school used a *science specialist* who served the entire school by having classes come in and out of a science laboratory for instruction on a regular rotating basis. The third school used the science specialist as a *science resource teacher* to supplement to the classroom teacher by providing whole class instruction and also working with teachers as the regular teacher planned science instruction. Demographics of the three case study schools are shown in Table 17.1.

A variety of data were collected at each school, including observations of schools, classroom teachers, and science specialists; interviews with principals, classroom teachers, and science specialists; a review of school promotional materials, including school Web sites; and a quantitative survey administered to teachers in the participating schools.

Observations were done with the intent of noting the visibility of science in the school, as well sampling the quantity and quality of science in-

TABLE 17.1
Demographics of Participating Schools

Science Program Model	School Name	Enrollment	Racial/Ethnic Breakdown	% Receiving Free/Reduced Lunch
Science Specialist	Peach Street	804	76% African American; 2% multiracial; 2% Hispanic; 3% Asian; 17% Euro-American	31%
Science Resource Teacher	Harper	305	84% African American; 6% multiracial; 10% Euro-American	64%
Traditional Science Program	Bismuth	740	62% African American, 10% Hispanic, 28% Euro-American	66%

struction. Formal school observations were done after the school year had already been in operation for several months. Observations were made of the entry to the school, hallways, and the surrounding school environment for evidence of science instruction. Field notes were made of any student work in hallways and bulletin boards of science-related content, as well as the presence or absence of science-related resources, such as nature trails, greenhouses, butterfly gardens, or weather stations. Two days of instruction were observed in two different randomly selected fourth-grade classrooms for each school in order to develop a snapshot of the instructional day. We believed it was important to observe a full day of instruction in order to capture formal science instruction as well as informal science teaching that might take place during an integrated lesson or in a teachable moment that might take place during recess or during instruction on other nonscience topics. In addition, one lesson by each of the science specialists was observed.

Each regular fourth-grade classroom teacher was interviewed concerning her teaching experience, comfort with teaching science, perspectives on science instruction, and the presence of science in the school. The two science specialists were interviewed; the questions addressed issues such as their backgrounds, their perspectives on teaching science, their perceptions of their role in the school, and their interactions with teachers and parents in the school. The principal in each of the schools was also interviewed regarding her philosophy of science instruction, perception of the science instruction occurring in the school, and support for science.

In addition to school and teacher observations and interviews, promotional literature (school brochures, and the school and district Web sites) from the school was analyzed for references to science and to the importance placed in science by the school. Putting all the different evidence together allows us to situate the science program models in real school contexts as we consider the pros and cons of the science specialist.

The Science Instructor Model: Peach Street Elementary School. Some schools use a science specialist as the school's science instructor who teaches science to classes across different teachers and grade levels. Peach Street Elementary School used this model, and as seen in this case study, the model allowed students to have approximately one science lesson a week in the science laboratory with a motivated teacher who provided quality science lessons.

Peach Street is a year-round K–5 elementary school that serves a racially and economically diverse student body. The principal, who was previously a science teacher, has been there for 7 years. The school is a relatively new, large school situated in a middle-income neighborhood. The grounds are large, with two large parking areas and two playgrounds with fields in between. Outside of the school is a nature trail that runs through the woods behind the school, created and maintained by the school's parent organization. There is also a fenced-off science area with rabbits and chickens running loose; this area can be seen through the windows of one of the hallways in the school. Teachers are able to use the animal pen, although few other than the science specialist take the opportunity.

Peach Street espouses a philosophy of student-centered instruction that uses an integrated approach to highlight students' connections to learning. Its Web site says Peach Street "is committed to creating a continuous year-round learning environment that is holistic and hands-on in its approach." Science is not explicitly mentioned in the school's literature, although partnerships exist with a local science museum and a planetarium. The hallways have displays of science activities outside the individual classrooms: volcano drawings, coastal habitats, and space explorations. The art teacher has been integrating science into his work through products involving spiders and bats. The school holds an annual Science Spectacular Day in which every class in the school invites a local scientist to make presentations to children throughout the day.

Ms. Dartmouth, the science specialist at Peach Street, was originally an art teacher who became involved in science through being trained in a new science curriculum adopted by the school system. Her principal asked her if she would be willing to become a science specialist for the school and she agreed.

The model of science instruction Ms. Dartmouth follows is that of a science specialist who provides instruction in her classroom once every 6 days to every class in the school. She has her own classroom at Peach

Street known as the "science lab." The classroom is the size of a regular classroom and contains tables for students, animals in cages, and supplies and science resources for her and other teachers. Classes come to the "lab" for science instruction and, for the most part, neither teachers nor assistants stay with their students; this is considered a planning period for the regular classroom teachers. Ms. Dartmouth sees herself as a "science leader" in the school. "My other role is to get children (throughout the school) excited about teaching science." Teachers in the school express respect for Ms. Dartmouth and appreciate her work coordinating science and providing them with science materials.

In one of the lessons observed, Ms. Dartmouth presented the topic of simple machines for fourth graders. On the tables were materials to make levers, a fulcrum, a board, and a spring balance, plus various weights to be lifted. Ms. Dartmouth began the lesson by asking students what they already knew about levers. She then ensured that they knew how to use the materials on their desks and gave the students 10 minutes to explore with the materials. The students constructed a lever and measured the force that was required to lift each weight while using the lever. At the end of the lesson, Ms. Dartmouth led a brief discussion of the conclusions the students had drawn from their explorations and let the students know that the following week they would use the same set-up with different weights and would collect and graph the data.

Despite the existence of a science specialist and her science instruction, regular classroom teachers at Peach Street Elementary are also held responsible for science instruction by the principal. In her interview, the principal said, "Every teacher ought to teach science." Teachers are expected to teach science 45 minutes a day in 2-week blocks, alternating between science and social studies. Actual implementation of this expectation, however, appeared to be dependent on the scheduling of the individual teacher.

Ms. Cooper, a fourth-grade teacher who was observed and interviewed, said that she was least comfortable teaching science; however, her students had just completed a unit on animals where they had chosen and researched particular animals. The two observations in her classroom, done a month and half apart, showed no science being taught. The lessons observed (language arts, mathematics, and social studies), were all teacher-directed and utilized the textbook as the center of the lesson.

Ms. French was the other randomly selected fourth-grade teacher whose class was observed at Peach Street Elementary School. Her style of teaching differed greatly from Ms. Cooper's. The class day began with the students writing about their memories of their first day of school. When one boy finished, she asked him to read the story aloud to another student, because "Sometimes the brain works faster than the hand." Her children then completed a social studies activity; in this case, they went to the computer lab where they read the assignment on the web and answered questions about the web-based articles. This was followed by mathematics in which the students were assigned to one of two groups where they worked on equivalent fractions, primarily through a game format. Ms. French asked the students why it was important to learn equivalent fractions, "If anybody says, so you can pass the End-of-Grade test, I am going to scream!" The morning concluded as the students set up for a science lesson to be continued after lunch. On their tables, the students placed a vacuum, a computer, an alarm clock, a VCR, a television, a timer, and a fan. The science lesson assignment was to take these complex machines apart and "Pay attention to the simple machines used to make this machine." After lunch, the students broke into groups and spent 45 minutes finding the simple machines in their more complex ones, recording their predictions of the number and kinds of simple machines and the final data on a chart. After the exploration time, the students shared their results with the other groups, prompted by questions from the teacher.

Despite the expectations held of the teachers, there is some concern at Peach Street that teachers may pull back from teaching science because of the existence of the science specialist. The principal pointed out, "With science resource teachers, teachers may not feel that they have the need to teach science." Nevertheless, she felt the risk was worth it, "We are doing a better job of teaching science by having expectations of the teachers as well as having (Ms. Dartmouth)."

The principal supported the model of the science specialist model used at Peach Street. She shared her belief that science is an important part of the curriculum: "I believe everything that has had a profound impact on our lives has a science perspective. It is extremely important. I don't ever want to see science in the elementary (school) compromised. I want every teacher to teach science. I don't want any child to miss out on science." Her support sustained the science specialist model even when a small group of parents questioned why she did not use the science specialist funds to fund a second language teacher.

The Science Instructor Model as implemented at Peach Street Elementary School resulted in considerable presence for science in the school and buildings. At this school, science had a place in the science laboratory and there was leadership for science-related events. The teachers tended to provide quality lessons to students and the principal provided support for the science program and the work of the science specialist.

The Resource Model: Harper Elementary School.   Science specialists can also take the role of a science resource teacher whose role is to provide leadership for the school's science program rather than providing the majority of the instruction to children. As seen in the case study of Harper Elementary School, this instructional model keeps the responsibility for teaching science in the hands of the classroom teacher but provides assistance in the form of curriculum and materials support, professional development, assistance with planning, and leadership for the overall science program.

Harper Elementary School, established in 1965, has historically been a school serving almost exclusively low-income African American students. Five years ago, it became a magnet school and as a result of becoming a magnet, the school's population has changed somewhat, with a reduction in the percent receiving free and reduced lunch from 95% to 64% and an increase in the Euro-American enrollment from 1% to 10%. The school is located in a historically African American section of town near subsidized housing projects. Although the school is small, the grounds are spacious, with two playgrounds and large fields. A nature trail winds through the trees in the back within which is a wooden stage with log benches, a bird blind, and a tracking box. Outside of the school, there is also a small weather station, a wetland area with picnic tables, a greenhouse, and garden plots scattered around the school.

Harper is an arts school whose literature emphasizes the role of student creation. Of the three schools observed, this school had the most evidence of science being integrated into other subjects. For example, in the primary hallway, teachers had displays of bones, life cycles of human beings and other animals, and descriptions of some of the physical processes involved in creating mummies in ancient Egypt. Other school specialists at Harper Elementary School also integrate science into their area: For example, the art teacher did a lesson in which students made animals that used the principles of simple machines to move.

The school has numerous schoolwide science activities, including a Science Fun Day, a natural areas clean up program, a recycling program, presentations by visiting scientists, and a variety of family nights such as Bat Night and Astronomy Night. All of these events are coordinated by the science specialist, who experiences significant support from many of the parents in the school. When the number of resource positions was cut, the group of Harper teachers and parents made the decision to keep the science resource position. This decision was significant because Harper is known as, and sells itself as, an arts magnet school. The end result was that the parents voted to keep the science position and reduce one of its arts positions to half time.

Ms. Tremain, the science specialist at Harper, has a master's degree in science education. She worked as a science educator for a museum and was a middle school science educator until she joined Harper's staff as the science specialist. She is currently a half-time science specialist and a half-time (information) technology specialist for the school. She has her own very crowded classroom; the computer laboratory takes up half and the other half serves as a science laboratory. One wall of the science lab is filled with commercial science kits and kits she has made herself. There are also many teacher and student resource books, as well as manipulatives and consumable materials that have been purchased over time. Teachers may sign out any of the materials for their own use. The teachers and the principal feel that the lab provides the resources necessary for science instruction to happen in the classrooms.

Ms. Tremain sees her role as a support person and a facilitator, "It was never my intention to teach all the science in the school." She teaches classes; however, there is no set schedule. Instead, she works out a schedule with individual teachers. For some teachers, she goes into their classroom and assists with a lesson. For others, she takes the students and works with them in her science lab. Teachers can choose whether or not to have her focus on science or on information technology. In cases where the primary focus is technology, she attempts to integrate science content into the technology lesson, such as researching animals or hurricanes on the Internet.

Ms. Tremain was observed teaching a lesson to fourth graders that was focused on the creation of rockets that they planned to launch on field day. The students worked in groups, following written directions on the creation of the rocket, guided through the process by Ms. Tremain. The children were divided into groups that were responsible for reading the directions and making the components of the rockets. The activity integrated mathematics by involving substantial measuring. In addition, the subject was integrated with regular classroom instruction on the solar system.

The science teaching that occurs in regular classrooms in Harper Elementary School is often integrated with other subject areas, an approach supported by the principal: "I like the way we do it here, it is integrated." Ms. Pritchard, one of the randomly selected fourth-grade teachers, exemplified this philosophy. In her interview, she said that she did not do a "science lesson that wasn't integrated." The observation in her classroom bore this out. A schedule that represented instruction at Harper Elementary included the following: Students began the day with silent reading. This was followed by a class lesson on measurement and volume equivalencies, with a worksheet. Midmorning, students attended various special activities: They went to physical education in the gym, had a special artist-in-residence experience, and visited the book fair. Upon returning to the classroom, students worked in groups while they finished their measurement worksheets. The final activity was preparation for a play on the solar system, followed by the performance of the play.

A primary focus of the observed instruction of Ms. Brown, the other fourth-grade teacher selected for observation, was preparation for the math and reading tests that occur at the end of the year. One of the entire observed mornings from 8:00 AM to 12:00 noon was spent reading passages with multiple choice tests and reviewing the mathematics skills that would be on the test, including place value and expanded notation. On another day, the teacher gave science assignments out of the science textbook.

Both the classroom teachers and principals recognized that science instruction does not happen as frequently as they would like. The principal said she thought the students were getting about 30 minutes a week of science-focused instruction, "which is not a lot." One of the teachers said there was "no science lesson that wasn't integrated." Both teachers and the principal said that time spent on science was reduced by the state's standardized testing program in reading and mathematics. Harper has a history of low achievement on the tests and the teachers shared that they feel significant pressure to increase their students' reading and math scores. "My kids almost don't get any science instruction. We concentrate on language arts and math due to testing" (Ms. Pritchard).

The Science Resource Teacher model as implemented at Harper Elementary School provided leadership for teachers' development of skills in science teaching, leadership for science-related events in the school, and management of the science materials. There was quality instruction within the resource teacher's instruction, but other teachers in the school tended to place science secondary to other subjects. The schoolwide emphasis on the arts, integrating the curriculum, and a focus on raising test scores worked against making science a school priority.

The Classroom Teacher Model: Bismuth Elementary School. In the traditional elementary program each teacher teaches a self-contained classroom of students and teaches all the curricular subjects. This model has the advantage of allowing the teacher to know each student well and to modify instruction for individual students. The traditional elementary program structure can make considerable demands on the single teacher to do all the planning and teaching for a wide range of subjects. One of the benefits of the traditional program structure is the considerable flexibility teachers have to do creative planning and interdisciplinary instruction. However, as seen in the case study of Bismuth Elementary School, there is potential in the traditional elementary program for some subjects to be left out of the day's instructional program or to be put aside because of competing demands. The traditional model also makes it difficult for each teacher to have access to the materials and equipment that they need to teach science.

Bismuth Elementary School includes a large building located on a very busy street in an area of town that has witnessed significant business growth. This school is also on a large plot of land with plenty of room for parking and for playgrounds. Outside of the school, there is an unused gardening area. The school has changed much in the past several years. Initially, the school served rural Euro-American students. Bismuth's demographics changed dramatically with an influx of African American students from the inner city when the county and city school systems merged. In response to this, many of the Euro-American students fled to the religious school located across the street from Bismuth. Bismuth has also had seven principals in the last 6 years, a factor that has contributed to confusion over the vision for the school.

The focus at Bismuth is on reading, writing, and mathematics through "direct instruction, bench-mark testing, and criterion-reference testing. This knowledge is most beneficial when they apply it to directive instruction from the teacher" (school Web site). The only reference to science occurs in the discussion of the gifted program in which the focus in K–3 is on science enrichment.

As you walk in the front door of Bismuth Elementary School, you see flags of different countries. The school mission statement shows the importance placed on diversity in the school: "Bismuth Elementary School will provide a safe, accepting, and challenging environment where students develop creative and critical thinking, interact with and influence others positively and respect diversity of backgrounds and cultures." Many of the displays on the walls outside of classrooms reflect a priority placed on writing such as: "Welcome to the World of Books," "Falling into Writing," "Thanksgiving," "Illustrate the Main Idea," "Columbus Day," "Writing," "Pumpkins," "Fall," "Inside Topics to Write About," and "Favorite Authors." One of the classes on another hall displayed students' writing assignments about spiders. Bismuth has a writing specialist but no science specialist. There was very little evidence of science instruction; science kits were noticed in one of the classrooms and several of the classes displayed work on spiders and there was also one example of a picture made of macaroni entitled, "What Plants Need." There were no schoolwide science activities reported by the teachers or the principal.

Without a science specialist, teachers at Bismuth are expected to teach science for 45 minutes a day. In practice, this may or may not happen because science is often scheduled at the end of the day. One teacher said, "We're supposed to teach science for 45 minutes every day. Sometimes social studies runs over though." Teacher instruction of science is characterized by the principal as "all over the board," based on the comfort level of the teachers. The teachers observed and interviewed indicated that science was the subject they felt least comfortable teaching. They also noted that they did not have any science supplies unless they purchased them with their own money. Nevertheless, science lessons were observed in both classrooms.

Ms. Farragutt was the third teacher this year for the fourth-grade class that was observed. Consequently, she felt a strong need to be very firm and strict, in order to control student behavior. When asked about her philosophy of science, she stated, "Students need discipline. They need rules and need to know who's in control." Her classroom reflected this need to maintain order. The first lesson of the day was a mathematics lesson on line graphs. The content of the question was science-related, "For a science project, Roland keeps the record of the temperature at 1:00 p. M. each day for two weeks in October." The children then had to graph the hypothetical weather data. Students worked individually and when one student was seen helping another, she was told to "Let [S.] do her own work." The teacher did all activities in teacher-centered whole group formats. Divergent thinking was generally not encouraged. One student said that she had solved the problem differently, "I did it a different way. I did it like that." The teacher responded, "What do you mean you did it a different way? Do you have those dates?" After the student had answered no, the teacher went on, "Then it ain't right. I told you to write the dates."

The next lesson of the day was on making nouns plural, followed by a spelling lesson, and then the students worked on writing and publishing their own stories. After lunch, the students had a social studies lesson using the test book. The final 15 minutes of the day were spent on a science lesson, "The Powers of Observation," an activity in which students made different observations of candles. The teacher began the lesson by passing out materials and going over the observations they had recorded the day before: "White, round, fat, made out of wax." For the last minute of the lesson before the final school dismissal, the students began measuring their candles in groups, with the teacher prompting them to measure it different ways.

Mrs. White, the other fourth-grade teacher observed at Bismuth, was also teacher-centered in her instruction, an approach consistent with the school's philosophy. On one of the days observed, Mrs. White began the day with a test followed by a reading lesson where students silently read books of their own choosing. She followed this with a 90-minute, whole class lesson based on a weekly reading periodical. Students broke for lunch and upon return to the classroom worked for an hour on mathematics out of their "End-of-Grade Test Notebooks" in preparation for the official test that would be given in two months. The lesson's objective was to review the concept of elapsed time and to answer multiple choice questions that were similar in format to the end-of-grade test. For the next hour of the day, students went out of the room for instruction in art and music.

When they returned to the classroom, students who still had assignments to complete worked at their desks; the remaining students went to the back of the room to do a science lesson on solutions. She began with a review of the idea of mixtures, which one student described as "It's when you mix something up and you can separate it." She then passed out an activity sheet on solutions where the children were asked to predict what would happen when they mixed various substances together in a baby food jar: sugar and water, sand and water, salt and water, water and vinegar. The teacher then led them step-by-step in following the directions on the sheet. The students waited between each step for the next directions by the teacher. When one group spilled its water on the floor, the group had to return its supplies and watch the other groups finish the activity. The teacher then asked the students what they had observed in their different experiments. The students turned in their sheets, cleaned up, and the lesson was completed quickly to give time for the social studies lesson.

The principal, who was new as of the interview, is supportive of the importance of science at the elementary school. "Science is a way to whet student appetites. It is an explanation of how things occur." The principal indicated that she would like to develop a long-term staff development goal in which the teachers would receive a full year of staff development centered around a subject area; she admitted, however, that a full year of science staff development would likely be 4 or 5 years down the road.

The Classroom Teacher Model of science instruction suffered from a lack of leadership and attention at Bismuth Elementary School. The focus was on teaching reading and mathematics with little emphasis placed on science. There were no schoolwide science events and virtually no leadership for science program improvement. Although the model has the potential for every classroom teacher to be a quality science teacher, the lack of focus on science within the full elementary curricula meant that Bismuth teachers placed their energy on other subjects.

ELEMENTARY SCIENCE MODELS: EVIDENCE OF EFFECTIVENESS

Looking at additional sources of data across different schools provides further evidence of the congruence between the different science program models and science education reform. Science program surveys were administered to all teachers at the 3 case study schools, as well as 9 additional elementary schools for a sample of 11 schools. The survey was developed by the National Science Teachers Association for schools to use to assess their instructional practices, resources, school, and district support.[1] The surveys were analyzed according to program type (those that have science specialists and traditional science programs). The mean response for all 70 questions for schools with science specialists was 3.48 ($SD = 0.40$) on a scale of 1 (Not achieved in practice) to 5 (Excellent, is actually in practice), whereas the mean response for schools without science specialists was 2.89 ($SD = 0.48$). The two different types of science programs were compared with a proportional odds model chi square and Hochberg's adjustment (Hochberg, 1988). The results showed that in schools with science specialists, teachers rated their school's science program significantly higher on every survey item than schools without a science specialist.

An examination of differences in item means for schools with science specialists and those without reveals that schools with specialists tend to be rated by teachers as having a program more congruent with current reform efforts (see Table 17.2). For example, teachers at schools with specialists rated their school's program higher than the ratings of teachers from schools without specialists for having "all elements of the program reflect a commitment to the intellectual values of the scientific community and to the scientific method of inquiry." The mean for the item, "Science is typically taught by a science specialist in our school" (3.37 out of a scale of 5) could be reflective of the idea that, even in schools in which science specialists are present, science is seen as a significant responsibility for the regular classroom teacher.

A content analysis of the questions in the survey indicates that the largest differences between schools with specialists and without specialists occurred in questions that dealt with leadership for the school's science program and the involvement of parents and members of the community (Table 17.2). For example, of the 20 survey items with the largest differences in means, 5 items addressed administrative support and 4 items focused on parent and community involvement in the science program. Schools with specialists were more likely to agree that administrators "show leadership by demonstrating a positive attitude toward science," "show leadership by encouraging and supporting the development of ef-

---

[1]The survey instrument "NSTA Assessment," was developed by the National Science Teachers Association (NSTA, 1989) and addresses issues such as instructional practices, resource availability, and school and district support. The items were first published in 1975 for high schools and were reviewed by school administrators, science teachers, science consultants, state science supervisors, and college science educators. In 1978, these guidelines were revised for elementary school programs and were published in 1989. Validity of the instrument for use with elementary science programs was established in a study of five schools by Jackson (1991).

TABLE 17.2
Survey Items with Largest Mean Differences for Schools
With and Without Science Specialists

Survey Item	Means (SD)	
	Without Specialist	With Specialist
Science is taught by a science specialist in our school.	1.52 (0.95)	3.37 (1.27)
A science supervisor/consultant/key leader is employed by the school or has release time to coordinate the science program.	2.11 (1.30)	3.83 (1.19)
Science program makes provisions for students with a special interest or aptitude for science.	2.40 (1.07)	3.48 (0.99)
Administrators show leadership in science by demonstrating a positive attitude toward science.	2.63 (1.15)	3.68 (1.11)
Parents are made aware of our school's science program by parent–teacher meetings, open houses, and by publicity in school or community news media.	2.71 (1.28)	3.70 (1.09)
All interested parties, such as teachers, parents, and representatives from the community, have the opportunity to become involved in the design and evaluation of the school's science program.	2.56 (1.17)	3.45 (1.11)
I have attended one or more science-related seminars or conferences in the last 3 years.	2.34 (1.46)	3.22 (1.51)
All elements of the science program reflect a commitment to the ideal that all students can learn science.	3.20 (1.15)	4.07 (0.91)
Science is taught at least 100 minutes in K–2; at least 150 minutes in grades 3 and 4; and at least 200 minutes in grades 5 and 6.	2.34 (1.27)	3.18 (1.28)
Local scientists, science educators, and leaders in business and industry assist the science program with field trips, class presentations, and science materials.	2.57 (1.17)	3.41 (1.25)
The emphasis on developing science literacy, science knowledge, positive attitudes toward science, and the abilities and habits of mind needed for science, is appropriate for the elementary school program.	3.13 (1.04)	3.92 (0.98)
Administrators show leadership by encouraging and supporting the development of effective science teaching and assessment practices.	2.63 (1.17)	3.42 (1.16)
The administration encourages teachers to attend professional growth activities, such as institutes, courses, and conferences.	2.86 (1.28)	3.65 (1.22)
Instructional technology is used appropriately in science classes.	2.61 (1.08)	3.38 (0.92)
Administrators show leadership in science by visiting classrooms when science is being taught.	2.45 (1.12)	3.22 (1.28)
All elements of the program reflect a commitment to the intellectual values of the scientific community and the scientific method of inquiry.	2.74 (0.98)	3.50 (0.87)

(Continued)

TABLE 17.2
(Continued)

| | Means (SD) | |
Survey Item	Without Specialist	With Specialist
The school's budget includes sufficient allocations of funds for all aspects of the science program (e.g., supplies, professional development, and field trips).	2.30 (1.07)	3.02 (1.22)
I work with colleagues, within and across disciplines and across grade levels, to plan the science program.	2.59 (1.25)	3.30 (1.27)
I identify and use resources outside the classroom, such as science specialists, science centers, industries, and the physical environment.	3.07 (1.25)	3.77 (1.00)
Class activities and experiences are consistent with contemporary research on conditions for student learning.	2.81 (1.05)	3.51 (0.90)

Note. The response scale ranged from 1 (Not achieved in practice) to 5 (Excellent, is actually in practice). $N = 298$ teacher surveys from 11 schools. There were significant differences ($p < .05$) between schools with resource teachers and those without resource teachers for all items when analyzed with the proportional odds model chi square and Hochberg's adjustment.

fective science teaching and assessment practices," and "encourag[e] teachers to attend professional growth activities such as institutes, extension courses, and conferences."

The schools with science specialists reported more community involvement through "local scientists, science educators, and leaders assist the science program with field trips, class presentations, and science materials" and "all interested parties, such as teachers, parents, and representatives from the community have the opportunity to become involved in the design and evaluation of the school's science program."

In addition, schools with specialists had higher means for questions that involved additional time commitment on behalf of individuals: providing experiences to students with an extra interest in science, planning for science activities with other teachers, and pursuing improvement in science content knowledge. In contrast, questions that dealt with instructional strategies that could be used in any subject area had smaller differences between means. For example, there were smaller differences between schools with specialists and those without for questions about using graphs and charts, concept maps, and technology.

In summary, for schools in this study with science specialists, the teacher assessment results suggest that these schools had more leadership for science, more community involvement in the science program, and teachers in the schools with specialists tended to involve students with special interests and aptitude in science, worked with colleagues across grade levels to plan the science program, and had teachers more involved in professional development in science. Over time, these program components could make important contributions to the quality of science instruction and the culture of science education in these schools.

DISCUSSION

The three models described in this chapter represent the most common models of science instructional programs for elementary schools. Although many more models may exist, these three provide some archetypal variation. The first model, the Science Instructor Model exemplified by Peach Street Elementary School, had a science specialist whose primary responsibility was to deliver science instruction to all the classes in the school. The specialist had her own room in the school, the "science lab," in which she collected resources that teachers could use. In the instructor model, time for working with other teachers is limited due to the heavy teaching load of the specialist. The trade-off in this school was that the teachers in the school were freed for instructional planning while the students were in their science class with the science specialist. The Science Instructor Model has the potential to make a significant impact on students because the science instructor teaches all the students in the school year after year. Not only do students get instruction from a qualified and motivated teacher, they receive this instruction year after year with the science teacher in the science laboratory.

The second model, exemplified by Harper Elementary School, is the Resource Teacher model. In this model, the school has a science specialist who functions in a more flexible manner. She has her own classroom, also called the "science lab," in which she has collected a large number of resources to share with other teachers. Her more flexible schedule means that she negotiates with teachers concerning what they would like her to do. In some cases, she teaches classes in her room, functioning more like the Science Instructor Model. In other situations, she may work with teachers in their classrooms to help them teach a specific subject. This flexibility has also meant that the teacher at Harper gets pulled in many different directions, taking over as half-time technology coordinator and writing grants for the school. In both models involving science specialists, the regular classroom teacher is still expected to teach science.

The third model, present at Bismuth, is the Classroom Teacher Model. In this model, the school has no science specialist; instead, each classroom teacher is the sole provider of science instruction. Thus, teachers take responsibility for obtaining their own resources and for planning their own lessons.

Observations of these three case study schools, as well as the surveys of nine other schools, show that the science specialists can make a difference but this impact is influenced by the other goals and reform initiatives as discussed in the sections that follow.

A Stronger Science Presence

As seen in the cases of Harper and Peach Street, the two elementary schools with science specialists, there was a much stronger visible presence of science in the school. Both schools had outside science areas, such as a nature trail, which had been either started or greatly enhanced by the science specialist. Both schools had a science laboratory in the school with materials and resources for teachers and both had schoolwide science activities for students and parents. In addition, the walls of both of these schools showed more student work related to science. In contrast, Bismuth Elementary, with the classroom teacher model, had no nature trail or other physical locations explicitly dedicated to science-related activities, nor were there any schoolwide science activities. There was little obvious science instruction, as indicated on the walls of the most publicly traveled spaces in the school. Bismuth, however, did have a significant presence for writing, something that can be perhaps accounted for by the existence of a writing specialist in their school. This gives credence to the idea that specialists in the school, in whatever discipline, may result in an increased physical presence of that discipline in the school. The assessment data from the survey of 11 schools provides additional evidence that science has an increased presence in schools with science specialists. Teachers from schools with specialists indicated that their school's science program placed emphasis on science literacy and had a commitment to the intellectual values of the scientific community and the scientific method of inquiry.

The presence of a science specialist can contribute to increased support for science in the school among the parents and local community. Results from the survey showed that schools with science specialists were more likely to indicate a higher level of community involvement in the school's science program and a higher level of administrative support for science. This was also found in the case study schools. With Harper School, the school governance committee voted to continue support for a science specialist at the expense of not hiring an arts specialist.

We speculate that the increased presence of science is due to the fact that the specialist views her role as one of promoting science and she therefore makes this a physical reality. Another possible reason is that science specialists tend to feel passionately about their subject area and are therefore willing to spend time outside of the regular school day increasing science opportunities in the school.

Standards-Based Instruction

The science instruction presented by the science specialists more closely resembles the science instruction promoted by the *National Science Education Standards*, a finding consistent with Schwartz et al. (2000). For example, Mrs. Dartmouth's lesson involved an investigation launched by the teacher (determining how levers reduce the force needed to lift an object), followed by exploration by the students. She provided general directions and then let the students explore for 15 minutes. The inquiry approach is, of course, supported by recent research (cf. National Science Resources Center, 1997) for fostering divergent thinking and creativity (Hodson, 1993).

Mrs. Tremain's lesson had less opportunity for creative exploration, although the hands-on activity (building rockets) was of high interest to the students. In addition, the students were active participants throughout the activity, integrating other subjects (e.g., mathematics or reading skills) as appropriate. It should be noted that both of these lessons were part of a longer series of activities.

Science instruction provided by classroom teachers in some cases was of as high quality as those lessons provided by science specialists. However, the survey with the larger sample of schools provided additional evidence that schools with science specialists may provide higher quality instruction in science—as seen in the higher scores for items that indicate the school's science program makes provisions for students with special interests in science, as well as the larger scores on items related to teachers' preparation in science and participation in science staff development. The survey data also showed schools with science resource teachers appear to do more instruction in hard to teach topics like physical science, nature of science, and history of science.

An Island of Its Own

In many cases, the teachers' classroom is considered their private fiefdom, impervious to most outside influences (Vesilind & Jones, 1998). The existence of a science specialist in the school is no exception. Teachers, like Ms. French, who already had an interest in and the knowledge to teach science, are likely to use the extra resources provided by the science specialist to teach science. Other teachers, like those at Harper Elementary and Peach Street Elementary Schools, will do the amount and type of science instruction they feel comfortable doing. In some cases, teachers who are uncomfortable teaching science and are in schools with science specialists may use the science specialist as an excuse not to teach science. More data must be collected, however, to make any sort of definitive conclusion.

Strong Interest Required

Science specialists bring enthusiasm and interest to the school's science program. Although the science specialists studied had very different backgrounds, they both had strong interests in science. Because of their interest in science, the specialists studied appear to pursue further staff development opportunities to improve their knowledge of science instruction and content.

Costs

One of the concerns about using the science specialist model is the potential cost. Schools that use specialists must fund an additional teacher position and provide classroom-laboratory space. As seen in Peach Street Elementary School, the trade-off for this expense is that the science specialist releases the classroom teacher from instruction and allows the teacher to have instructional planning time.

Locating qualified elementary science specialists can also be problematic. With the current shortages of science teachers at all levels, it can be difficult to provide the large numbers of teachers with science expertise needed to implement the science specialists model on a large scale. It is also problematic to find teachers who have expertise and experience teaching elementary school as well as having competence in science (Swartz, 1987).

For some, the Science Specialist Model conflicts with the belief that children need to be taught by a single significant adult who can know them well and can guide all of their learning. The use of a science specialist moves the program toward the middle and high school model, which has students changing classes and having a variety of teachers with different specialties.

Competing Reforms

As seen in the cases presented, other efforts to reform the school program can conflict with efforts to improve the science program. The state in which this study was done has a high stakes testing program that commences with third-grade students taking standardized tests in reading and mathematics. These tests have various consequences connected to the performance on the tests. Permeating discussions with all the teachers and principals was a sense that the testing program has diminished the time spent on science. Teachers felt much more pressure to teach reading and mathematics; the principals also felt substantial pressure to ensure that their schools were able to do well on the tests. Our classroom observations also found a high focus on testing in the day-to-day instruction that was observed. It is interesting that the one teacher who had a high quality science lesson was also the one who attempted to downplay the importance of testing.

In addition, in one case study, the science specialist's role was expanded to focus on another area of instruction—information technology, thereby reducing the amount of time spent on science. Thus, reforms that should not necessarily be in conflict with one another often end in shifting priorities in these schools, with the end result of less time spent on science instruction.

SCIENCE SPECIALIST TEACHER PREPARATION AND DEVELOPMENT

It is possible that elementary teachers may work in schools that follow any of the three science program models. Thus, teacher preparation programs should be structured to prepare student teachers with information about the different possible models of science instruction they may encounter. Although most undergraduate programs prepare teachers with only the initial skills needed to teach successfully, elementary teachers need to be able to collaborate with resource teachers, special science instructors, or teach science in a self-contained classroom setting.

The pros and cons of each model and the implications of the model for instructional practice should be explicitly discussed with beginning teachers. An elementary teacher who works in a school with a Science Instructor Model will need to know that this model can provide students with high quality science instruction, but it will not substitute entirely for classroom instruction in science. With both the Science Instructor and Science Resource Teacher Models, general elementary teachers will need to learn the importance of planning with the science specialist to maximize the connection between what happens in the lab and what happens in the classroom. The data show that this sort of coordination happened more frequently at Harper than at Peace Street, but it is key to effective teaching in both situations.

The Classroom Teacher Model at Bismuth shows that elementary teachers may not have the expertise, time, or resources to establish a strong science program in a school without a science specialist. This is also what the survey data suggest. Yet, this is exactly the model that is followed in the majority of schools. Therefore, elementary teachers need to be provided with the science skills and confidence necessary to effectively teach science in the self-contained classroom.

There is evidence that the science specialist has the potential to provide significant leadership for the elementary school's science program. Science specialists could benefit from advanced education programs that could help them develop skills in managing science programs (maintaining science materials, writing grants, balancing budgets), establishing community support and involvement, supervising and mentoring other teachers, increasing depth in science content knowledge, and developing skills providing staff development for other teachers. A traditional elementary teacher education program may not include these skills and competencies.

The *Standards for Science Teacher Preparation* (Gilbert, 2001) recommended that elementary science specialists should have mastered "conceptual content . . . balanced among life, earth/space, physical and environmental sciences, including natural resources" (NSTA, 1998, pp. 5–6). In addition, these standards stated that teachers should be knowledgeable about content (concepts, relationships, processes, and applications), the nature of science, inquiry, contexts of science, skills in teaching science, curriculum, social contexts, managing an environment for learning science, and professional practice.

As leaders within the school for the school's science program, science specialists should also possess leadership skills such those identified in the National Science Teacher's Position Statement (NSTA, 2003). These leadership skills include skills in the broad areas of science teaching and learning, professional development, science curriculum, and assessment.

NSTA recommends that leaders be able to implement inquiry instruction, use teaching styles that encourage constructivist approaches such as differentiated learning and cooperative learning, utilize student self-assessment, use effective communication with parents, and build principals' ability to recognize standards-based science instruction. Within the area of professional development, NSTA argues for leaders to be able to facilitate meetings effectively, involve teachers in decision making, use disaggregated student achievement data, promote collaboration and partnerships among policy-makers and universities, and mentor new teachers. As leaders of curriculum, NSTA suggests that science leaders be able to design and align curriculum, as well as to select developmentally appropriate and standards-based curriculum. For assessment, science leaders should be able to use a variety of qualitative and quantitative assessments, use assessments appropriate for diverse learners, and use assessment data to inform instructional practice.

In order for an elementary science specialist to meet NSTA's *Standards* and the leadership skills described earlier, it is likely that an elementary teacher would need advanced graduate preparation. A master's program that could lead to a specialist licensure could include the following:

- Advanced courses in science content domains
- Leadership skills development
- Strategies for management of materials and laboratories
- Advanced training in instructional technology
- Research on professional development for teachers
- Science research internship
- Advanced science methods
- Preparation to teach science to students with special needs
- Internship with a science coordinator at the regional or state level
- Research on effective school reform and program development

Special preparation for becoming a science specialist could also take the form of a district-based professional development program. However, due to the limited number of specialists likely to be found in a single school system, a statewide or regional program may be more effective in meeting the need for professional development for science specialists.

Few states offer certification or teacher licensure specifically for science specialists. Utah (Utah State Office of Education, 2003) offers elementary science specialty licensure at three levels: Elementary Science, Emphasis; Elementary Science Specialist; Advanced Elementary Science Specialist. The advanced elementary science specialist requires the following:

- three university/college science courses
- two university science courses with labs (including two of the three areas of science)
- one science field course
- one course in science assessment
- one course in elementary science teaching methods
- a portfolio that demonstrates professional growth

This model of licensure appears to prepare individuals with strong preparation in content but appears to have less emphasis on the leadership skills promoted by NSTA. As seen in the schools observed, the jobs of the elementary science resource teacher and the elementary science specialist are demanding and require a myriad of skills. There is a need for specialized teacher education programs that can enable science specialists to serve as outstanding teachers of science, leaders within the school and district, mentors to other teachers, as well as serve as the spokesperson for promoting science within the school community.

CONCLUSIONS

Three different models for science instruction have been discussed: two that depend on a science specialist, and one that relies exclusively on the classroom teacher. The cases and survey data presented are exploratory and thus any conclusions that can be drawn must, of necessity, be considered preliminary and should be explored in further research.

Overall, a science specialist shapes the school's science program in a variety of important ways. The specialist heightens the presence of science in the school's community. This presence is both physical (in the form of numerous resources, such as gardens, greenhouses, science materials, and nature trails) and in terms of leadership that impacts the science program (through inquiry lessons, high interest in science, a school emphasis on science literacy, increased planning for science across grades, as well as more teacher professional development).

The cost to a school to have a science specialist is relatively high. The school must fund a teacher position for the specialist, as well as free a classroom for a science laboratory. Schools typically find a variety of ways to fund the position, including increasing class sizes across the school, using special resource teacher funds, or obtaining parent organization support. For teachers at Peach Street Elementary School, the cost of the specialist had the additional benefit of providing the regular teachers with a planning period.

The two specialist models we observed also played different roles in the overall school culture. It is likely that in the Science Instructor Model the science specialist could teach a class of first graders a lesson and then 45 minutes later teach an entirely different lesson to a group of fifth graders. These teachers often have multiple preparations each day and must serve as quick-change-artists as they move independent sets of materials in and out quickly for different classes. For elementary teachers who want to teach science all day, the energy costs may be worth the opportunity to teach their chosen subject. This model also allows the science specialist to teach all the students in the school year after year, allowing the science specialist to plan for the development of multiyear goals and long-term science literacy development.

The Resource Teacher Model has the potential to provide more school-wide leadership for science but costs more to implement. This model is highly dependent on the skills of the teacher specialist who must break into the isolated classroom teacher's world to collaborate in the planning and teaching of science. In addition, because this model does not require the science specialist to teach a full load of classes, the potential exists for other teachers to question the workload of the resource teacher (a situation observed in a local elementary school outside of this study).

The traditional Classroom Teacher Model appears to include pockets of high quality science instruction, as seen in several of the lessons that were observed. However, the model can be problematic because science is dependent on the interests and dedication of each independent classroom teacher. This model also lacks the efforts of a science leader who can coordinate staff development, science materials management, school and community science events, and maintain resources such as nature trails or gardens. Furthermore, the costs and benefits of the three models should be considered in light of the impact and benefits to the whole school. The Science Instructor Model has the potential to provide high quality science instruction to every child in the school on a consistent schedule. The Resource Model could provide high quality instruction through direct instruction to children and through collaboration with classroom teachers. Both specialist models add leadership to the school community for science instruction, materials management, and schoolwide events.

It is possible for the traditional Classroom Teacher Model to provide quality instruction, but the model is dependent on the leadership of the principal and the interests of each classroom teacher. Given the competing demands made on teachers, consistent high quality instruction for science is more difficult to maintain across time. As indicated in these case studies, science instruction can be adversely affected by reforms that run in competition with one to the other.

The science specialist in either the Science Resource Teacher or Science Instructor Model has the potential to make a huge impact on the quality of science instruction in elementary schools. The science specialist job is complex and can provide significant advantages to the elementary school program as Ms. Tremain, one of our case study science specialist teachers noted: "The advantages [of the science specialist] are bringing support to the teachers . . . helping them see things as possible that they might not have seen as possible, helping with those events, coordinating things . . . maintaining materials, helping them find resources when they need them, and then modeling and demonstrating activities."

The challenge is for science teacher educators to create professional development and teacher education programs that can prepare both generalist teachers of science and science specialists for their respective demanding and complex roles, especially because it appears that some form of school leadership in elementary science enriches the science education experience of students.

ACKNOWLEDGMENTS

Appreciation is extended to Chris Wiesen from the Odum Institute for Research in Social Science for statistical analysis advice and to the teachers and principals who kindly participated in this study.

REFERENCES

Abell, S. K. (1990). The case for the elementary science specialist. *School Science and Mathematics, 90*, 291–301.

Blank, R., Kim, J., & Smithson, J. (2000). Survey results of urban school classroom practices in mathematics and science: 1999 report. In *How reform works: An evaluative study of National Science Foundation's Urban Systemic Initiatives.* Study Monograph No. 2. Norwood, MA: Systemic Research, Inc.

Gilbert, S. (2001). A continuum of standards for science teachers and teaching. In J. Rhoton & P. Bowers (Eds.), *Issues in science education: Professional development, planning and design* (pp. 43–58). Arlington, VA: National Science Teachers Association Press.

Goodlad, J. I. (1984). *A place called school.* New York: McGraw-Hill.

Hochberg, Y. (1988). A sharper Bonferroni procedure for multiple tests of significance. *Biometrica, 75*, 800–803.

Hodson, D. (1993). Teaching and learning about science: Considerations in the philosophy and sociology of science. In D. Edwards, E. Scanlon, & D. West (Eds.), *Teaching, learning and assessment in science education* (pp. 5–32). London: Chapman.

Hounshell, P. B. (1987). Elementary science specialist? Definitely! *Science and Children, 24*(4), 20, 157.

Jackson, M. (1991). *An evaluation of an instrument assessing elementary school science programs in North Carolina.* Unpublished doctoral dissertation, NCSU, Raleigh, NC.

Jones, D., Hardin, B., Chapman, L., Yarbrough, T., & Davis, M. (1999, November). The impact of high stakes testing on teachers and students. *Phi Delta Kappan, 81*(3), 199–203.

National Research Council (NRC). (1996). *National science education standards.* Washington, DC: National Academic Press.

National Science Resources Center. (1997). *Science for all children: A guide to improving elementary science education in your district.* Washington, DC: National Academy Press.

The National Science Teachers Association (NSTA). (1989). *Guidelines for self assessment-elementary school science.* Washington, DC: Author.

The National Science Teachers Association (NSTA). (1998). *Standards for science teacher preparation.* Retrieved April 14, 2003, from http://www.nsta.org/main/pdfs/nsta98standards.pdf

The National Science Teachers Association (NSTA). (2003). *Leadership in science education: NSTA position statement.* Retrieved April 14, 2003, from http://www.nsta.org/159&psid=36

Schoenberger, M., & Russell, T. (1986). Elementary science as a little added frill: A report of two studies. *Science Education, 70*, 519–538.

Schwartz, R. S., Lederman, N. G., & Abd-El-Khalick, F. (2000). Achieving the reforms vision: The effectiveness of specialists-led elementary science program. *School Science and Mathematics, 100*, 181–193.

Swartz, C. E. (1987). Elementary science specialists? We should know better. . . . *Science and Children, 24*(4), 20, 157.

Tilgner, P. J. (1990). Avoiding science in the elementary school. *Science Education, 74*, 421–431.

Utah State Office of Education. (2003). *Checklist of minimum requirements for elementary science specialty*. Retrieved April 13, 2003, from http://www.usoe.k12.ut.us/curr/Science/core/endorsements/elendscien.htm

Vesilind, E., & Jones, M. G. (1998). Gardens or graveyards: Science reform and school culture. *Journal of Research in Science Teaching, 35*(7), 757–775.

Weiss, I., Banilower, E., McMahon, K. C., & Smith, P. S. (2001). *Report of the 2000 national survey of science and mathematics*. Chapel Hill, NC: Horizon Research, Inc.

Williams, D. (1990). Making a case for the science specialist. *Science and Children, 27*(4), 30–32.

18

Future Directions in Elementary Science Teacher Education: Conclusion

Ken Appleton
Central Queensland University

Given that our only guide to the future is our past, and perhaps some wishful and even fanciful thinking, this chapter draws on much of what goes before. It naturally reflects my own biases and interests, although I have tried to guard against that.

POLITICAL AND CULTURAL INFLUENCES

The main constraints on, and opportunities for, elementary science teacher education come from the political and cultural environments in which it operates. These differ considerably from country to country, although there are some remarkable similarities. For instance, the current worldwide trend to outcomes or standards-based education, in some instances accompanied by extensive testing programs, has the potential to provide benefits or major disasters. I see benefits in the form of greater innovation as science teacher educators attempt to help prospective and practicing teachers come to grips with the cultural contexts and education systems in which they work. I see disasters in the making where schools respond, for instance, to harsh testing regimes in misguided attempts to improve test scores (see Diane's story in chap. 10). Unfortunately, these same measures threaten the education of the very children that the tests were supposedly designed to benefit. They also drive good teachers from the system. It is significant that the majority of innovations outlined in this book are framed by standards-driven "reforms" in the United States. In some other countries, such as Australia and New Zealand, there is a more measured approach to accountability in education systems, and reforms are driven more by concerns to improve practice—although they are not immune from overzealous and good intentioned legislators who deliver their own notion of "reforms" that embody more ignorance than wisdom.

It will be some years yet before there is a realization that some of the so-called reforms have been counterproductive, so elementary science teacher education will continue to be framed by these events. The innovations driven by the reforms, however, have the potential to lead to major changes in elementary science teacher education. For instance, the respective attempts to introduce prospective teachers to inquiry by van Zee, and Akerson and Roth McDuffie via related but different routes are providing new pathways for others to follow. Similarly, the notion of a standards-infused, integrated curriculum puts a different spin on teaching to standards. Different perspectives on how to deal with the overemphasis on literacy in some school systems are provided in these chapters, and in others such as that by Prain and Hand, and provide some ways forward in the difficult context in which teachers work. These issues do raise the question of whether or not prospective teachers should be trained in the art of political lobbying.

Other issues arising from these political considerations include system priorities and resources. Resources tend to follow stated systemic priorities. If the standards emphasized by a system are literacy and numeracy, and are linked to high stakes testing, then resources will tend to be allocated to these areas by districts and schools. Even if standards in science are emphasized within a system, resources will tend to be allocated to teacher actions and programs that elevate test scores rather than those that actually benefit student learning. I make this distinction between test scores and student learning deliberately, because there is growing evidence that test scores, even the widely accepted Third International Mathematics and Science Study (TIMSS) scores, do not measure student knowledge well. For instance, Harlow and Jones (2004) explained how they replicated a selection of TIMSS science items in a test with grade eight students, and supplemented the test with other strategies to find out how the students were interpreting the test items, and what they knew about the science content. They concluded that just 13% of the items actually measured what they purported to.

Political and cultural conditions also influence the nature and type of preservice and inservice teacher education. For instance, there is little incentive for teachers to take formal inservice courses in some states of Australia, where some education systems do not to require higher degree qualifications for promotion and advancement. There is pressure for universities to offer their existing courses via distance education in order to service teachers located outside their normal area. Boone's chapter highlights the complexities of doing this in a way comparable to face-to-face teaching. In the longer term, alternative pedagogies may be developed for use with information technologies, which are also acceptable to students. Until then, pressures to simulate face-to-face teaching via technology will continue.

Informal professional development can only be offered where there is funding and demand. Funding availability depends on political priorities. Further, the type of professional development that is effective in generating transformative teacher change in science teaching tends to involve intensive mentoring and sustained classroom contact (e.g., Appleton, 2003; Peers, Diezmann, & Watters, 2003). This is both costly and time consuming, and so is not likely to be appealing politically.

In preservice programs, current moves toward alternate licensure, standards, and more school-based programs are only just beginning to generate changes. The naïve notion that content knowledge required for teaching in the elementary school can be acquired during preservice teacher education in all subject areas, and that content knowledge per se is adequate for teaching science is shaping some preservice programs, and is driving them in a counterproductive direction. The chapter on science PCK provides some insights into the complexity of teacher knowledge, and recent research shows that science content knowledge, of itself, is inadequate for helping neophyte science teachers (e.g., Skamp & Mueller, 2001). Abell's analysis of field experiences shows the possible directions that might emerge as school-based programs that focus on the real-life context of teaching are promoted and funded.

DIRECTIONS FROM RESEARCH

For a futures perspective, however, it is to the more theoretical chapters that we must turn, because in these we see how science teacher education can potentially be transformed because of our greater understanding of what science teacher education is about. For instance, our developing understandings of the learning process guide not only our notions of how to teach children science, but also how we teach prospective and practicing teachers. The emergence of a greater emphasis on the cultural context of science teacher education and social-constructivist views of learning have paved the way for more effective teaching based on the use of real-life, situated teacher education. Similarly, our understanding of the nature of science teachers' knowledge, and the particular needs of elementary school teachers in acquiring that knowledge, provides directions for future innovations.

Both of these fields, despite recent progress in research and in conceptualization, require further research. In particular, the practical outworking of our theoretical frameworks need to be tested and evaluated. University learning environments do not always lend themselves to different approaches to learning and teaching, so there is a particular challenge for elementary science teacher educators to further innovate and push the boundaries of what is possible, based on our growing understandings.

The content of preservice teacher education programs in science will be influenced considerably by both research and by political pressures. For instance, the current interest and emphasis on pedagogical content knowledge has the potential to reshape aspects of preservice courses. Our growing understanding of the nature of PCK will provide impetus for a greater integration of content, pedagogy, and context knowledge acquired in authentic educational contexts, to aid construction of science PCK. However, the trend toward using PCK as a means of accreditation and licensure will push courses toward teaching "hints and tips" and the like, which more readily suit written tests that purport to measure science PCK.

If the need for neophyte teachers to develop science PCK is taken as a guiding principle for the development of preservice programs and science methods courses, then the difference between PCK as an epistemological construct and pedagogy for enhancing the development of PCK needs to be clear. For instance, a naïve interpretation of the epistemological position that science PCK is derived, in part, from general pedagogy, content knowledge, and context knowledge might be to teach each of these, in turn, in different courses. We must remember that an imposed epistemological construct does not necessarily imply that learning occurs in the same way as our construct is supposedly organized—I believe that learning is much more holistic, episodic, and complex than this.

Also of concern is the trend toward "alternative licensure" of elementary science teachers, where the notion is that somehow people with life experiences and noneducation degrees need little else in order to teach. This denies the validity of the essential contributory areas of knowledge to the development of science PCK, and assumes that all that is necessary is content knowledge and/or life experience. The consequences of this fallacy will emerge in time, but not before many well-meaning people "sink or swim" in efforts to become teachers, and innumerable students suffer the consequences. I speak from observing a similar move in my own state some 30 years ago, when there was a major shortage of teachers. Short-term, intensive teacher preparation programs were devised to fast track those with suitable qualifications and life experiences into elementary teaching. Few survived in the workplace more than a year, and the "initiative" died within a short time.

Another aspect needing further research is the use of science specialists in elementary schools. Although the Jones and Edmunds chapter provides our first real indication that learning gains accrue from use of science specialists, further evidence is needed, and explorations of the cost-benefits of specialist science teachers compared to generalists, explored. This is not a common direction for science education research, but it is necessary for schools and school systems to evaluate whether the increased costs are outweighed by learning gains in students.

The place of technology with respect to science teaching and science teacher education has been clarified by Jones, and presents particular challenges to science teacher educators in jurisdictions where science and technology are treated as the one curriculum area, or worse, where technology is omitted altogether. Further, confusion about information technology and technology itself abounds. A major issue is the extent to which science teacher educators should also be technology teacher educators. This often happens in institutions by default because of staffing issues or misunderstandings about the nature of technology. Can one person become an expert in both fields? With technology still very much an emergent area, this will become an issue with which many more science teacher educators will have to grapple. In particular, what areas of knowledge are appropriate for technology teacher educators?

A further challenge, outlined by Flick, is how to prepare those who would be elementary science teacher educators. As he explains in chapter 2, the demands of the job are varied and extensive. How can a person be prepared in several cognate areas at an appropriate level, as well as gain sufficient experience in elementary teaching? Further, there is an emerging need for teacher educators to be politically astute so they can maneuver their way within state and national legislatures: Teacher educators can no longer afford to be academics who merely teach and engage in research. The current and future educational contexts make it necessary for incisive research to be conducted, and for the findings to be reported in public forums as well as science education conferences. It is noteworthy that much of what is written in this book highlights how teacher educators tend to be reactive rather than proactive. That is, we tend to react to political and cultural circumstances and adjust what we do to fit. This is a natural reaction to external pressures, but there is little evidence of proactivity in trying to create different cultural and political circumstances. This is not to say that some do not try, but we have not been particularly successful in most political arenas.

For those of us engaged in elementary science teacher education, it is most rewarding when we see prospective and practicing teachers freed from the structures of their own limited perceptions of science and science teaching, and begin to engage their students in exciting and meaningful science. Collectively, we know a lot about how to be effective in making that happen. The challenge for the future is to know how to create the conditions that allow us to work with teachers for their own effective professional development and the betterment of their students' science learning.

REFERENCES

Appleton, K. (2003, July). *Pathways in professional development in primary science: Extending science PCK.* Paper presented at the annual conference of the Australasian Science Education Research Association, Melbourne, Australia.

Harlow, A., & Jones, A. (2004). Why students answer TIMSS science test items the way they do. *Research in Science.Education, 34,* 221–238.

Peers, C. E., Diezmann, C. M., & Watters, J. J. (2003). Supports and concerns for teacher professional growth during the implementation of a science curriculum innovation. *Research in Science Education, 33,* 89–110.

Skamp, K., & Mueller, A. (2001). A longitudinal study of the influences of primary and secondary school, university and practicum on student teachers' images of effective primary science practice. *International Journal of Science Education, 23,* 227–245.

Author Index

Subject Index